T0291675

CAMBRIDGE LIBRARY COLLECTION

Books of enduring scholarly value

Travel and Exploration

The history of travel writing dates back to the Bible, Caesar, the Vikings and the Crusaders, and its many themes include war, trade, science and recreation. Explorers from Columbus to Cook charted lands not previously visited by Western travellers, and were followed by merchants, missionaries, and colonists, who wrote accounts of their experiences. The development of steam power in the nineteenth century provided opportunities for increasing numbers of 'ordinary' people to travel further, more economically, and more safely, and resulted in great enthusiasm for travel writing among the reading public. Works included in this series range from first-hand descriptions of previously unrecorded places, to literary accounts of the strange habits of foreigners, to examples of the burgeoning numbers of guidebooks produced to satisfy the needs of a new kind of traveller - the tourist.

The Duab of Turkestan

W. Rickmer Rickmers (1873–1965) was a German explorer and mountaineer who visited and explored central Asia five times between 1894 and 1906. This book provides an account of his travels in the area he calls Turkestan, which incorporates modern Uzbekistan, Tajikistan and south-west Kazakhstan, and was first published in 1913. The region, which contains the ancient cities of Samarkand and Bukhara, had not been previously described in so much detail by a western European traveller. Rickmers includes accounts of both these historic cities as well as describing the social life of the indigenous people, with a comprehensive survey of the geography of the region. Richly illustrated with 207 maps and photographs, this volume provides an insight into the everyday life of the area before the upheavals of the Soviet era.

Cambridge University Press has long been a pioneer in the reissuing of out-of-print titles from its own backlist, producing digital reprints of books that are still sought after by scholars and students but could not be reprinted economically using traditional technology. The Cambridge Library Collection extends this activity to a wider range of books which are still of importance to researchers and professionals, either for the source material they contain, or as landmarks in the history of their academic discipline.

Drawing from the world-renowned collections in the Cambridge University Library, and guided by the advice of experts in each subject area, Cambridge University Press is using state-of-the-art scanning machines in its own Printing House to capture the content of each book selected for inclusion. The files are processed to give a consistently clear, crisp image, and the books finished to the high quality standard for which the Press is recognised around the world. The latest print-on-demand technology ensures that the books will remain available indefinitely, and that orders for single or multiple copies can quickly be supplied.

The Cambridge Library Collection will bring back to life books of enduring scholarly value (including out-of-copyright works originally issued by other publishers) across a wide range of disciplines in the humanities and social sciences and in science and technology.

The Duab of Turkestan

A Physiographic Sketch and Account of Some Travels

W. Rickmer Rickmers

CAMBRIDGE UNIVERSITY PRESS

Cambridge, New York, Melbourne, Madrid, Cape Town, Singapore,
São Paolo, Delhi, Dubai, Tokyo, Mexico City

Published in the United States of America by Cambridge University Press, New York

www.cambridge.org
Information on this title: www.cambridge.org/9781108010665

This edition first published 1913
This digitally printed version 2010

ISBN 978-1-108-01066-5 Paperback

THE DUAB OF TURKESTAN

Shakhzinde at Samarkand.

THE DUAB OF TURKESTAN

A PHYSIOGRAPHIC SKETCH AND ACCOUNT OF SOME TRAVELS

BY

W. RICKMER RICKMERS

WITH 207 MAPS, DIAGRAMS AND OTHER ILLUSTRATIONS

Cambridge:
at the University Press
1913

CAMBRIDGE UNIVERSITY PRESS
London: FETTER LANE, E.C.
C. F. CLAY, Manager

Edinburgh: 100, PRINCES STREET
Berlin: A. ASHER AND CO.
Leipzig: F. A. BROCKHAUS
Chicago: THE UNIVERSITY OF CHICAGO PRESS
Bombay and Calcutta: MACMILLAN AND CO., Ltd.

PREFACE

THIS book is an attempt at combining a record of exploration with the teaching of a little elementary physiography. Facts newly observed and theories of my own are also woven into the general description of a characteristic region. May the difficulties of the experiment in some measure reconcile the general reader, desirous of hearing something interesting or amusing, and the critical scholar in search of information. The style of the book sufficiently accounts for the omission of an incubus of footnotes and literary references. I have absorbed the writings of others, not quoted them. The bibliographical list at the end will, I hope, form the necessary acknowledgment of intellectual property. Scientific discussion has been reserved for the Appendix, where I have laid stress on the intricacies of many problems, also drawing attention to various questions to be cleared up by future travellers. The student of physical geography for whom this book has chiefly been written need not despair when faced by frequent doubts. Scepticism tempered with enthusiasm is the attitude of true science.

I express my deepest gratitude to the Council of the Royal Geographical Society and to the Syndics of the University Press who, by their munificent help, have both enabled me to publish this work. Nor can I forget that invaluable guide behind the scenes, the Corrector.

My sincere thanks are due to the Imperial Russian Government and its political representatives, as well as to His Royal Highness the Amir of Bokhara, for permission to travel in their dominions and for the invaluable assistance they always gave me. I also include many kind friends and helpers in Russian Turkestan and the Caucasus.

For advice in various matters I am indebted to Dr J. Scott Keltie, Secretary of the R.G.S., as also to Prof. U. Dammer, Berlin; Prof. H. v. Ficker, Graz (photographic work); Prof. A. Grünwedel, Berlin; Dr H. Hoek, Freiburg i. B.; Prof. F. Machacek, Vienna, and others. The D. Oe. Alpenverein and the Geographical Societies of Berlin, Bremen and London have allowed me to use my contributions to their journals (*Geogr. Journal*, Dec. 1899 and Oct. 1907; *Deutsche Geogr. Bll.* XXII. 2; *Ztschr. d. Ges. f. Erdk.*, Berlin, 1907; *Ztschr. d. D. & Oe. Alpenvereins*, 1902, 1907 and 1908). Professor von Ficker gave leave to draw freely upon his *Meteorologie von West Turkestan.* His sister Frau H. Sild was the most unselfish travel companion one can imagine and I owe much to her kindness, while Mr Douglas Carruthers proved a thorough sportsman and true friend under very trying circumstances. The dead, alas, can hear me no more. The late Albrecht von Krafft was more to me than I can say, and Aemilius Hacker who perished in an avalanche last winter, was likewise associated with one of my first visits to Bokhara.

W. R. RICKMERS.

INNSBRUCK.
November 1912.

NOTE

The metrical system, correct pronunciation, and other matters are explained in the glossaries at the end.

CONTENTS

LIST OF ILLUSTRATIONS

Photos Nos. 4, 5, 6, 7, 11, 25, 26, 47, 48, 82, 144, 195, are by Frau H. Sild.

All photographic negatives of above, mostly full plate size, are at the Royal Geographical Society, London.

INDEX TO A FEW IMPORTANT SUBJECTS IN THE ILLUSTRATIONS

CHAPTER I

THE DUAB OF TURKESTAN

To begin with, some apology is needed for a name which the reader has probably not seen before. "The Duab" as applied to Turkestan is an innovation which I have chosen for practical reasons, the names in general use being either too sweepingly vague or too restricted. I wanted to circumscribe a field for systematic research which might at the same time serve as generally representative far beyond its narrower boundaries. What is known as "Turkestan" is an atmosphere, there is no better word for it, but I also wanted a locality and I defy any one to outline on the map a country called "Turkestan." Nor would that be desirable, for a real atmosphere cannot be imprisoned within visible walls. My idea was to have a compact laboratory or natural park of definite shape but intensely saturated with an atmosphere extending far beyond the border and gradually losing its intensity as it merges into other climes.

The Duab of Turkestan is the land between the two rivers (*du*, two; *ab*, water; analogy, Panjab), between the Amu-darya and the Sir-darya or Oxus and Jaxartes. Its outline has the charm of simplicity. From the Wakhjir source of the Oxus we follow the river to the Sea of Aral; thence round the northern shore to the mouth of the Sir and up this river to where (under the name of the Narin) it breaks through the Ferghana mountains. The watershed connecting this point with the source of the Oxus completes the circuit. This area which I call the Duab, contains everything that is typical of and common to various overlapping or subdivided conceptions such as Turkestan, Western, Russian, Chinese, or Eastern Turkestan, Central, Middle or Inner Asia, Turan, Iran, Transcaspia, Transoxiana, Bokhara, Kashgaria, Tarim, Aralocaspia, etc.

Here we have the Pamirs and all the most important ranges radiating from them to the west, with their glaciers, fauna and flora. Between these rivers—they themselves being the all-embracing distinction, the beginning and end of the whole—we find a complete collection of typical phenomena from the summits of the mountains to the cities of the plains and to the salt in the Sea of Aral. Turn to what you like, to climatology, geology, zoology, botany, to the human races, to political economy and history, to the Russian conquest, administration and present commercial development, search every department of science as applied to Turkestan, and you will find all the central and leading facts or types included in the Duab.

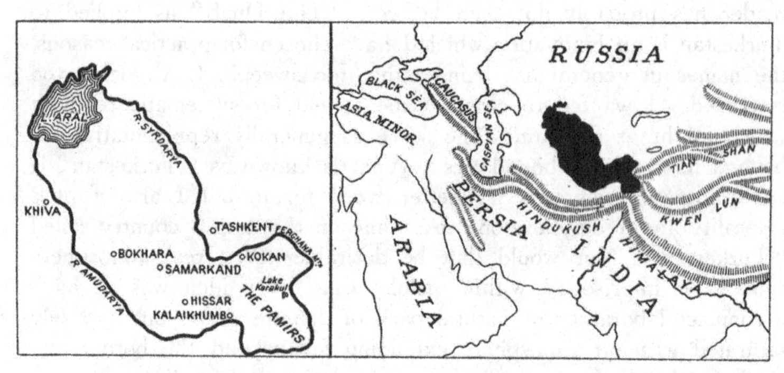

1. The Duab of Turkestan.

The Duab in this sense is, of course, not an indigenous name but an artificial, scientific term like the Alps, the only difference being that it has been made instead of having grown historically out of topographic misunderstandings and literary generalisations. Apart from purely political dominions nearly all the geographical names of Inner Asia denoting large districts were created by European travellers and writers. Peoples who have no maps use only two kinds of names: definite, special names for every topographic feature, however small, which they own or visit regularly; and vague names for foreign lands, such as anything across the mountains, to the east, across the sea or river, belonging to other tribes, or any appellation chance or fancy may beget. Only European science and politics with their statistics and classifications have produced an abundance of terms of definite international value. The origin of the name of a large district is still very often traceable to a small place or to local usage. Turkestan was only a town before it became one of the greatest expressions in modern history and geography.

A sharp boundary is always something arbitrary, just as classified nature compared to living nature. But the more artificial a system the

2. Turrets and Bastions of Conglomerate (Yakhsu).

clearer and simpler it becomes, as witness political frontiers. So-called genetic systems are often a contradiction in terms. There is continuous shifting from special to general or vague, from abstract or artificial to concrete or applied. The "Germans" were perhaps a small tribe before "German" attained a wide, racial sense, and now "Germany" is a strictly defined political area, although it often lends its name to more vaguely distributed features of central Europe. A book on the "Birds of Germany" might just as well be called French or Swiss or Austrian birds, for the bulk remains the same. The national name simply stands for each people's own glass case containing a but slightly varied collection from the common ground.

The Duab of Turkestan is a strictly topographic boundary enclosing a typical assortment of natural things. What is within *is* Duabic; anything without—but within Asia—*may* be duabic (compare "Alpine" and "alpine"). The geometrical definition is limited but the application extends to many physical characteristics widely distributed over Asia. The Duab is a place with the atmosphere of Turkestan.

3. Gossips.

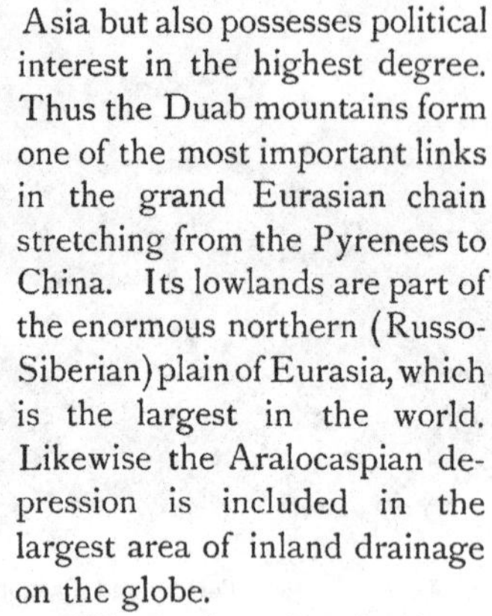

The Pamirs, that massive nucleus which covers a surface equal to that of Ireland, is not only an orographical centre of Asia but also possesses political interest in the highest degree. Thus the Duab mountains form one of the most important links in the grand Eurasian chain stretching from the Pyrenees to China. Its lowlands are part of the enormous northern (Russo-Siberian) plain of Eurasia, which is the largest in the world. Likewise the Aralocaspian depression is included in the largest area of inland drainage on the globe.

The total surface is about 250,000 square miles (595,000 sq. km.) peopled by 5,000,000 inhabitants, who are however very unequally distributed.

Many historians ascribe to Turkestan and especially to the Duab regions a preponderant influence on the origin of the Aryan peoples. The province of Samarkand and the khanate of Bokhara almost correspond to Sogdiana, to which Bactriana joined on in the south. Alexander the Great took Samarkand in 329 B.C. Huns and Turks conquered these regions during the sixth century A.D., the Arabs in the eighth, and Genghis Khan established his empire during the thirteenth century.

After the death of Tamerlane (1405) his dominions split up into many small states, the playground of incessant wars, until Russia gradually incorporated the Duab in the course of the nineteenth century.

Politically the whole belongs to Russia, the independence of Bokhara being nominal only. The Duab borders on China and Afghanistan, while the frontiers of the British and Persian Empires are not far off.

Ethnographically we find here all the races and tribes, settlers as well as nomads, of Middle Asia, or their equivalents. Close neighbours and occasional visitors are the "national" Afghans and Persians. By a national people I mean, of course, that typical, modern combination to which the name applies in every sense, being Afghan, Persian and so forth not only by descent but also in features, language, religion, dress, habits and, very often, politically. From a genetic point of view, i.e. racially and historically, there are practically no boundaries in continental Asia.

4. An Araba.

Mongol and Aryan elements have about an equal share in the population. In enumerating the various races I shall follow the German savant Schwarz who has written the best general work on Russian Turkestan. To begin with, however, I shall have to say a few words anent the name "Sart," which is to be found in nearly every book. In common parlance and every day life the Russians use "Sart" in much the same way as British colonists would speak of "niggers." It is applied to all and sundry "natives" whose dress does not single them out at once (Jews, Turkmen, Kirghiz) or who are not evidently foreigners (Europeans, Afghans, Chinese, Hindu, etc.). "Sart" is the "ordinary native" of Bokhara, Samarkand and Ferghana. For this reason no importance whatever can be attached to this description unless the speaker or writer clearly specifies his meaning. The origin of the term is local. It has been used for centuries by the Duab peoples to designate the peasants of the plains and the dwellers in the towns, that is to say, all those who are not nomads, mountain tribes, foreigners or Jews. This is the second and true application. Thirdly: Sart is also an ethnological distinction to which scholars have given a specific value. Schwarz says that the word Sart has an ethnological meaning applied to a people which are neither pure Uzbek, nor Kirghiz, nor Tajik. Their language is Turki and their type is as different from that of the Mongoloid Uzbeks and Kirghiz as from that of the Indo-germanic Tajik, being perfectly independent and peculiar. They are a mixture resulting from the intermingling of the Indo-germanic aborigines with all the races which have lived and ruled in Turkestan during the last two thousand years. Thus "Sart" has three possible meanings. I shall only use it in the second sense of typical, settled native, having chosen "Uzbeko-Sart" to express the scientific definition.

5. Ladies of the Harem.

Numerically and economically the Uzbeko-Sarts are the preponderating element, forming, as they do, the majority of cultivators and traders. But they are also the worst portion, especially in the towns. On the whole they are cowardly and cruel, liars and cheats, and given to every kind of vice. Outwardly they pose as dignified gentlemen and perfervid Muhammadans. The Tajiks are far better, while the nomads (Kirghiz and Turkomans) and mountain peoples (Galchas) most nearly conform to our ethical ideas. We must not forget, however, that among children of nature, especially among nomads, mere thieving cannot be called a fault of character. The same applies to curiosity and fits of temper, while so-called lying can generally be traced either to indifference or to the eastern love of exaggeration. Likewise, we should take a lenient view of boorishness and suspicion towards strangers at a first meeting and of the reluctance to give information, all primarily due to the instinct of self-preservation, alas! but too well justified. It is only when hypocrisy and cowardice are added to the faults of the child, the warrior and the peasant, that the tribe becomes an undesirable one, despised both by general and inter-racial consensus of opinion.

The Tajiks are also a mixture but more Aryan in every way, having preserved themselves more or less from the last Mongol (Uzbek) invasion. They speak a Persian dialect.

Pure, original, nomadic Uzbeks are very few in number, but divided into many sibs of high lineage, some of which have furnished rulers for the thrones of Bokhara and Kokan. Being the only pure remains of the Mongol conquerors they form a kind of racial aristocracy. Some writers have used the name Uzbek in the sense of Uzbeko-Sart.

6. Children of Bokhara.

The great nomads of Asia are the Kirghiz, and they are well represented in the Duab. Their origin points back to some near relationship with the Uzbeks, for their family names and language show close similarity. Of their mode of life I shall speak in the course of my narrative.

In Transcaspia the noble race of the Turkomans (Turkmen) replaces the Kirghiz as the great nomadic nation. Sections of some of their easternmost tribes live on the right bank of the Amu-darya. Their name is sufficient indication of their origin. They are the Beduins of Middle Asia and among the proudest, if poorest of the nomad tribes. Of horses they only keep a few, but these are the best in Turkestan, the Turkoman treating his steed like a child, even valuing it above his goods and wives.

Among the mountains, in the valleys of the Zarafshan and Yagnob, in Darwaz and Wakhan, we find the Galchas or Indo-germanic Pamir peoples in whom the students of Aryan origins take such an ardent interest.

Of Arabs a few sporadic occurrences have been traced. They are probably the remnants of Arabs settled by Timur.

Jews of a pure and ancient type are found in every town. According to their own tradition these native Jews are descendants of those brought here by Assyrian and Babylonian kings. They are everywhere called "Bokhara Jews," probably because their strongest colony and intellectual centre is at Bokhara, where also their original mode of life is most typically preserved. Their daily language is Persian.

Of Mongolian varieties represented in the country one may also mention the Karakalpaks, Kuramas, Kipchaks, Kalmucks, Targauts and the Tatars from Kazan and Siberia. Two kinds of Gipsies are also found.

7. Among the Zarafshan Mountains; Rest on the Glacier near Akhun Peak.

Amongst visitors we meet Persians, Afghans, mostly traders or political refugees, and the Indians from the Panjab. These latter, called Multani by the natives, mostly hail from Shikarpur and are detested all over Turkestan as the worst of usurers. From what I saw of their houses they are also the dirtiest of all the various races. Russia has banned them from districts under her administration and will probably soon give them notice to quit Bokhara.

Last, not least, we have the Russians, the conquerors, rulers and colonists of to-day.

A noteworthy fact in connection with the Duabic population is the minority of the female element, there being only from 80 to 90 women for every 100 men.

Leaving aside the Christianity of the Europeans and a few mysterious Pamir sects, the whole of the Duab is under the sway of

Islam. The nomads only profess their religion nominally, but the settlers, and especially the Bokharan subjects, may be counted among the most fanatical Moslems in the world. In Bokhara, however, religious fervour is the work of a few leading spirits as is best shown by the condition of things across the Russian border where spiritual tyranny is hardly noticeable. Here one never hears of quarrels between Christians and Mussulmen, the population being exceedingly tolerant on both sides. The Russian administration has strictly forbidden all proselytising even on the part of the Orthodox Church. This wisdom has excellent results, contrasting favourably with the questionable effect of missionary work in other countries.

The geography of pathology can be dismissed very briefly. Natives suffer chiefly from malaria, inflammation of the eyes, small-pox and an occasional epidemic of cholera. Characteristic diseases, though not numerically important, are leprosy, a thread worm called rishta (*Filaria medinensis*) and the "Sart complaint," namely a kind of Aleppo boil. Even in the cities which are so filthy according to our notions, the dry climate seems to prevent the frequent occurrence of decimating epidemics. The steppes and deserts are practically free from contagious or infectious diseases, their inhabitants mostly dying of various complications, accidents, starvation, neglect or old age.

8. In a Market Place at Bokhara.

Europeans can easily preserve their health by the simplest elementary precautions. The disease to which they are most liable is malaria, more than half the foreign population suffering regularly from it. It is, however, relatively harmless death being hardly ever immediately traceable to it. At times, of course, it must prove a predisposing cause, owing to its weakening effect upon the human body. Next in order come, intestinal catarrhs, mostly due to gross carelessness in eating or drinking, internal inflammations, rheumatism and veneric diseases. Briefly, the Duab climate may be described as a healthy one for European constitutions.

The fauna is palæarctic with a few immigrants from the south, such as the tiger and the cobra. Wild sheep are among the most characteristic and typical denizens of the mountains all over Asia, and they are ubiquitous in the Duab. Wherever the ground rises above dead level forming hills however small, wild sheep are sure to occur. Red marmots and birds of prey, especially the bearded vulture (lämmergeier), are most familiar to the traveller in the mountains, while in the plains one hardly ever sees anything more exciting than a troop of gazelle. Sturgeon, carp and waller are plentiful in the great rivers, while *Scaphirhynchus* is a curious fish for which the Amu-darya is remarkable. Insects are the most eager in claiming our attention especially in the

9. Evening—Samarkand from Afrosiab.

shape of flies and mosquitos and the usual vermin with its irrepressible claim upon our companionship. Scorpions, tarantulas and others of that ilk abound in the earthy districts but nobody really fears them. The only dangerous spider is a little black animal called Karakurt by the Kirghiz. Its sting is said to be deadly to cattle and great losses are ascribed to it in a few localities.

The flora is that of the deserts, steppes, cultivated lands and alpine mountains of the hotter temperate zone. The absence of the spruce (fir) is worth mentioning. Carruthers found the first spruce just outside the Duab in the extreme western end of the Tianshan, so that here we observe a sharp boundary between two important botanical districts.

As to communications we shall have plenty of opportunity for studying the native caravan roads and mountain paths during our travels. The Russians developed a good system of post roads immediately after the occupation of the country, many of which are now superseded by railway lines traversing nearly all the more important centres of cultivation. The last great undertaking will be the projected railway *via* Aulie-ata, Vierny and Semipalatinsk, which is to connect the Siberian and Turkestan systems. Steamers run on the Oxus and on the Sea of Aral.

10. Picturesque Disorder.

It goes without saying that the advent of steam and electricity have revolutionised trade. Before the Russian occupation the imports from India *via* Afghanistan or Persia were very important, the risk incurred from nomadic robbers on the northern routes having much to do with this. Now things have changed and Russia supplies almost everything. Indian trade declines swiftly, save for green tea without which the natives cannot exist and which is not obtainable elsewhere. A few muslin goods and a small quantity of opium hardly weigh in the balance.

We must also remember that there is no duty on Russian wares, not even in Bokhara. The staple articles are printed cotton and the products of metallurgy, sugar, sewing cotton, aniline dyes, china, glass, and, of course, the usual motley of ironmongery, cheap fancy goods and useful oddments. The bulk of the export trade is concerned with raw cotton, silk and wool. Specialities are carpets, lucerne seed (alfalfa), wormwood seed (santonin), karakul skins, dried raisins, pistachios and walnut wood. To these might also be added curios or antiquities, faked as well as genuine.

That the empire of the Tsar knows how to rule its subject peoples is beyond dispute. One great secret of success is the principle of non-interference with religions and habits, even missionary propaganda being forbidden. Nor are special efforts made at "improving" the intellectual or moral standard of the native, beyond giving him the possibility of a better education. Very much, however, is done in the way of irrigation, roads and other economic works, which after all form the basis of intellectual progress.

Much nonsense is still talked about so-called Russian conditions. Naturally this country is very different from ours, but why excite ourselves about its internal affairs which are really as stable and as reasonable as those of many western states? Nobody is obliged to go there. That a traveller, a foreigner, will easily get into scrapes is simply an outcome of ignorance on both sides. Curiosity, suspicion and so forth are easily excited when strangers meet, but hospitality is writ large on the balance sheet. With good manners and the exercise of a little tact one will get along very well in Russia as elsewhere. Customs examination and the passing of aliens is far worse in the United States of America and "prohibition" a hundred times worse; the political police is as good as that of Prussia and military secrets are guarded as jealously here as anywhere, while the discipline of the church is entirely restricted to the members of the Orthodox faith. Life in Russia is free, easy, unconventional; there is less theory and more humanity of the best kind as well as of the worst. One must take one's chance as everywhere in this world and the chances of the individual are certainly greater than in many better known regions.

From the explorer's point of view travel in the Duab cannot be called difficult. One can get almost anywhere and inhabitants are found in the remotest corners, while with the exception of the civilised centres murderous attacks are practically unheard of.

Such is the Duab, a section of Asia, from the Pamirs to Lake Aral. Here we have glaciers, foothills, deserts, steppes and landlocked seas. We have fertile countries; we have all the peoples and remnants of peoples that made one of the greatest histories of the world; we have Bokhara and Samarkand. We have the conquest, the new era promising to the land a magnificent future of industry and commerce, for the Duab is the Egypt of Russia.

Around all this, fed by a thousand veins, are the two great rivers, the beginning and the end, the life and content of the whole, from which they are born and in which they die, which they lovingly encircle as a harmonious entity, self-contained and grand.

11. A Public Meeting.

CHAPTER II

THE PHYSICAL FEATURES OF THE DUAB

THERE is a great problem called the "desiccation of Inner Asia," and it is intimately bound up with the riddle of mankind. It seems that great areas have become drier and drier even within historical times, and the Duab lies within the sphere of this influence. Thus the study of its deserts, mountains and glaciers and the questions relating to the great ice age are endowed with peculiar interest. Our exact knowledge being still very meagre we are obliged to do a little careful speculation now and again in trying to arrive at some idea. In the course of this book I shall not show undue hesitation in following the line of deductive reasoning, whenever interesting questions arise, which cannot as yet be surrounded and overpowered by a sufficient array of facts and figures. For this purpose I shall make a liberal use of the direct method consisting in the comparison of large features and striking phenomena with familiar European, and especially Alpine, conditions, for the Alps are the fundamental standard of comparison, whether openly avowed or unconsciously traditional, of all research concerned with mountains, valleys or glaciers.

In physical geography or physiography, which is the study of the life of the earth and of the expression of her features, a definite meaning has been attached to Central Asia by Richthofen. He defines it as the region hemmed in by the Altai, the Pamirs, Himalaya and the watershed towards the great Chinese rivers, or, in other words, Chinese Turkestan and Tibet. Western Turkestan and Iran he calls the peripheral districts. The rivers of Central Asia have no outlet to the sea, nor has the Aralocaspian area of inland drainage of which the Duab forms a completely representative part.

Although the shape of the ground is the initial cause, the question of imprisoned rivers is not quite so simple as appears at first sight. If there were more rainfall many of the interior basins would easily fill up and overflow, there being no very considerable difference of level between their bottom and the lowest point of the rim. Thus the problem is one of action and reaction between surface and climate. Australia is almost entirely a desert in spite of being surrounded by a

12. A Bridge in the Mountains.

vast ocean. Its neighbour, New Zealand, on the contrary, boasts of a moist climate, but then it has high mountains with snow and ice. This shows us that latitude being equal, mountains are the great decisive factor of climate. Upon their height, extent and position the distribution of water and life largely depends. The vapours of the torrid zone are condensed on the southern slopes of the Himalaya, thus forming a stagnation of rain clouds to which India owes its fertility. As witnessed by frequent famines this rainfall becomes irregular towards the South of India, agriculture being endowed with a speculative element as distinct from the certainty of artificial irrigation. Siberia's condition is partly due to latitude, partly to the mountain barrier of the Tianshan. But the effect of these barriers is more apparent in their negative influence upon Central Asia, which thus enclosed has become one of the driest regions of the world. If there was a more open sea north of Asia and the Tianshan non-existent, the highlands of Tibet would probably be covered with huge glaciers and the depression of Chinese Turkestan occupied by a lake. Mountains have been called the skeleton of the land, being guiding lines to the eye, but we can extend the analogy to the backbone which is also a nerve centre and as such the ruling influence in the development and vitality of a continent. The mountains have been upheaved by the land and in return mould its outward features. From this it will be seen that the solution of Asiatic problems lies in the mountains and with them we shall chiefly concern ourselves.

The peripheral districts fully represented by the Duab are not so extreme as Central Asia which is a commanding type of its own, having no parallel in the world. Great lakes and rivers are found in the outer belt, while the swamps of Lopnor and Tsaidam are all that is left of Central Asian water circulation. Thus the Duab forms a transition.

With regard to latitude the Duab occupies about the same position as Italy between the 37th and 46th parallels. Lake Aral corresponds to the upper end of the Adriatic with Venice and Trieste. The 40th parallel, which passes close to Samarkand and Bokhara, traverses the middle of Sardinia and Spain, while the southern extremity of Sicily would coincide with the Afghan frontier of the Pamirs and the sharp bend of the Oxus.

Of the total surface the Alai-Pamirs or Duab mountains claim one-fourth or 56,000 square miles (plains 174,000 sq. miles, or 145,000 and 450,000 sq. km.) which is more or less the superficies of the Alps. Although not an isolated group this mountain system to which the name of Alai-Pamirs is frequently given, forms an individual branch of the Asiatic structure. It would not be complete without the Pamirs to which it has, however, no exclusive claim, the great nucleus of the "Roof of the World" being also attachable to any of the other chains or groups. It forms the common centre of all, being indivisible itself like the hub of a wheel and incomprehensible without at least one of the spokes. To the eye which arranges and connects the lines of the map the Pamirs form the focus of radiation. But viewed in the light of geological history this centre is one of effect not of cause. The solid bulk of the pillar of Asia owes its origin to the meeting and intersection of the long mountain folds.

Where the crests of these waves crossed each other in their tumultuous upheaval a kind of compact eddy, a labyrinthic surf was formed. But the reader should not look upon this picture as anything more than a simple metaphor, symbolising one of the most complicated

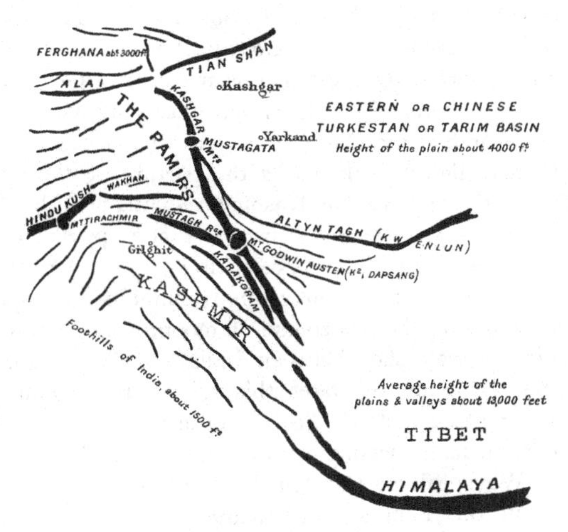

13. The Orographical Centre of Asia.

convulsions of the earth's crust. Maybe another century of patient exploration and surveys will pass before we are able to unravel these mysterious interlacings in a definite and scientific manner. In parenthesis I may mention that the high Eurasian mountains, the Alps, the Caucasus and the Himalaya are, geologically speaking, very young, inasmuch as the folding up responsible for their present elevation happened during the tertiary age which is the latest of the three great epochs of geological history and the last during which considerable uplifts took place.

The Pamirs form one half of this area of the flocking together of the ridges. As any map of Asia will convince the student, this region covers nearly the whole of Kashmir, Baltistan, Chitral and the north-east corner of Afghanistan. Fig. 13 is a diagram giving a general idea of this orographical centre of Asia. It is a dense array of wrinkles and crinkles bent in various directions. Here the solid wave of Himalaya breaks up into its northerly and westerly branches. Whoever has watched a stormy sea will remember that the highest waves are not continuous but move in steps or echelons. Likewise the folds of earth never run smoothly throughout but are interrupted and begun afresh. It seems to me that Mount Godwin Austen or K2 (28,200 ft.; 8620 m.) is the important signpost in this maze, that, poetically speaking, it is the true gatherer of the folds and deserving as none other to be called the monarch of the Old World. As well as can be done by a single point, this mighty peak marks the parting of the main lines. Towards the north we have the Kashgar mountains with Mustagata (25,800 ft.; 7860 m.), while the Mustagh range appears to lead us to Hindukush, where Tirachmir (25,400 ft.; 7750 m.) rises into the clouds. The triangle of these three kingly summits encompasses the most ponderous and terrible concourse of mountains in creation. The Hindukush line or watershed between India and Turkestan, between the Indus and the Oxus, separates this remarkable region into two districts of different character. To the south we have what, for want of a better general name, we might call "The Mustaghs," to the north the Pamirs. We speak of the Mustaghs meaning many mountains and ridges; but the Pamirs mean many valleys, their names thus expressing their salient and distinctive features. The Mustaghs have gigantic crests and pinnacles with the Hispar and Baltoro glaciers sunk between them, whereas the Pamirs are very broad valleys with gently sloping sides, the divides between them being relatively low. Of the probable causes of this condition of things I shall speak later. Here I only wish to present a preliminary survey by the aid of Figs. 13 and 20, drawing attention to a few important facts which one ought to keep in mind.

One thing is quite certain, namely, that the Pamirs are on the northern side of the climatic divide traced by the sweeping curves of the Himalayan system. That the line of the highest ranges does not always coincide with the watershed need not trouble us here. Although transgressive tapping is an interesting problem in itself and one which may throw light on the physical history, it does not make much difference in the present state of affairs, the quantities of water in question not being sufficient to play a decisive role in the weather conditions on either side of the rim.

The simple topographic position is this. The main axis dividing this region from southern Asia follows the Himalaya-Mustagh-Hindukush line. From the plains of India a long and very high ascent leads to the gaps or passes of this wall, but the drop down the other side to Tibet and to the Pamirs (i.e. the floors of the valley plains) is something like 11,000 feet less and also very much shorter by horizontal distance. Tibet and the Pamirs are from 12,000 to 13,000 feet high (3600—4000 m.); Lhassa and Pamirski Post are both 11,800 feet (3600 m.) above sea level. Thence from both these pedestals comes a step of 8000—9000 feet down to the real oven and central depression of Asia, to the Tarim or Lop Basin (Chinese Turkestan), while towards the north and another thousand feet lower down, after crossing the rise of the Alai, we find the open valley of Ferghana in the Duab. Finally, the Tianshan and its branches form the northern outline and barrier of the whole.

Thus it will be seen that the Pamirs are a replica of Tibet, that the Alai corresponds to the Altintagh and Ferghana to the Tarim basin. But there is this difference, that the Pamirs and Ferghana are open to the west, which gives them a better climatic chance. We observe therefore that near Mount Godwin Austen the great ridge, which until then serves a double purpose, branches north and west. The transverse chain of the Kashgar margin shuts off the central saucer which is closed all round; the Hindukush branch becoming the frontier of the Mustaghs against Western Turkestan, that is to say, against the peripheral or duabic regions. In this angle or wedge lie the Pamirs, and it will now be readily understood why they form an intermediate type between the Asiatic extremes of Tibet and the alpine or humid sides of Himalaya, between innermost and outermost Eurasia. The fact that they are neither a pure plateau nor a deeply riven mountain group appears to be another point in favour of my suggestion.

Let us go back, then, to a bird's-eye view of Asia as any good atlas will show it. Below us is the great, solid octopus of the Mustagh-Pamirs sending out its long, spare tentacles towards the east, gripping the expanses of Tibet, Lop and Mongolia. Its western side, however, looks more like a frayed edge of short but thick fringes, the further geological continuations leading to the Alps not being apparent at first sight. To this outer fringe, but deeply rooted to the heart of the continent at Mount Godwin Austen, belongs the Duab of Turkestan.

I have been speaking of the Pamirs and Tibet as "pedestals," and I must make a digression here in order to explain a point which must be fully understood. In comparing countries or masses of land the height of the summits alone does not give a sufficient indication of the general character. In judging a mountain chain or a district full of peaks and ridges we must also consider how far the highest points are apart and how deeply the notches or valleys between them are cut down. In other words, we must find in each case the common and

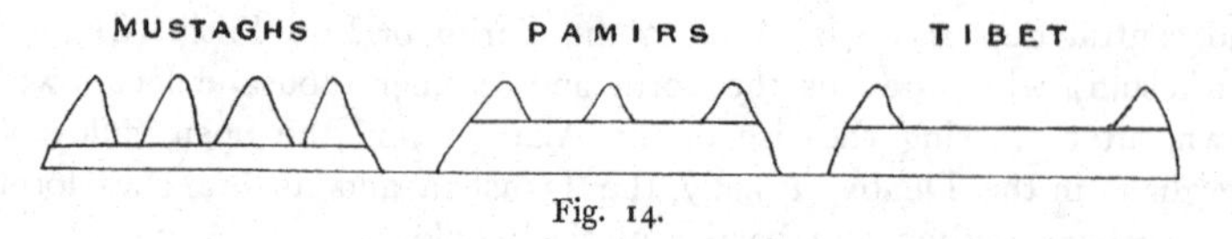

Fig. 14.

uncut platform on which the rises of the ground are built up. Likewise, in estimating the relative strength of two saws, we must also consider the size of the blades without the teeth. The common base of all land on earth is of course that imaginary floor formed by the continuation of sea level under the continents.

The diagram, Fig. 14, shows at a glance what is meant. Roughly we may assume that the average height of the crests and summits of the Pamirs, the Mustaghs and Tibet is about the same. The Mustagh valleys are, however, much deeper; the summits stand more boldly and steeply on a lower pedestal. The Pamir ridges do not rise so high above the ideal platform, while in Tibet there are even wide plains between the crests. Tibet is a plateau. The Pamirs are about the limit of what may still legitimately be called a plateau, and such I shall always call them—that is to say, the whole mass—since now the reader understands the term and will not mistake it for a table land or table mountain. I consider that plateau is a term applicable to any relatively broad feature of the earth where the inequalities are not higher above their common base than this common or high level is above the third level or base of comparison. This latter may either be a neighbouring plain or the absolute base, namely, sea level. On this last distinction it depends whether the plateau is one for the eye, by direct impression in nature, or on the map and by geographical reasoning. Nor need a plateau be very horizontal, for the name of summit, plateau or glacier plateau is given to many places in the Alps, the crossing of which requires a good deal of climbing. Thus the plateau of physiography has a wide but still very graphic sense. Every table land is a plateau but not *vice versa.*

A "pamir" and the Pamirs stand in the same relation to each other as an "alp" and the Alps. The great geographical name has been

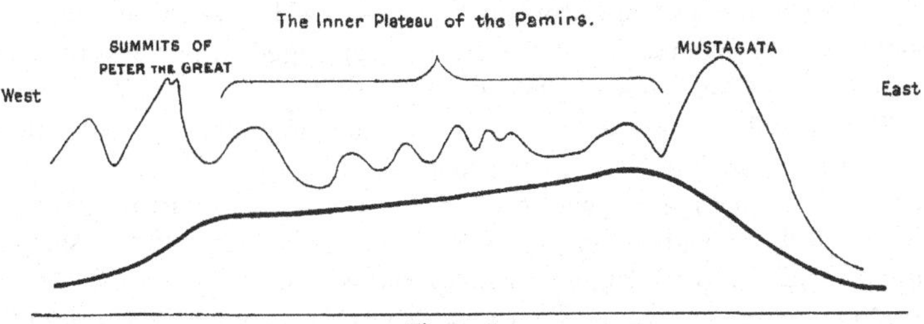

Fig. 15.

derived from that given by the inhabitants to their pasture grounds. The Pamirian plateau (see Fig. 20) is composed of a group of broad shallow troughs, treeless and desolate. Such a valley the natives call a pamir (Figs. 138, 160, 171). As the map shows, the Pamirs proper or interior Pamirs are surrounded by a semi-circular rim of higher mountains which exceed the greatest elevations of the inside by about three thousand feet (1000 m.). With the exception of Hindukush this crater-like rim nowhere forms a watershed but is breached by the rivers from the plateau. Within this arc all is genuine Pamir. But there are also a few typical pamir valleys just outside the fringe or between it and the north-south watershed, such as the Alai or Kizilsu valley, the upper reaches of the Gez river, the Tagarma plain and the Tagdumbash Pamir. These might be called the marginal or Outer Pamirs as distinct from the central mass of the Inner Pamirs.

On the whole the Pamirs are inclined from east to west and a valley running east-west will therefore reproduce the elevation or outline of the slightly tilted plateau as suggested by Fig. 15, where the heavy line represents the floor of such a valley tributary to the Amu-darya. On the other side a shorter valley leads past Mustagata and down the more sudden drop to the Tarim. The pamir character ceases where the valley assumes a steeper gradient, where the river becomes more torrential, where the surrounding peaks appear higher and wilder because we see them from a deep gorge. Here then the scenery begins to look more "alpine" as opposed to "pamirian." This contrast is intensified by the coincidence which places the highest or marginal clusters of summits alongside of the steepest gradient of the rivers, thus magnifying still more the relative height, as shown in the diagram. Owing to shortness of the drop on the eastern side there would be no room for real pamirs if they were not arranged transversally, as exemplified by the Tagdumbash and others.

The traveller, after passing the plains and foothills of the Duab, meets with great difficulties in the long, deep valleys which guide him eastward. Sometimes he has to wend his way through narrow gorges where hardly enough room is left for passing between the rushing cataract and the overhanging rock. But once he has overcome this world of fissures and turmoil, once he has stepped upon the threshold of the Pamirs, his troubles are ended, for a haven of peaceful surface receives him. Nobody can properly appreciate the Pamirs who has not ridden through the weary canyons or over the cruel stone and glacier passes of the Asiatic Alps. After surmounting the last and steepest ascent of the road—which may be a good many miles—we gradually emerge upon a broad plain-like valley the sides of which gently slope towards the lines of peaks which, although bold and snow-capped, accompany us at respectful distance on either side.

The bottoms of these flat valleys or shallow troughs are between 12,000 and 14,000 feet above sea level, while the dividing ridges are from 3000 to 6000 feet higher. Small lakes are formed by nearly every Pamir river, but the largest of all, Great Karakul (12,400 ft.; 3780 m.) has no outlet. The presence of this salt lake is but a further indication of the general plateau character of the Pamirs.

If we keep in mind that the eastern section of the Alai is the ultimate northern rim of the Pamirs—the Alai valley being one of the outer Pamirs—then the mountain ranges of the Duab present themselves as a fraying and drawing out of the upper plateau towards the west. The rivers show it clearly enough. All the right affluents of the Oxus, including the Zarafshan, are a pouring out from the inside of the crescent, whereas the flank of the Alai chains can only develop a few short streams intended for left-hand tributaries of the Sir-darya.

The Duab side of the Amu-darya is, of course, only one-half of its basin or drainage area, but it is the more important half as regards the

16. The Upper Ten and the Populace; Types of Karategin.

quantity of water. The surface of the complement, namely, northern Afghanistan and Transcaspia, is much larger but of its rivers five do not reach the main channel, a fate which is only shared by two on the left bank, one of them being the Zarafshan, which by itself does almost more useful work for humanity before it dies than all the Afghan affluents of the Oxus taken together. The reason for this is the general tendency of draining towards the south-west which we observe in the western fan-tails alike of the Tianshan, the Alai and Hindukush.

The Herirud or Tejen, the Oxus and the Sir-darya also share the same peculiar knack of turning north round the ends of their mountains. The Murgab river, intended for a left affluent of the Oxus, was the mother of the once famous oasis of Merv.

On the other side we have one half of the Sir-darya system. But this river is already strong before entering our territory. I dismiss it here with a few words because its landscapes are the same as those of the Oxus. Ferghana has nothing to distinguish it from the oases of the Zarafshan, being inhabited by the same people and subject to the same methods of cultivation, which are those of the whole of Middle Asia. The Narin or original river of the Sir rises in the Tianshan, and although thus partaking of another climatic region, its headwaters of the left or south side have pamirs of their own in the shape of the Aksai plateau, lately visited and described by Carruthers.

The common goal of the two historic streams is Lake Aral, a sheet of salt water covering 25,000 square miles (65,000 sq. km.). Its surface is 160 feet (48 m.) above sea level and its greatest depth 223 feet (68 m.).

From this landlocked sea to the foothills the land between the rivers is occupied by the sandy desert Kizilkum which thus claims nearly one half of the whole surface of the Duab. Some of it, but little, is steppe, and where the steppe skirts the mountains, which it generally does, it has been irrigated. Narrow strips of cultivated ground are also found along the banks and in the deltas of the streams which are able to traverse the arid plains. A few scattered oases are hardly worth mentioning, nor do the Kirghiz make themselves very conspicuous in this most desolate country.

It will greatly simplify the reader's general conception of the Duab if he looks upon the whole of it as one tract of desert and steppe, whether flat or mountainous, low or high, and upon fields and gardens as nothing but oases content with a small portion of the entire surface. On a liberal estimate, all patches of cultivation with villages and cities occupy a tenth of the total area. Here live more than half of the population or an average of a hundred to the square mile. This leaves ten inhabitants per square mile to the rest. But here again a certain concentration is brought about by the nomads of the steppes and the peasants of mountain valleys, so that 100,000 square miles must be satisfied with two or three little dots of humanity each, which number is further reduced to one or almost nothing over the 50,000 square miles or so of nearly absolute desert.

The term "desert" has such a vague meaning, at least in the scientific mind so prone to sweeping generalisations, that it is useless to attempt a strict definition. Nowadays so much is certain that there is no kind of ground or soil and no shape of *terra firma* enjoying a special privilege. Desert is ubiquitous, being a state rather than a shape. Absolute absence of vegetable life is the ideal condition of the desert, and unless we wish to split hairs we must concede so much to our traditional picture that this barren condition is due to drought and location in a sunny place. But that is all. As readily as one admits that the glaciers or naked rocks above the snow line and in the polar regions are deserts by poetical courtesy only, for direct names suggest themselves more easily, as strongly must one destroy the popular fancy that the desert is always a sandy plain. As a matter of fact, it is associated with any and every kind of soil, rock, plain, valley and mountain below the snow line and between 45 degrees north and south of the equator.

In the Duab, from the Pamirs down to the Sea of Aral, we find every variety: deserts of the mountains, hills and plains, deserts of rock, sand, shingle, clay and salt; for throughout this region a dry and sunny climate prevails (compare Figs. 42, 95, 196, 197). The lowlands are desiccated because the water from the glaciers cannot quench the thirst of 100,000 square miles; the slopes are bare owing to their inability to retain the rainfall.

While the idea of the desert is for ever bound up with that of the great Sahara, the original of the steppe has come to us from the Great Steppe of the Kirghiz which unfolds itself in that vast space between the salt lakes (Caspian, Aral, Balkhash) and a boundary indicated by a line from the lower Volga to the swamps of Western Siberia (Map I). Here the steppe signifies an expanse of grass land inhabited by nomads. It is luxurious under the rains of spring, but gradually withers up as summer advances. During three-fourths of the year it is dry and shrivelled, sere and yellow or greyish. The characteristic trait which has been picked out as the leading one of a steppe in the geographical sense is the patchiness of the vegetation or, in plain language, its moth-eaten and mangy aspect. The grasses and herbs grow singly or in tufts, leaving dusty earth between them (Figs. 120, 123, 124, 156, 159, 171, 188, 189).

In its typical sense the name of the steppe ought to be restricted to grass land with a vernal flora and affording some sort of pasturage.

17. A Rich Man's House and Courtyard (Karategin).

But geographers have acquired a habit of using it biologically, identifying it with any kind of straggling, poor or dry vegetation, even scattered trees being admitted into some subdivisions of the term. Thus, while the various kinds of desert are classified according to their minerals and surface formation, the addition of a new element, namely, vegetation, determines the higher order of the steppe. Hearing of a sand-steppe, we know that it must be a sandy place or perhaps a sea-shore with a thin sprinkling of weeds or shrubs; an astragalus steppe must be a colony of isolated tragacanth (milk vetch) bushes. Third in order come what may be called the forms of full life represented by the

18. Mount Akhun on the Zarafshan Glacier.

familiar trinity of forest, field and meadow (marsh). These hide the soil so completely that we describe them by their plants or groupings or their usefulness to man. The negative influence of a dry climate rules the desert; the relation of plants to the soil gives the cue to the definition of the various steppes. There is thus a genetic meaning included in these names, a suggestion of cause. Deserts reflect the killing power of the sun; steppes express the degree of hospitality offered by the soil without regard to climate. But all other vegetable associations are seen in the light of the life which the individuals lead with each other, it being left to a special explanation in each case how soil, climate or cultivation are responsible for the social form. For practical purposes the steppe may pass as an intermediate class between the desert and full vegetation. The transitions from desert to typical or standard steppe (i.e. the Kirghiz Steppe) are very numerous, while those from steppe to forest, jungle, marsh, meadow and swamp are innumerable. Practically no desert is absolutely dead and the sand steppe or desert steppe is called a desert by most people. Thus much is left to personal opinion and tact in the selection of the terms.

I have seen very typical steppe in North Germany. Some of the low Westphalian lime hills are covered with poor, dry grass and studded with broom and juniper bushes. The grass downs on the chalk heights along the English coast are practically steppes. Were it not for the frequency of rains and slow evaporation the grey-green carpet would not last. In both instances the high lying and porous soil cannot retain the atmospherical water for long, being, at the same time, out of reach of subterranean irrigation afforded by the ground water which plays such an important part as an invisible reservoir in the lowlands of temperate zones. The famous lawn of Old England was mostly steppe at the close of the great drought of 1911. The drier forms of the heath, as apart from moorland, may also be looked upon as steppes. Heather, juniper, broom, buckthorn, hard grasses, by reason of their foliage and habits of growth, are xerophilous or drought-loving plants, are makers of the steppe.

Steppes generally surround the rocky or sandy deserts. The great steppes are to the north of Kizilkum and Lake Aral, where the soil is more clayey and higher latitude affords a moderate rainfall. To the south a narrow belt is interposed between the mountains and the sands. Thus the Duab has very little of the typical, lowland pasture steppe and what there is of it occurs in small patches, the rest having been turned into fields and gardens with the help of the water from the hills. The steppe is the nearest and most natural object of irrigation, just as a marshy moorland offers itself to amelioration by drainage. But high up, close to the alpine limits of vegetation, the Duab boasts of the great pasture steppes of the Pamirs, the counterpart of the

Russo-Siberian plains and harbouring the same nomadic people, the Kirghiz.

The whole of the Duab, therefore, whether mountain or plain, slope or level, is a country of deserts, steppes and desert-steppe. Forests and perennial meadows are practically non-existent and all agriculture is an oasis. Lastly, there are the high or alpine regions with their rocks, snowfields and glaciers claiming the rest of the surface.

The eternal snows are the desert of cold, the opposite extreme to the desert of heat; but while the torrid sands are rest, stagnation, death, the towering ice is a promise of work and energy, is life to be. The "yellow belt" lies between the high latitudes and the high altitudes, shaded off by steppes and fertile ground in either direction (Map I). From the Arctic Circle cold winds keep the atmosphere alive by setting up a circulation and the warm vapours of the tropics are condensed into rain. The mountains hold fast the clouds and send down rivers to work on the land. The North Pole and the Pamirs are like two magnetic centres forcing into manifold currents the even flow of solar energy. Both overshadow the continental plain between, both have produced glaciers, moraines, climates, winds, rivers; and where their forces meet or neutralise each other we see exhaustion and dead refuse. The mighty southward influence has painted its successive areas with a broad and sweeping brush; the mountains, repeating the same process but over a shorter distance and mostly downwards, have produced quicker contrasts in their gradations. Thus the stagnating dump of the desert lies nearer to the mountain glaciers than to the Arctic ice cap, but the life stages, whether wide or narrow, are analogous and in the same order in both directions—desert, desert-steppe, steppe, cultivation and civilisation, mountain or Arctic steppe, alpine or Arctic flora, ice. It is as if the short vertical zones were projected in proportion on to the thousands of horizontal miles towards the north. From the two cold radiators life goes increasing to a maximum which reveals itself in the prosperous districts on each side. Thence a falling off towards the common minimum of the yellow death. The nearer we approach this centre, the further away we go from the beneficent shadow of the mountains, the more barren the Duab becomes. We must understand, however, that the simplicity of this arrangement, although the ultimate expression of a dominating cause, is really the result of complicated processes. The configuration of the ground gives direction to the general climate, but this in its turn reacts upon the destiny of the land. In the mountains ice is stored and there is more rain, but as the mountains become lower in the course of ages the glaciers and condensation will decrease. The further we go away from the mountains the less water, because there is less rainfall and because the rivers are used up and evaporate very quickly. To this must be added the accumulation of waste matter freed from all soil dust by the wind, or impregnated with salt which evaporation has left behind. Finally, the bare and sandy desert increases the heat of the air by violent reflection, thus further impeding condensation of vapour into rain. So one thing begets another, leading to a cumulative succession of detrimental effects. The flow from the mountains, after having brought forth fertile soil and water, bringing them to a happy union, must at last be exhausted by the manifold reactions which it has set up against itself. The perfect symmetry of the phenomena we owe to the regular build of the land and its interior drainage, to the great continentality and its simple climate.

The Duab is exceedingly dry, the precipitation (rain and snow) in the plains averaging 6 inches (145 mm.) per year, and from 10 to 15 inches (250 to 380 mm.) in the mountains. In the British Isles the rainfall is about 40 inches, but as much as 80 inches in some parts of the Cumberland and Scotch mountains; moreover it is well spread over the seasons. This dryness is productive of sharp contrasts in every way—contrasts of temperature between day and night, summer and winter, sun and shade, glacier and desert; contrasts between fertility and desolation, aridity and inundations; contrasts of colour, shape and life. In short, there is very little mugginess of weather or softness of perspective. In the plains a yearly extreme of temperature of 100 degrees is not rare, as for example between 10° F. (– 12·6° C.) in winter and 110° F. (42° C.) in summer.

Not only is there little rain but also bad distribution, much of it coming at a time when it cannot help agriculture, and next to nothing falling in summer. What water is stored in the bowels of the mountains seems to be an almost negligible quantity judging from the rarity of springs. Thus the greater portion of the water sustaining

19. Among the Yakhsu Conglomerates.

the life of wild or domesticated vegetation during summer is supplied through the intermediary of cold storage. Hence it comes that the high mountains and glaciers of the Duab assume a preponderating role in all questions connected with the quantity and activity of the water. Without them a doubling and trebling of the rainfall, the distribution remaining the same, would leave the country even more arid than it is now. We have here a great contrast between drought and a well-developed system of glaciers. Such extremes facilitate measurement and the separate estimation of the values of individual factors. Owing to their conspicuous share in the welfare of the country, as well as owing to their singular and well-defined influence, the Duab glaciers are eloquent historical witnesses to and barometers of destiny. A study of their present aspect and an inquiry into the records of their past will yield answers to many problems connected with the glacial periods and the climate or desiccation of Inner Asia.

Twenty-five years ago W. Geiger published his book on the Pamir Regions, in which he covers the same ground to which the name of Alai-Pamirs is now usually given and which is identical with the Duab mountains. Geiger's is still the best general survey of the major lines and features of this district. I have followed his classifications and the diagram of Fig. 20 is based on his map. Since then exploration has advanced in detail without, however, changing the now well-established construction of the fundamental skeleton. The whole country has been mapped by Russian military topographers on the scale of 1 : 420,000 or about 6 miles to the inch. This labour was completed towards the end of the last century and must be looked upon as a great achievement considering the promptness of its execution. Peaks and glaciers are merely hinted at and correct heights are few, but all rivers, villages, roads and important passes are shown, so that one can travel by this map, using it as a framework for detailed investigation.

Nearly all information we owe to Russian travellers whose names form a long and glorious list, especially since the conquest of Turkestan. Nothing is more exciting for the climber than a perusal of their reports, which constantly refer to long mountain chains, mysterious glaciers and an army of mighty summits of 20,000 feet (6000 m.) and more. But most of these things were seen from the valleys or easy passes, for few penetrated to these realms of ice so treacherous to the foot and eye

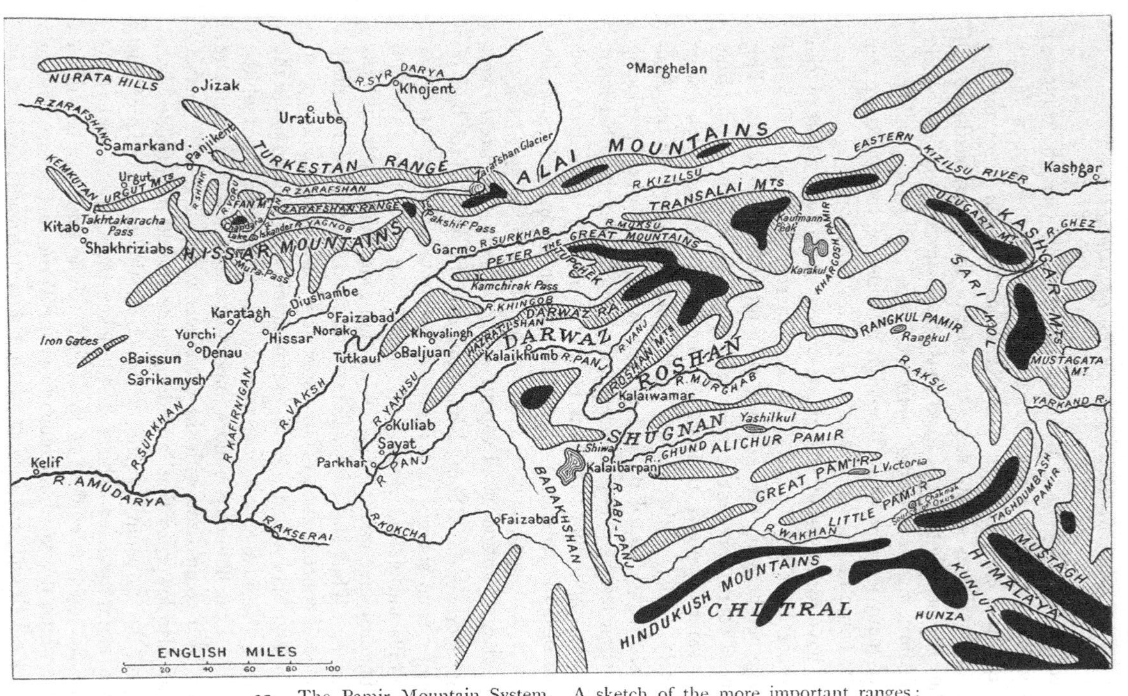

20. The Pamir Mountain System. A sketch of the more important ranges; the highest elevations are shown in black.

of the sons of vast plains. Mushketov is the only one who traversed a high glacier pass, while Sven Hedin attempted the ascent of Mustagata. But the late Albrecht von Krafft and myself were the first trained mountaineers to lay siege to some of the inner bastions. Of late the Russian botanist Lipski has done much excellent work, published in a massive volume. He has visited and photographed the snouts of many ice streams. The high altitudes of the Duab are still a treasure house for the alpinist in quest of fresh surroundings.

As I have already said, the greatest elevations form a sort of rim to the Inner Pamirs. Towards the west there is a constant decrease of average height along the ranges leading into the Duab plains. This does not, of course, prevent certain knots, such as the Fan group, from rising considerably above their general environment like watch-towers in a fortress wall. Although the earth space occupied by the Alai-Pamirs is, roughly, equal to that of the Alps, their water-producing capacity is far less as plainly shown by the rivers. On the other hand, we must remember that the Oxus is almost entirely dependent upon the favour of high mountains, that is to say, upon permanent snowfields and glaciers, and upon rain or snow falling close to the upper regions where the larger portion of all vapour is condensed. The great rivers of Middle Europe, the Rhone, Rhine and Danube, receive the bulk of their volume from sources and from the precipitation in lower levels or in the hills and plains. Important streams like the Weser and Elbe or the Thames have nothing to do with high mountains at all, still less with glaciers. Thus the dry climate of the Duab expresses itself in a very marked concentration of atmospheric water in the alpine region. Owing to their enormous height, these mountains are able to make the best use of the modicum of moisture which reaches the heart of the continent. That humidity is more important than a lower temperature is shown by a comparison with the Caucasus, which is a little further north but also lower, so that latitude is compensated by altitude; but, thanks to the proximity of the Black Sea, the ice mantle of the Caucasus is far grander, whiter and thicker than that of the Alai-Pamirs. So much is evident at first sight. The same can be said of the Alps, which do not differ much from the Western Caucasus, being three and a half degrees further north but correspondingly lower in height (Peter the Great 39° Lat. N.; Elbrus 43°; Aletsch-Bernese

Oberland 46·30°). The Eastern Alps are slightly drier than the central and western portions, but in the Caucasus this phenomenon of continental influence is illustrated in a striking manner. The snow line is much higher in the Eastern Caucasus than in the western half and the aspect of the landscape correspondingly drier. After having made due allowance for the solar climate or latitude we thus find that owing to the dryness of the atmospheric climate the Duab wears a lesser coat of eternal snow than its majestic elevation entitles it to.

21. A Castle of Mud (Denau).

The number of glaciers in the Alps is 1155 and that of the Duab, unknown at present, will not prove much inferior, judging from what I have seen and read so far. Number alone is no indication of power, but sufficient proof of the fact that the mass as a whole is well above the level of glacier formation. The total surface and volume of snow and ice cannot be calculated before the country has been mapped like Switzerland in an exact manner and on a large scale. The Zarafshan glacier is so far the longest of the Duab, measuring 14 miles or 23 km. (Aletsch 17 miles).

A substance which we meet at every step and which plays as important a part as water in the household of Middle Asia is the loess (pronounced "lus" as in 'bus or lust). It is a soil, it is *the* soil of steppe and field, of all plant life worth speaking of. The name, meaning "loose," hails from Germany, where the loess of the Rhine valley was first described by geologists. It is a friable, yellow clay containing lime and a large percentage of sand. Its characteristic quality is porosity, for it is traversed by a close network of capillaries which give a kind of spongy appearance to the mass. These fine tubes are the negatives of grass roots and to them hang the history and theory of the yellow earth. Unlike other soil, it cannot have been spread by water or formed by local disintegration. Against that speak its porosity—for wet loam becomes dense—the land snails which it contains, and the fact that it reaches far up into the mountains, not merely filling up depressions but coating hill and vale with an even layer which preserves the contour of the underlying surface. Geographers now mostly agree that loess has been sifted and deposited by wind. It is assumed that the enormous heaps of mountain refuse left behind by the glaciers of the different diluvial or ice ages furnished the necessary quantity of first material on which other agencies then set to work. Strong winds blew the fine dust out of the moraines, carrying it far away. Water helped by laying bare successive portions of the *débris* to the action of the wind and also by transporting mud and sand down to a lower level, there again offering it to the wind. Sand, being heaviest, was sorted out last.

Now it is clear that an almost impalpable dust is at the mercy of every breath of air and cannot form deposits unless held fast in some way. Here vegetation came to the rescue, or, stating the case more narrowly, the post-glacial steppe, for loess and steppe go together as readily as sand and desert. Where grass had once gained a footing it behaved like a carpet, retaining the dust that came sweeping across it. As the air-borne powder rose—say a tenth of an inch per year—the blades of grass pushed upwards or from time to time established a new storey by their seeds. Thus the vegetation grew, leaving behind its roots, which rotted in due course, thus accounting for the slender veins which traverse the matrix raised by thousands of patient generations of the steppe. Together with the impress of the plants,

22. A Loess Cliff at Samarkand.

the shells or bones of animals are found embedded in the earth on which they lived and died. Most plentiful are certain snails, while the skeletons of rodents or gazelles are less common.

That in a wild state and under a dry climate loess and steppe are almost identical terms, especially on higher ground, has been conclusively shown by the occurrence in the loess deposits of Germany of the bones of typical mammals of the steppe. From this we may conclude that a spell of drought everywhere followed immediately

23. A River Valley in the Loess.

upon the glacial period. But this dry interval did not last long in Europe, being relieved by more modern conditions, as one is led to infer from the lesser thickness of this formation represented by the belt of straggling loess hills from Hungary to France. It steadily increases towards the east until we reach China, the loess country *par excellence* where the face of earth is smothered under a yellow load sometimes many thousands of feet in thickness, cut up by vertical canyons and precipices and swarming with a population that live in it as well as on it.

In the Duab the colour and outline of the loess dominate the hills, the fringe of plains nearest the mountains and all the broader valleys. In the outer and flattest lowlands it has been washed and kneaded into clay. Here much of it is also saturated with salts dissolved out of the highlands and left behind by evaporation, thus leading to the formation of salt steppe, mineral desert or salt swamps. Wherever loess rests on a base, however slightly elevated above its surroundings, it can preserve its original porous structure, that is to say, remain original and genuine loess. Being porous, it sucks up rain water like a sponge, allowing it to drain away below. This explains why under ordinary conditions it will only support steppe which thrives under the wetting of spring but quickly withers and goes to sleep as summer advances with rapid strides. To be productive of exuberant life the yellow earth must be thoroughly soaked, a process to which the peasant devotes much labour and ingenuity. It is the sole secret of agriculture, for where loess and water are brought together under a sunny sky, they will return a thousandfold the seed entrusted to their care. Wherever loess occurs it is a giver and harbourer of life. The vernal steppe is green with succulent grass and-ablaze with lilies and tulips, while myriads of insects, reptiles and birds issue forth from cracks or holes to greet spring with revelry and love. Man, by the toil of his hands, lengthens the harvest and, moulding the earth, raises himself a shelter, humbly thankful to his yellow mother's sacred and ever fertile womb.

In the outer belt and lower level of the mountains the loess reigns supreme, impressing its character upon many thousands of square miles of landscape. It almost entirely obscures the geological history of the foothills, where outcrops of solid rock are scarcely to be found. Wherever it is disturbed it reveals its tendency to vertical cleavage by steep cliffs or narrow gorges (see Figs. 22, 23, 24, 188). Whoever has not seen loess can hardly form an adequate idea of its sway, though it may have created a sufficiently vivid impression as a geological theory. But whosoever has seen, smelt, breathed and felt loess, to him it is Middle Asia, a fragrance, a colour, a sensation; to him it is the East which he hears a-calling.

At the present time, when the bulk of the loess has already been brought together, its renewal is chiefly by a circulation comparable to that of water, though within a narrower limit, reaching neither so far up nor so low down. The grandest and widest circulation on our planet is that of the air. All others are subject to it according to the

24. Loess Canyon in the Steppe near Guzar.

weight and mutability of matter. Every succeeding and smaller return depends on those which involve and support it. Dissolved into its minutest particles, water floats in the air which tore it from the bosom of the sea. Wind shifts it as clouds and in the shape of mist it hides the view. Clay "evaporates" on the great plains. Rasped off the surface by violent gusts, it rises to the heights in the embrace of ascending whirls and storms. The drifting dust sometimes attacks us like a shower of rain, or on a hot, still day, enshrouds us like a thick and yellow fog, which slowly settles down upon the ground. Nevé is porous water; loess is spongy clay, is the solid earth-sea of the lowlands raised to a height and deposited in a lower level than the water changed to snow. But the two unite when the mountain snows melt, flowing together in a hurrying line of muddy water, a frothy, seething turmoil. Then they slow down, settling calmly and separating from each other, seeking their own level of plain and sea. Three elements they are, three states and three circulations—air, water, earth—all driven by the fourth, by the fire of the sun.

25. A Double Cutting in the Loess; formed by the Tread of Caravans.

This much I wanted to say as a general introduction to the physical aspect of the Duab and its place among the surrounding features of Asia. For this region I can truthfully claim the same distinction as that possessed by New Zealand, namely, as a collection of geographical models. "The unending variety, the striking freshness, and diagrammatic clearness of almost every conceivable geographical feature have been dwelt upon by many writers..." (*Geographical Journal*, November, 1910). Curiously enough the paper on the Waitaki basin from which these words are culled also contains three photographs that might have been taken on the Zarafshan glacier and on the Pamirs. Can there perhaps be some relationship between these distant lands, so widely apart in their disposition—the one continental, the other oceanic? Or shall we assume that it is merely a question of different causes leading to similar effects? Be that as it may, one thing is certain, that the Duab is equally a natural and plastic diagram of the fundamental laws of physiography. Some day a comparison between these antipodes may shed new light on great theories. Here and there our vision is directed towards the mountains as if an answer was expected from them, and to the Duab mountains I am pledged.

26. Autumn's Bounty.

The mountaineer whom one leads to the city gates of Bokhara the Holy will not express enthusiasm about the landscape. True, the fruit gardens are lovely, steppe and desert have their melancholy beauty, but can the pedestrian for ever dream in the shade or patiently perambulate into endless distance? The waters are sluggish and however far we may stride with dogged obstinacy our view point remains ever the same, the middle of infinity.

Whoever has tasted of the glacier spring, on him is the spell. His is a peculiar state of the mind: nature in the plains can never satisfy him for long. Are they not all beautiful—the heath, the peat moor, the seashore and the forest—full of moods and sentiment, of peace or sullen menace or tender memories? But hardly have we become conscious of deep gratitude towards them, before there awakes that old feeling which calls and drives us to the heights. On gentle planes progress is but a suffering of distance, a burthen of time; on the mountains it is work, and when work is best it is a battle. Such is the secret of ice and rock. On the level we feel an oppression, a stagnation of the blood, there being too many things near us and alongside. We strive to escape from this equality and treadmill, we aspire to freedom. The eye thirsts after the glitter of the high snows and the challenge of the jagged crest; our muscles are ever spoiling for a climb.

He who from the dirty roofs sweeps the horizon with inquiring glance may never know that south-west of him the greatest mountains of the world have sent out their furthest tentacles and that one-half of the Khanate of Bokhara is the home of mountain peasants. Below him, between the fields and roads, slimy ditches contain the last expiring drops of the Zarafshan, a river highly born from a glacier of Alai. And he thinks of his journey over the weary plains of Russia and over the interminable sands of Transcaspia. But then he enters the train once more and next morning, from the minar of Ullugbeg at Samarkand, he salaams to the vision wonderful to Hazrat Sultan on rocky throne, with icy crown. To such an one it must indeed appear that the chief end of rails is a mountain.

CHAPTER III

THE ZARAFSHAN

THE Zarafshan (Zerafshan, Zeravshan) (see Map II), the Polytimetus of the ancients, the nourisher of Sogdiana, is *the* river of the Duab. It is the Strewer of Gold, the Picture of Life, the River Symbolic.

What the Duab is for Middle Asia as a representative of type, that the Zarafshan is for the Duab, a summary of its features. Along its course the whole panorama from mountain snows to desert sands is unfolded before us. I shall therefore use it as a thread on which to string the first part of my description.

The Zarafshan is the very essence of life to Samarkand and Bokhara. Springing from the Alai mountains it runs for two hundred miles through a ravine and then for two hundred more in open country, ultimately losing itself in the plains without reaching its destination, the Oxus.

Let us issue forth from the busy streets and crowded bazars of the noble city of Bokhara. Through the massive gate we pass and through the silent graveyards where the dead lie in tombs of brick ; we walk along the shady avenues outside, where hostelries and tea-houses are filled with the din of caravans. Gradually the rows of shops and houses break up and we pass between the interminable mud walls of vast gardens, with their mulberry trees and vines. Through many villages we travel, around us the thick abundance of a fertile soil, till, at last, the clusters of dark foliage open out to the gaps of a distant view. The trees are rare and lonely in the last yellow wheat fields, the canals and runlets vanish one by one, losing themselves in swampy pools and clumps of huge reeds. Drier and barer becomes the ground ; it turns into steppe with fissures in the scorched soil and the scantiest of stunted growth. At last, after many miles, we feel a crunching under foot: sand. Here we may still discover somewhere a darker tint upon the ground, a spot of evanescent moisture. We touch it; it is a faint humidity which fades away in the burning breeze as we spread the sand upon the palm. Beyond is sand, rising in dunes, which retire into the hazy distance like an ocean of yellow waves. The last blush of moisture on the confines of utter aridity is one of the very last drops of water oozing from the last life pulse of the dying Zarafshan. The last sigh of a wonder-working slave who has given his last to make a paradise for man. Once that drop was ice among the great peaks whence the river came; that very same drop may be one of the snow flakes which perhaps a hundred years ago alighted on the highest point of the valley, on the divide of the Zarafshan Pass, fifteen thousand feet above the plain. This is the end, that was the beginning, and between them is the life-time and the work of a drop of water; between them are generations of men.

We have stood here in the desert; we have stood at the top where the ice-fall thunders, and we have gone along the line. Follow me now to watch again the progress of the water, the magic that shapes the landscape and its destiny.

The map (Map II) shows the whole course of the river with a strip of country on either side. To the village of Karakul, the last point where one can speak of an individual stream, the entire length is 400 miles (640 km.), of which one-half belongs to the mountainous section from the glacier to Panjikent. The drop over this distance is 8500 feet or 21 feet to the mile (4 m. per km.) on an average. The various gradients are distributed over the whole distance as follows:

Glacier to Panjikent 31 feet per mile.
Panjikent to Karakul 13 feet.
Glacier to Pakshif 25 feet.
Pakshif to Varziminar 30 feet.
Varziminar to Panjikent 36 feet.
Panjikent to Samarkand 20 feet.
Samarkand to Karakul 11 feet.
Frontier to Karakul 8 feet.
Bokhara plains 3 feet.
Karakul to Oxus (would be) 1½ feet.

From these figures we see that the river is very evenly graded and that the steepest part is between Varziminar and Panjikent, not in the highest reaches as most people would expect.

Any large stream rising in the mountains can generally be divided into three characteristic sections: the torrential track among the wild upper valleys where the drop usually exceeds 50 feet per mile; the valley track in the wider, open basins and rarely more than 10 feet per mile; finally the plain track of the lowlands where the descent is only a few inches to the mile[1] (Mississippi 3 in. p.m.). Thus the Zarafshan has only a valley track, if we take the figures just mentioned as the absolutes of the fully-developed, typical river. Relatively to its upper course it has a plain track which cannot, however, be so gentle as that of a stream reaching the sea because the Zarafshan is (or was) the tributary in the middle course of a larger river, the Oxus. A torrential track it does not possess, not even comparatively to the lower reaches, the whole trough being flat-bottomed and evenly graded, a condition the reasons for which we shall discuss later. Practically it has only valley and plain track. Being neither too steep nor too level the general slope of the Zarafshan is just right for irrigation.

Along this line of 400 miles which might be called a text book valley, all the typical phenomena of the country are arranged in successive order.

At Charjui the railway traveller from the west reaches the banks of the Amu-darya. Here the Oxus, spanned by a fine steel bridge, is over a mile wide, its whirling, muddy waters forming many channels between the multitude of ever shifting shoals and sandbanks. Navigation is exceedingly difficult and uncertain, the steamers from Kungrad (Khiva) to Patta-hissar taking weeks where days would suffice on the Volga or Danube. Father Oxus is said to pass 63,500 cubic feet (1800 cb.m.) of water per second under the bridge, or one sixth of what the Danube, the second largest river of Europe, discharges into the Black Sea. The length of the Amu is 1500 miles (2400 km.) and during summer, when no rain falls in the plains and very little in the mountains below the snow line, it is practically a glacier stream, an impression also conveyed by the slaty colour of its troubled waters.

[1] H. R. Mill, *The Realm of Nature.*

Being thus chiefly a son of alpine heights it does not lend itself to irrigation as readily as the Nile whose floods are more regular. Practically all irrigation is on the left bank at Charjui as well as at Khiva. This is due to the fact that there is a secular shifting of the whole river from west to east, so that it presses towards the right, digging a new bed there and leaving its left bank behind. The cause may be found in prevailing winds from the west which force the current to undercut its right side, while a few savants ascribe great importance to the law of Baer which attributes deviation to the rotation of the earth.

In very ancient times the Zarafshan joined the trunk somewhere near Charjui, but if the reports of ancient geographers have been rightly interpreted this was already the condition of things 2000 years ago. As the missing interval is not great in any case, measuring only 15 miles (24 km.) to the last branch and 20 to the largest swamp, an idea is conveyed of the long periods with which we have to reckon even within the modern age of geological history. Though we know nothing more, we know at least this much that the utmost possible maximum is these 20 miles in 2000 years or a mile per century. This is the maximum rate at which the desert has encroached upon the cultivated land and at which the oases shrink like drying drops on a hearth-plate.

Before reaching Karakul the railway crosses 12 miles of sand hills. This narrow strip is the Sundukli Desert which connects the Kizilkum advancing from the north and the Transcaspian Karakum crossing the Oxus in a broad front. Here then the forces of the enemy have joined hands by breaking through the last line of defences. They have already surrounded and are beginning to cut off the isolated outpost of Karakul, the next to fall before the slow but relentless onslaught. The demands of irrigation alone cannot be made solely responsible for the Zarafshan's sad fate. In winter especially there would be quite enough water left to form a brave little stream fighting its way to Abraham's bosom. But on top of exhaustion from overwork came the westerly winds blowing a heavy load of sand right into its teeth. Thus the Zarafshan succumbed, scattering itself in the vain attempt at finding a loophole through the creeping fiend.

Here we have a good opportunity of watching the forms of the wandering sand, for sand moves by waves or hills commonly called dunes. In Europe these are best developed along the shores of the Baltic and on the coast of Brittany where they will gain from 10 to 20 feet in a year unless kept within bounds by defensive works. In the dry climate of the Duab the pure, yellow sand is nearly always able to assume the ideal shape of the travelling dune, namely that of the barkhan or crescent shaped sand hill. As the word already betrays, it is a half-moon. Its open side with the tapering horns and steep inner wall reminds one of part of a volcanic crater. It is the lee-side. To windward there is a long gentle slope, up which the quartz grains are drifted, presently to fall over the edge of the leeward precipice. Thus the barkhan is continually built forward, being taken to pieces at the back and added to in front. The same happens to the other billows that in serried ranks heave through vastness like a fallow host.

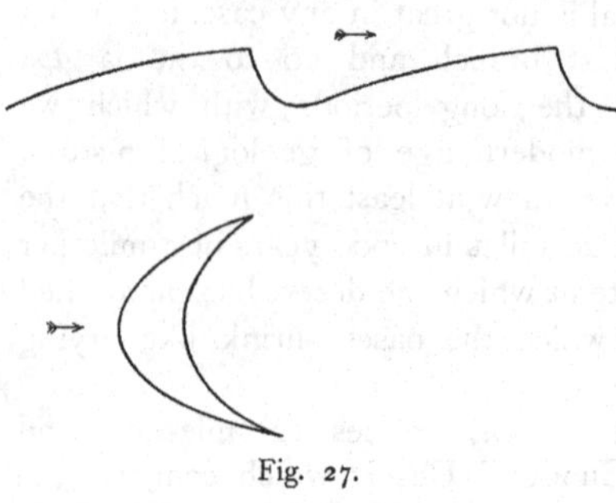
Fig. 27.

As each moves forward it robs the rear one of support, thus maintaining the steepness of the lee-side and saving the trough from being filled up. As shown by the ripples on sand, on snow, on the sea-shore and on water, wind beats with rhythmic vibration upon the yielding surface. Engendered by the rebounds of first impact and the inequalities of the ground this motion is made stronger and deeper by the mounds which it has raised itself or found before. Like a gathering swing this action and reaction grow apace in the sifted substance spread evenly over the open plain. This continues until the average maximum size of dunes is reached, locally determined by gravity and other factors. Thus the air heaps up the waves of sand—the negatives so to speak—which in their turn suck in and throw out the wind-waves guiding their energy into longer undulation. In this fashion atmosphere and lithosphere excite each other, swaying to and fro, fitting crest into hollow, curve into counter-curve.

Were the storms to blow from one quarter only the golden ocean would ride as fast as plague and locusts. But they change with the seasons and gain exceeds loss by a few feet in the direction of the most persistent winds. Thus the desert as a whole travels a short distance each year in the direction of the prevailing air currents. When a powerful river like the Amu-darya is to be crossed only a little sand is lifted over by the most violent blasts of a gale, the rest being swept away by the river into which it fell. A like fate awaits that portion which has safely reached the opposite bank but is partly driven back again.

The map also shows a salt lake, Dengiz-kul, 15 miles from the Oxus. This may be looked upon as the ultimate end of the Zarafshan, although like Makhan-kul and the swamps of the Karakul delta it is merely one of the many terminations which are only fed during late autumn, winter and early spring by the overflow of water not then required for cultivation.

At Farab which is our first railway station on the right bank the Russian government has established a very interesting nursery-garden for plants of the steppe and desert. In the early days enormous expense was caused by the sand drifts which continually overwhelmed the rails. The only durable preventive is a broad belt of vegetation on either side of the line. Experiments were made at Farab with local as well as foreign species in order to find out the shrubs and grasses which would grow quickest and settle down most firmly. After long and patient researches success was attained and the steel-road protected. A strip of hardy weeds and bushes was reared on either side of the line, while draconic laws punish all damage done to this "horizontal hedge" by man or straying cattle. Some stretches have still to be provided with the living belt; the existing one has to be widened and occasional gaps must be filled in. The Farab nursery is therefore kept at work raising seedlings to be planted out on the restless flanks. There are a great number of plants that will grow well on the dryest sand, forming the thatched desert or sand-steppe (Figs. 42, 43) wherever they have obtained a hold. The difficulty, of course, consists in the first footing which, once securely established, condemns the barkhans to comparative quiescence. As we shall see later on, during our visit to Makhan-kul (Chap. IV), the desert nearest the swamps and fields of the Zarafshan is mostly covered with vegetation, which explains why progress is slow here although not entirely suppressed. The worst barkhans are in the open, and those menacing the railway are chiefly in the great Karakum of Transcaspia, between Merv and Charjui.

The plants of the steppe and sand have only a short season, namely, spring, for summer to them is as useless as winter. They are protected against drought and evaporation. Excessive transpiration is checked by a reduction of foliage, the leaves being turned into almost invisible scales which lie close to the stem, or a large portion of the leaves are changed into thorns which at the same time afford protection against

28. A Karavansarai in the Steppe (Yakasarai).

animals. Others again have hairy or woolly coats as a shield from sun and air. Many of them withdraw their vegetative parts and a store of nourishment into bulbs or thick roots which lie dormant during the dry season and winter. Therefore bulbous flowers, having a ready fund of building material, sprout with amazing rapidity, gaily carpeting the ground in a few days, thus being able to make the best of a short season. In other countries, such as America, this kind of vegetation makes a speciality of succulent stems or leaves protected by tough skin and acting like water reservoirs. To these belong the cacti and stone-crops. Another group of frugal vegetation is called halophytes or salt-loving plants which also play a great part in the Duab, and which, as the name implies, are able to thrive upon soil impregnated with more alkali, sodium, and magnesium chlorides than ordinary herbs can bear.

One of the most notable families of the brave colonists of arid soil is the genus *Astragalus* (tragacanth, milk vetch), of which no less than 1200 different species people the steppes and semi-deserts of the temperate zones, mostly in the Old World. Besides them the following are most frequently found in dry places of the plains and mountains, for as I have said before, the conditions of the steppe and desert are everywhere and only increasing cold sets a limit to the height at which lowland species can be found: *Artemisia* (wormwood), whereof one kind in the plains supplies santonin, the remedy against tapeworm, and another, in the Pamirs, provides the Kirghiz with fuel; *Ephedra*, *Tamarix*, *Eleagnus*, and many varieties of buckthorns, hawthorns, barberries, caper shrubs, and many other tough or thorny fighters with hard or leathery limbs. Of grasses *Stipa* is well to the front, even near the snow line. Rhubarb and leeks also climb high. A well-known legume of the plains is *Alhagi camelorum*, the camel thorn, which exudes a sort of manna. One of the species useful to man is the Saxaul (*Haloxylon ammodendron*) which only grows in sand and may almost be called a tree. Its trunk takes a hundred years before attaining the thickness of a man's leg, and provides very heavy and brittle firewood which burns with a hot glow, leaving but few ashes.

Proceeding further on our journey, we enter the oasis of Karakul. Here, for the last time, the Zarafshan has been drained into many channels conducting water to fields and gardens. But the district is not purely cultivated land like the plains of Bokhara and Samarkand, as the lesser number of villages already indicates. Swamps, steppe and sand hills share a large part of the surface, which is thus marked out as a mixture of reclaimed land and wilderness, or, rather, land being reclaimed by the desert.

While the large centres of dense cultivation seamed with innumerable channels and ditches may be called intermediate or artificial deltas, the fan of Karakul may be defined as a lost or landlocked delta. A delta forms in a lake or sea because the current of the entering river is checked. The mudflats built up of the sediment which is dropped in the still water have so little slope that the stream is obliged to spread out fan-like into many branches or distributaries. Obeying the same fundamental law, a stream must split up when its force is weakened by a diminution of its water and by an opposing army of sand hills blocking the way in every direction. Nor could very much be done with this last blood of the Zarafshan come to the end of a life pulse of healthy circulation. I imagine that by now, after having passed Samarkand and Bokhara, the water must be pretty well "salted up." This salinity is not noticeable to our taste, but one cannot allow it to become stronger. It is a weak solution which can still keep the salts moving in suspense, and which must be passed on before getting so concentrated as to form a deposit on the slightest provocation. Irrigation requires plenty of water owing to great losses by evaporation, and because part of it must be kept flowing, must be passed *through* the fields, for if it were always left to the dregs it would but clog the soil with an accumulation of noxious minerals. When new steppe of the lowlands is reclaimed the action of the first water brought on to it consists in washing out the bad salts. Subsequently their accumulation must be prevented. Therefore irrigation is always coupled with an irreducible minimum of drainage. Stagnation is death to all things, or, expressing the case more correctly, every kind of life, organic or inorganic, every physical condition on earth depends on some sort of circulation for the maintenance of its individuality, character, existence. When the circulation upon which it depends sinks below a certain minimum another form of life or surface steps in. Thus our food plants are superseded by salt plants or sand

29. A Forsaken Monument.

grasses or reeds; fields by steppe, desert and marshes, when the circulation of air and water and the rotation of the soil fall below or exceed a certain limit. Likewise the barkhans cannot live without wind, and the formation of any special kind of mud at the bottom of the ocean will be interrupted if a change takes place on land, in the currents or in the animal life of the sea. The salt lakes, salt steppes, swamps and sand deserts are evidences of the great stagnation of those geographical forces and turnovers which at the present time are favourable to agriculture and civilisation.

Presently we come to a neck which separates Karakul from the oasis of Bokhara. A close scrutiny of the map reveals on each side of the railway a few of those contour lines which are used to indicate differences of level. Within a wide plain—if we except the rarer instance of a depression below the general surface—they must necessarily mean greater height, a rise of the ground. Here, then, we have the edge of a plateau, only a few feet high but sufficient to prevent the adduction of water from the Zarafshan by gravity alone. The bucket-wheel or chigir is not much in vogue in this particular neighbourhood, probably because there is only just enough water to irrigate the lowest districts. This higher ground then prevented cultivation, a fact of which the desert took advantage by closing in upon the neck on both sides. To the north a valley is sketched out by the opposition of two contour lines, and a lost branch of the river coming from Yakatut proves this shallow trench in the low plateau to be a "wadi" or dry bed, only used during certain seasons of abundance by the Gujeli-darya. These flat shelves are tables of clay and loess-clay, getting lower and lower as we recede from the mountains. Into them the rivers old and new, periodic or permanent, have cut valleys, sometimes narrow but generally very wide, to which the sharp, vertical edges stand in the relation of banks. As we approach the foothills these flat tablelands are more and more composed of true loess, forming cliffs along the rim that overlooks the bed of the stream (Fig. 23). Such elevated blocks are appropriately named high steppe, of which a tongue projects near the station of Kuyu-mazar further up-stream, and which, towards the middle and as it is continued southwards, changes into hill steppe, where the loess mantle repeats the ups and downs of the underlying country, instead of acting as a shelf-like fringe.

The reason is, of course, that the loess cannot bury big hills under a complete level, but only drown their lowest ribs and bights in an equalising plane. The peninsula of high steppe just mentioned rises to 1800 feet towards the centre where a ridge lies hidden, one of the last spurs of the Kemkutan Mountains near Samarkand.

Near Yakatut the Zarafshan sends off two branches in a northerly direction. Of the one we have already spoken, while the other goes to the swamps and reed-beds of Makhan-kul, where also a few villages manage to keep alive. I visited this small oasis, in all respects the same as that of Karakul, and shall describe it in the fourth chapter.

Of the fertile garden of Bokhara only a portion found room on the map, but it is by far the larger portion. Its shape is that of an elongated parallelogram or lozenge, to which the main course of the river serves as a diagonal. By imagining the small missing triangle added at the top, one can easily reconstruct the whole.

It is here that the Zarafshan finally leaves behind the last faint antennae of the mountains. While the eastern side is bounded by the first rising of the ground towards greater and higher things, the westerly sides (S.W. to N.W.) are hemmed in by desert sands. The railway keeps far to the left, taking advantage of drier ground in order to avoid the wet maze of ditches and inundated fields. On the same side there is also a row of marshes and small lakes skirting the edge. This is where the irrigation water comes to a stop because it cannot run uphill. Here, then, the mighty Pamirs still cast the shadow of their grace upon the plains, bidding the desert halt and verdure to spring forth. The sands are torpid, the steppe is the restless shift of wandering cattle, but here is concentrated life swarming like a million ants.

From time immemorial humanity has employed the expedient of bringing water to dry land or retaining it there. The Moors in Spain were famous for their irrigation works and no country in the world from ancient Greece and Rome and Upper Italy to India, California, or even England or Switzerland is without some method of supplying water to a place which has none or not enough. But the classical example is Egypt with the flood of the Nile. Although there is no strict limit we can distinguish between irrigation in the narrower sense of the word—on steeper land, especially near mountains—and submersion. The great inundation of the Nile, owing to its regularity and duration, favours the latter method. By a system of dams, rather than canals, the work of countless generations of toilers, the sheet of the vast inundation is divided into fields. Overseers look to it that everybody gets his just share, after which the gap is closed and the water left standing on the allotment. This process is repeated as long as the flood lasts. Naturally, when the flood level begins to sink again the higher land and that furthest away from the Nile will be left dry first and the sooner this happens the more need there is for irrigation, which now steps in. Canals take the water to such fields as must have moisture during the rest of the season and a good deal is even lifted to higher terraces by water-wheels and modern pumps. The chief value of the Nile flood lies in the mud which it deposits on the fields, thus manuring and rejuvenating the soil.

The Duab system as a whole is only irrigation, though of course local inundations, especially of rice-fields, are produced by filling a plot completely. In such close neighbourhood to the mountains the slope of the surface, a relatively small quantity of water, and its spasmodic supply, do not favour the submersion of huge tracts under a shallow and even sheet. Canals, called "ariks," conduct and distribute the precious liquid wherever possible. The great rivers of Middle Asia cannot be used in the same way as the Nile. The Oxus has very irregular floods, being too much of a glacier stream, nor can these regions boast of dams and canals of such magnitude as those of Egypt, which are the heritage of a grander and more stable civilisation. Likewise, the Duab rivers are quenchers of thirst rather than fertilisers. Their deposit, still fresh from the rock, is more sandy or slaty, whereas the Nile carries a fine, slimy ooze. To the fields of the Duab manure is chiefly brought by wind in the shape of loess, or the peasant sometimes throws new earth on the impoverished soil. If an uncultivated piece of high and dry land does not happen to be near he takes a little fresh loess from the road or maybe uses the ruins of a clay-built house or garden wall, for he needs not much, the yellow dust being exceedingly fertile.

The lower Zarafshan valley cannot compete in richness with Lower Egypt from which it differs nine degrees in latitude. Constant irrigation and a mild climate ensure to the Nile delta an almost perennial agriculture, often yielding three crops during the year.

In Bokhara the Zarafshan is tapped by 43 principal ariks, whereof Shakhrud, the main feeder of the capital and its surroundings, is the most important. Their combined length is estimated at 600 miles (1000 km.). What we see on the map are only some of the larger canals, whereas the medium ditches and small conduits must measure at least 10,000 miles. Shakhrud, which looks in plan exactly like a plume, shows the greatest density of its feather-beard in the outskirts of the city, where it grooves the country with close, parallel furrows. The near town is the best market for all produce, while on the other hand the city magnates wish to have their pleasure gardens within easy distance. Nor can it be due to accident that the thickest bundle of arteries marks the situation of Shirbudun, the Amir's summer palace, whereby one is moved to meditate upon the influence of rank on physiography.

As water is an absolute necessity without which nothing could exist, neither capital nor labour, neither master nor slave, a just distribution is more or less ensured, and the irrigation service the least corrupted of all departments of the State, which is saying a good deal. Being the circulating wealth of the country, the flowing water is endowed with something like the sacredness of coinage protected by ancient laws and regulations. The officials and overseers are perforce the most honest, for apart from a certain awe and responsibility which must fill the dispensers of the holy life-blood, they are also the most closely watched by all and sundry members of the community. Just as the legendary Scotchman keeps count of every farthing not only in his own pocket but in everybody else's, so the Duab peasant weighs every drop of water in his own and his neighbour's field.

Great care is taken in suiting irrigation to a particular kind of crop and the hardest work of the farmer is that connected with the watering of his plantations. Rice must be half drowned in a muddy lake while a beautiful ornamental pattern of labyrinthine design represents the water-furrow winding in and out among the little conical heaps of a melon field. In some places, especially near Tashkent, high-lying ground is irrigated with the help of the chigir or water-wheel, which is exactly the same as that used in Egypt.

Of cereals the most important are wheat, rice (only in hot lowlands), jugara (i.e. sorghum), millet and barley, which latter climbs highest into the mountains. Rye and oats are almost unknown, except where grown at the instigation of Russians and for their consumption. Horses are fed on barley. While various peas and beans are grown, the most conspicuous vegetables are cucumbers and carrots. The yellow roots are among the most visible ingredients of the pillau or rice dish. Cucumbers, melons and arbuses (watermelons)—why the one should be called a vegetable and the other a fruit is a mystery of human mentation, at least from the Oriental's point of view—form the overwhelming majority of raw foods. Next in order comes the abundance of grapes, whereof a large part is, however, dried and consumed as raisins or exported. Onions are of course a *sine quâ non* of Eastern existence. Of spices and restricted plantations I might mention red pepper, capers, caraway, sesam (oil), poppy (opium), and various odds and ends. Sunflowers are often grown and chiefly for the commoner of the Russians among whom the eating or rather the chewing of the seeds and subsequent spitting out of the husk is a favourite pastime. This I can quite well imagine, for the balancing of the thin, hard seed within the cavity of the mouth, the paring of the dry husk with the edge of the teeth, and its subsequent ejection must be a matter of considerable difficulty and patient practice, the essence of all true sport. Wherever a party of Russian peasants have been sitting for some time, at railway stations or on landing stages, the ground is sure to be littered with the glistening husks. If our symbolists only knew to what base use their emblem-flower is here put!

The best fruit is the grape, of which the natives distinguish a great number of varieties. Muhammadans use them for eating only, but the Russians are now making a fair amount of strong wine out of them.

Amongst other fruit we find apricots, peaches, apples, figs, almonds and pistachios. But apples and peaches are generally too watery and insipid for our taste. Apricots are very good and aromatic in the mountain valleys. Mulberries are of course indispensable for the silkworm. Flax and hemp are also cultivated but not for the sake of the filament, all textiles of the country (unless imported) being made of cotton, silk or wool. Even ropes are made of hair. Flax is used for oil, and hemp for a narcotic, that is to say, hashish or bhang. Cotton, of which several American sorts have been introduced with success, is perhaps the most important product of the region, and Russia is already in a position to cover a third of her demand within the confines of the empire. For filling his teacup, the native has to look to India, but the tobacco for his beloved chilim (water-pipe) can be grown at home. Lastly, lucerne deserves some notice, for besides being a nourishing food for cattle, its seeds form a valuable article of export.

There is no forestry in a land of steppes, and much building wood is now brought by rail after having floated down the Volga and being taken thence by steamers to Krasnovodsk. Locally timber is raised in the gardens from poplars planted in large quantities. This tree grows very fast and is a real boon in these poorly wooded regions. Plantations of young poplars are a familiar sight, their closely serried, slender stems reminding one of bamboos.

Everything grows out of the loess, even houses, and the simile of man's origin from earth is here understood to its full import. Owing to its richness in lime, loess mud hardens very well and walls keep a long time, especially in this climate.

Wherever there is loess there is its product—the Sart. Just as the houses, the men and traditions of Scotland have sprung from the hard, grey granite of the North, so the Sart and his character have risen from the yellow clay. He thrives where the sun shines and the water flows, but his progress and destiny are shaped by a few strong men or conquerors; he is a deposit moulded by active currents. His energy never goes beyond the mere upkeep of a life which allows him as many idle hours as possible, a philosophy in many respects wiser than the mad rush of the Occident. Take away the hand of a great ruler—the main ariks will run dry, the public buildings crumble, the Sart and his work return to what they were—dry mud. His wants are few and the envy of every explorer. Give him a horse, a pair of saddle-bags, a five-pound note, and he will travel two thousand miles in six months. In his baggage is a bed-quilt, a teapot and a chilim. He has two or three chapans (top coats in the shape of dressing gowns), which enable him to adapt himself to every temperature and which serve as coverlets at night. The chalma (turban) is his pillow and innumerable are the uses of the square cotton cloth or cummerbund which he wears around his waist; it is belt, purse, pocket, napkin, handkerchief, towel, table cloth, horse halter, prayer carpet, rope, or anything ingenuity and the necessity of the moment may suggest. Thus equipped, and with his wonderful talent for make-shifts, he is able to face any and every emergency, and let me say that there is no difficulty in this world which a Sart cannot overcome by waiting, if all else fails. If there be no bridge, he will wait till there is one; if life is a burden to him he waits till it is over. He can work when he must; he can work very hard and then rest with a vengeance; he can work permanently when it means sitting down and giving orders, for a Sart in power is a great oppressor. He really loves honestly, and for ever, one kind of work, namely, that which others do for him. I cannot help feeling, however, that this is but a sliding scale which fits humanity all over the world and that the differences between the races are merely slight shiftings of the scale along the basis of fundamental man.

Such are the people who live in the cities and villages, in the fruit gardens, vineyards and rice fields of the plains. They live on the bounty of the Zarafshan which has worked hard to collect the water of life in the mountains which are our goal.

Kermine (accent on the last letter) marks the situation of another neck or narrowing of the flood-plain of the Zarafshan. Here begins what is ordinarily called a valley, bounded on both sides by visible hills. Thus we can distinguish three sharply marked sections along the river's course—the narrow valley of the high mountains which ends fairly abruptly just below Panjikent; the basin of Samarkand sunk into the foothills; and the plains of Bokhara, which begin below Kermine. We may call them the upper, middle and lower course, respectively, separated by two distinct steps. The general slope of the middle course, the basin of Samarkand, is 20 feet or equal to the average grade of the river from source to end.

Soon after passing the environs of Bokhara the railway climbs a shelf of high steppe, the Malik-chul, as can be seen by the difference of level between the station and the city of Kermine. Beyond Katta-kurgan the line takes advantage of a dry nullah in this plateau. To the north the Samarkand basin is bounded by the Nurata and Aktau hills, the last outliers of the Turkestan range, while to the south are the Kemkutan mountains, a low branch of the Hissar range, called Hazrat-sultan in its western portion.

Judging from the appearance of the map, the district of Samarkand seems to be less densely populated than that of Bokhara, for the number of dots representing villages is inferior. But there may be other reasons for this discrepancy. Either the hamlets round Bokhara are smaller or the map is better, that particular section of the Russian survey being more recent by ten years. It would therefore not be safe

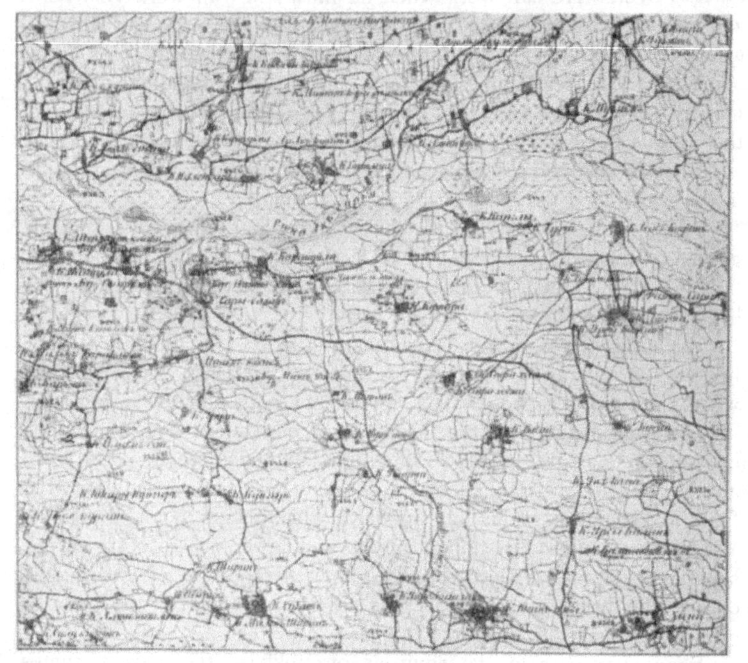

30. Part of the Samarkand Basin.

to indulge in conclusions as to greater fertility and the like. As the Bokharan authorities do not issue statistics, I have not been able to settle this question.

The small map of Fig. 30 represents a square inch of the Samarkand Valley cut from the larger sheet of the "Zarafshan from End to End." This little square, taken out near Mitan, shows the irrigation system over a surface of 34 square miles (88 sq. km.). The river is the

Ak-darya branch of the Zarafshan and there are 58 kishlaks (villages and hamlets). The heights are given in sazhens (fathoms) of 7 feet.

This rectangular cutting from the great oasis would contain about 12,000 inhabitants, which makes 350 for each of the 34 square miles (135 per sq. km.) included. This is of course an average, the figures rising to tenfold and more in and near the towns. As we must reckon that at least 2,000,000 people live in this area (Bokhara, Samarkand

31. A typical Village View in the watered Plains.

and Ferghana) it means that only 5700 square miles of the Duab (one fortieth of the entire surface) are thickly settled. In Belgium, which is an industrial country where crowding has practically no limits, the inhabitants number about 600 per square mile (230 per sq. km.).

Such an irrigated plain is a perfect maze of ariks and ditches, roads and paths crossing each other in every direction. In between are fields, groups of trees, villages, also bits of swamp and untilled steppe. Finding one's way is extremely difficult, even more so than on a glacier or among desert dunes, for steering by the compass is impossible owing to the multitude of irritating obstacles. Walking is out of the question if one leaves the main road and even then one must expect wet feet. During spring, when the water is highest, riding is nearly synonymous to wading when one tries to reach a village somewhere in the middle. The path runs hither and thither among the square plots, generally on top of the narrow ribs of earth or dikes which retain the water within the fields. The stumbling blocks of the smallest ariks are a trial by their number of about a hundred to the mile, while the medium ones afford surprising revelations as to depth. As there are no bridges the horse will sometimes be up to the shoulder in muddy water while the hind feet are still on the bank, his tail pointing heavenwards. Cuttings lead down to the bottom of the biggest ariks, but as many riders are continually crossing with dripping animals the descent to the ford is greasy and slippery. For a distance one travels through dust a foot deep and then again the road is turned into a lake by an escape of water from the fields. During summer these plains are sultry and decidedly unhealthy, forming a delightful breeding place for the malarial mosquito.

Occasionally one comes to a bit of higher ground or a kurgan (hillock) which has remained steppe, reminding one of the natural condition of the ground before the advent of man. But the river has also undergone a change from its wild state. What was once the flood-plain for its tempestuous and rambling currents during spring, when it blindly careered in the space bounded by hill slopes and the steep rim of the steppe—that is now a regulated system where small, local floodings will happen but where no dangerous inundation can any longer take place.

At Samarkand the Zarafshan splits into two arms—the Ak-darya and Kara-darya. They are natural river beds but their contents are dealt out by means of a barrage at Samarkand. The Dargom on the other hand is mainly artificial. It begins at Rakhmetabad below Panjikent, gradually cutting deeper and deeper into the steppe shelf until at Dargom Bridge (Map II) it runs in a canyon with vertical loess cliffs. How the position of dwellings is determined by water is very strikingly shown in some parts where villages are strung upon an arik as pearls upon a thread. The finely dotted halo surrounding the centre of Samarkand indicates the extent of interior garden country or suburbs. Within this circle there are no fields and no open view, the roads passing in between high walls behind which rise the poplars and elms. Their leafy tops flowing together form the ocean of thick verdure that sighs when the breezes play round the tomb of Tamerlane. Here all is one garden divided by mud walls into large and small properties reached by shady lanes (Fig. 32).

Every village has its name but many a one occurs again and again all over the Duab. Such are Juma, Denau, Agalik, Sari-assia, Mitan, Peishambe, and many others. They are general words recalling some peculiarity of situation or the first settler's happy thought, or may be that of his first caller.

32. In a Garden at Samarkand. Karagach and Poplar Trees.

The Zarafshan being a longitudinal river which has gathered strength between two parallel mountain chains, the irrigation system of its prime is like the network of arteries and capillaries of the human body, more especially as many of the veins flow together again. For the sake of comparison I have also drawn a plan which resembles the roots of a tree (Fig. 34). The Sokh river leaves the Turkestan (Alai) range at right angles intent upon joining the Jaxartes. Coming down such a short valley, it cannot have collected much water, which is at once tapped and used up as it issues from the mountain gate. Distribution is made easy by the fact that the Sokh has built out into the plains a vast, gently sloping cone or fan. Owing to this formation of the ground, the las water can drain away instead of forming swamps as in the flat delta of Karakul.

The Zarafshan below Panjikent feeds about a million people. If the Thames suddenly stopped flowing it would certainly be a tremendous calamity but one which can be borne, for nobody need perish. Were the Zarafshan to be engulfed by the earth the million must either die on their land or migrate, even under modern conditions. The railway might be able to keep them from starvation but not from thirst. The bulk of the population would throw itself into Ferghana, thus causing a general convulsion of the whole of Middle Asia from the Oxus to Siberia and Eastern Turkestan. The Russian troops might regulate this irresistible tide to a certain degree but stop it they would not, unless wishing to kill the natives outright. Even under the most

33. By the Tank in the Palace Yard at Diushambe.

favourable assumption some hundred thousand would succumb to starvation, thirst and murder before being supplied with food or reaching new land. Thus we can realise to some extent how the drying up of the Tarim basin may have been the cause of the migrations of Aryan and Mongol peoples. Of course the events leading to these ethnic tides have not been so catastrophic as just depicted for the sake of argument. But we can easily imagine a state of things gradually growing worse and worse until a climax was reached and a wave of unbearable unrest overwhelmed a region already crowded to suffocation. Then suddenly an electrical resolve shook the whole mass of the people, causing them to gather up their movable property in search of new homes.

We now come to the upper course of our river. The Alai mountains (see Fig. 20) split up into two chains called the Turkestan and Hissar ranges, and within their fork the Zarafshan Glacier is situated. The Hissar range again divides soon after the Pakshif pass by sending out a northerly branch known as the Zarafshan ridge. Between this and the Turkestan chain the famous river flows as in

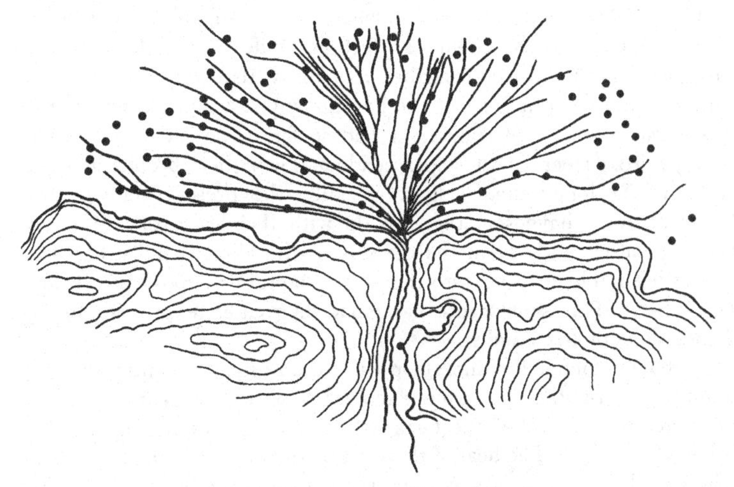

34. Sokh River entering Plains of Ferghana.

a trough of stone, giving what is perhaps the best example of a herring-bone valley in the whole world. Owing to this symmetry the Zarafshan possesses no proper torrential track distinguished by the antics of cataracts and waterfalls. From both sides short tributaries unite simultaneously so that we cannot follow any main feeder into steeper gullies. The great glacier, also formed at once by lateral affluents, has scooped out a broad valley during the ice age, afterwards providing an enormous quantity of morainic *débris* which were spread the length of the river. Thus the Zarafshan obtained a very even grade from end to end. Unlike streams of the branching tree system this herring-bone is very straight, the river going from start to finish almost as the crow flies. Hence, the effects of erosion and accumulation are concentrated along a shorter distance. All affluents support the main river singly instead of joining and helping each other from step to step. The only exception is the Fan river which debouches at Varziminar after having amassed a volume of water equal to that of the Zarafshan at the junction. This explains why the average fall is steepest after this point—36 feet per mile from Varziminar to Panjikent—the sudden increase of energy having led to a deeper erosion of the valley.

The gentlest grade of the Zarafshan's mountain course is between Pakshif and the glacier, being 25 feet per mile. In looking at this question of grade we must further remember that the Fan with its many feelers taps a nucleus of the best snow peaks and glaciers of the Hissar range while the main river had so far to be satisfied with the drainage from steep, short slopes sending down their brooks at right angles to the longitudinal ridges. Finally, the present state of the Zarafshan must have something to do with the desiccation of Inner Asia, as consequently upon the dryness of the climate the stream has not yet been able to cope with the glut of morainic deposit just below the glacier. This same cause has found its greatest expression in the phenomenon of the Pamirs.

Although confined within steep walls the Zarafshan continues to be bountiful. Irrigation, however, devolves upon the affluents which are conducted to the level of the terraces on either side of the main valley (Figs. 36, 79, 95). The heat of the sun thrown back from the bare rocks brings forth beautiful gardens wherever water can be had and cramped space will allow. Higher up where westerly winds are frequently condensed to rain there are even fields independent of artificial water supply. The same condition, by the way, is found in some districts on the northern slopes of the Alai and elsewhere, but always sporadically. Since time immemorial tax-gathering statesmen have been discerning in such a case, for the rain-sprinkled crops must pay one-fifth instead of the usual tithe, i.e. one-tenth, imposed on irrigated land.

The higher reaches of the Zarafshan are inhabited by the Galchas. Of them and of the many sights of the valley I shall have occasion to speak in future chapters.

35. On the Upper Zarafshan, The Mazar at Khudgif.

The Zarafshan is only a small river but it works very hard, carrying water at great speed and with unfailing energy. Its importance is enormous and the poetry of its name, which means the Strewer of Gold, is not to be traced to the few miserable grains of yellow metal found among its gravel, but should be taken as a symbol of the benign influence of the River of Rivers.

36. An Aqueduct; Upper Zarafshan.

The wars between the kingdoms and states of the Duab, especially those between Bokhara and Samarkand, were nearly always water-wars and chiefly around the ownership of the Zarafshan. Once the Russians had taken Samarkand, they controlled the waters of the Zarafshan, thereby holding 'twixt finger and thumb the vital artery of proud Bokhara.

CHAPTER IV

A VISIT TO MAKHAN KUL

WITH the decline of summer comes the bracing time in Samarkand. During October and November autumn has established itself. At breakfast time the thermometer will show between 32° and 41° F. (0° to 5° C.), rising to about 60 (15° C.) at two o'clock. The evenings are chilly again, being frequently followed by several degrees of frost at night. The trees are yellow, streaks of mist creep through the gardens in the morning and the air is crisp. That is the season for travelling and walking in comfort, for seeing the sights as a tourist.

Then, in December, comes winter with an occasional shower of rain or sudden rise of temperature, but on the whole clear and cold. Through the leafless branches distant views with snow-topped hills are revealed and hidden houses everywhere become visible among the recesses of the gardens. Now the bugle sounds for the gay chase after pheasant, duck and boar, across the crackling stubble-fields and through the rustling reeds. For us autumn and winter are favourable for roaming in the plains; now is the season for lusty deeds in the lowlands.

On the 1st of December 1907 two ladies, Carruthers and myself left Samarkand intent upon a shooting trip to Makhan-kul. According to the official calendar of the Russian railways it must already have been mid-winter, for our second class carriage was almost bursting with heat. The iron stove in the corner had been goaded to red wrath and was doing his level best to reduce to a sweltering stew the crowd packed into this boiler on wheels. Herein it succeeded with remarkable atmospheric effects. Knowing from the beginning that our modest hint at an open window would have been met with indignant protests from everybody, we tried to acclimatise ourselves, justly admitting that in this one respect we still were mollycoddles, that our training was sadly deficient in matters suffocatory and odoriferous. Nor could charitable souls understanding our plight—had there been any—have helped us, seeing that all the double windows were tightly screwed down.

After Carruthers had swooned once or twice we came to the conclusion that our hand baggage wanted re-arranging. When by common effort we had lifted a heavy gun-case nearly up to the rack it slipped, swayed frantically and getting beyond control went through the double plate glass with a terrible crash. Curiously enough, in spite of its great weight, we managed to save it from falling out of the

37. Ferry across the Zarafshan.

train. A hubbub followed, but everybody was compassionate, pitying us for being obliged to sleep near the influx of cold air caused by the regrettable accident. After having agreed to pay for the damage we sat down with a clear conscience, a breeze playing round our temples, and the happy feeling that the question of ventilation had been settled without disagreeable discussion. The conductor tried to keep out the chilly blasts by unscrewing a double window and fitting it into our gap, but wide cracks remained and supplied us with the modicum of oxygen necessary for the maintenance of our lives.

Late at night we arrived at Yakatut where the platform lay lonely and in inky darkness under the glittering sky of the steppe. Down in the blackness we could see a lantern twinkling like a lost star and hear how the luggage van disgorged the mountainous mass of our camping kit. Then the train moved again, softly rumbling into the vast stillness of the plain. After a time shadowy figures began to emerge from the gloom. One of these proved to be a messenger sent by our Armenian hosts, the owners of a small wine-press. Along the railway line the making of alcoholic grape juice is allowed in Bokhara, but on condition that nothing be sold to natives. Only Europeans and Christians enjoy the privilege of drunkenness within the frontiers of the Khanate, while outside these limits even freethinking Sarts may indulge in the liberty of the gullet.

The house where we were to pass the night was about ten minutes distant. With the help of two men we busied ourselves for an hour in dragging our paraphernalia to this place. Here a big, bare room was given us, the blank destitution of which was more agreeable to our mind than the threadbare trappings and furniture of semi-civilised inns. Spreading the tent canvas we camped on the floor and thus satisfied the hospitality of our friends by making use of the roof overhead.

Next morning we hired two camels and three asses, their owners, three men, accompanying us as servants. After buying a sack of bread and loading up, we follow in the wake of the pack animals. For the first two hours we pass through partly cultivated land, intersected by ariks and high dams, with groups of houses and trees. But behind us, and beyond the silvery line of the rails, there lies the southern horizon, a straight bar drawn across infinity, the horizon of the clay-steppe, as flat and hard as a billiard table. The low station buildings and the bold, angular shape of the water tower are like black holes sharply rent into the blue-grey sky. To east and west the telegraph poles, linking their wiry arms, deploy in an endless chain that vibrates unseen with the news of the world.

We tramp along a dusty road between dry fields and empty ditches. Here and there is a bit of untilled steppe or a drift of dirty sand mixed with much earth and dead plants ground to fine powder. Sometimes a pheasant rises from a clump of reeds, but our goal in the distance is his salvation.

The pastel-like shades of the landscape were exquisite against the blue of the sky. The bare trees shone at times a rich violet, the steppe itself was tawny coloured, broken here and there by patches of green. From amidst this coloured ocean arose at intervals kishlaks whose mud walls made yet another variety in the landscape. As the day drew on the distance became enveloped in a thick, soft haze of dust. Masses of faint, fleecy clouds appeared here and there, the light grew feebler, the colours still less determinate. Only the softly flowing Zarafshan still reflected the blue of the morning sky as it rolled on ever nearer its appointed doom.

38. Fording the Makhan-darya.

At Yakatut the Zarafshan sends out two branches, the Gujeli and Makhan daryas, and our destination was the triangular space between them (Map II). For this purpose we had first to cross to the northern bank of the main river, afterwards recrossing the right prong of the fork. This complicated route was imposed upon us by the fact that the Zarafshan is unfordable at this season even down here, near its end. Naturally the ferry connects the villages on both banks to the north of Yakatut, and not the desert and swamp on the right bank with the thin line of villages between the river and the railway.

We found the Zarafshan a very deep, narrow channel between high dams. Its waters were swift and slaty. The reason for this periodic swelling of the river must be sought in the plains, not in the mountains; in a decrease of demand and not in an increase of supply. The waters of the Zarafshan still keep circulating through a multitude of ditches but only for the wants of man and beast in the hundreds of villages. All fields are left dry and as it is they which claim the lion's share in summer the unused balance is very great. Hence it comes that in winter the districts of Makhan-kul and Karakul are glutted with water which knows not where to go, while such branches as the Makhan-darya and Gujeli-darya near Yakatut are changed from miserable, desiccated gutters into considerable streams.

The ferry-boat was a veritable Noah's Ark, large and roomy, built of rough hewn beams caulked with clay (Fig. 37). As we had no interpreter negociations with the ferry-man proved somewhat difficult, more especially as he seemed conscious of his importance. His business principle, undoubtedly a sound one, was cash down in advance, and as we did not know the local market value of his services we moved on uncertain ground. No one is by his calling so predisposed towards the gentle game of extortion as a ferry-man and I dare say that even Charon is not averse from relieving a soft sinner of more than the customary obolus. Being Occidentals, smitten as they all are, with the feverish madness of hurry at the sight of an obstacle or the prospect of losing time—by Allah, he is crazy indeed who pays good money for time, for air, for nothing—we paid silver where others probably give brass, whereon our friend with his long, stout pole punted us to the opposite bank. He took the whole of our caravan at once, there being about an inch to spare to the absolute load-line or certainty of getting swamped. Fresh customers were already waiting at the other side and opening a conversation by shouts before we had got halfway across. They were making their first offer to the bargee and we still heard their noisy bargaining when already well on the road. Nor could the return journey be begun in a hurry, for every time the leviathan had to be emptied of the water which spurted merrily through a dozen wide cracks.

At Malik, the last village on our way, we stopped for a while in order to complete our larder by the addition of ten melons. Great white and yellow dogs with cropped ears resented our intrusion, undoubtedly emboldened by the knowledge that stones are rarely handy on their native soil. Then we passed on and soon found ourselves eyeing the slow but steady current of the Makhan-darya.

Meanwhile the sun had risen high and was now shining hotly, through a thin fog of loess dust. A slight breeze from the north-west seemed to smell of the sea and gulls were flying above. Of humanity there was none but ourselves, standing on the shore, staring into the silent eddies, looking for a ford. After a long palaver our men decided upon a likely place. One of them stripped and partly wading, partly swimming led the camels across (Fig. 38). The backs of these long-legged animals were far enough out of even the deepest water as to ensure dry transport of passengers and luggage. After several journeys to and fro, the asses being swum across, the task was safely accomplished and our party landed on the delta of Makhan-kul. Here all signs of agriculture had vanished; flat steppe was under foot and a yellow rise on the horizon announced the nearness of sand hills, making a proud show of height against the dead level from which they sprang. Striking out in a westerly direction we wandered on, at random, though not aimlessly, in search of a lake which the name of the place (Makhan Lake) seemed to warrant. This our quest was more difficult than we had expected, for we did not discover our goal until the fifth day.

Meanwhile the sun was preparing to go down and we made haste to set up our first camp. Here stood weeds and low shrubs in splendid isolation, leaving between them naked soil, the upper layer of which formed a spongy, blistered mass produced by the efflorescence of salts, as proved by a slight whitening on top. Below a soft, brown earth was exposed as one scraped one's boot across. Water was hunted up in a river arm near by; stalks and roots of dead plants and the twigs of tamarisks served as fuel. At half past four the sunset was a glory to behold and a peace to cherish. Tent and flaring camp-fire stood against a transparent sky that was very light and ethereal, delicately tinted with the palest green gently dissolving into a thin and tender blue that arched the zenith like a crystal vase. Low down swam grey and purple clouds, the wondrous fish of a limpid sea. Suddenly against the sky-line a camel rider loomed, passed near us, mumbled greetings and then disappeared into the dusky shadows that gathered on the silent steppe. The night was fairly warm and windy.

Next morning Carruthers left very early in order to look for birds. Two hours later we started with the caravan and soon struck the river which we had agreed upon as a guiding line. Through a steppe of high grasses (Fig. 39) where we met some shepherds with their flocks,

39. Clay Steppe of High Weeds; Dunes in the Distance.

we reached a place where many branches of the river were busy reducing the surface to swamp. Continually we came upon mud, pools and glistening streaks, most of the water betraying a visible current in a westerly direction. Here we met Carruthers loaded with his booty: several hares, mallard and little birds of all sorts.

After a time the maze of straying water became hopeless and seeing nothing but reeds ahead we retraced our steps to the higher ground which seemed to skirt the inundated area. Keeping thus in close touch with the rising flood we hoped to find the lake without venturing too far into the labyrinth and endangering our line of retreat. Crossing a belt of sand-hills (Fig. 42) we selected a site for our camp where the last dry slopes abutted on a vast expanse of reeds. From here we intended exploring until a practical route for the caravan was made out. In vain we scanned the horizon from the highest sand heap. It was most tantalising not to be able to obtain a view and to see the flights and flocks of waterfowl travelling towards some object which they could see so well. We made energetic attempts in several directions, walking and wading, taking great care to remember our way back. But it was everywhere the same: an indescribable mixture of low sandbanks, bits of steppe, stretches of deep mud, pools, lakelets, broad ditches, reeds, all interlocked and dovetailed into each other. So here nothing could be done and we resolved to try elsewhere next day. We also felt too much cooped up in this locality where the tall reeds blocked the view and a good walk was only possible in an opposite direction from our goal, the existence of which we now sometimes doubted. Apart from this crowning desire which made us restless, we enjoyed the landscape very much, for there is a strange fascination in this wilderness.

Towards evening we were visited by some shepherds from the dry ground in the neighbourhood. We are here within the sphere of the famous Karakul breed of the fat-tailed sheep, which supply the so-called Astrakhan lambskins. It is, however, a legend that the lambs are taken while still unborn (except where the mother has died), for that entails the destruction of the pedigree ewes and thus the whole stock would have been annihilated long ago, almost as soon as it began. The curly fleece is taken from the lambs about a week old. Owing to the great value of the Karakul lambskins, which hardly ever go out of

40. Dunes of the Sand Steppe near Camp III.

fashion, many attempts at acclimatisation have been made in European countries, so far with but little success. This is probably due to the change of food which causes the animals to lose the fine, close curls of their wool. At present a new experiment is in progress in Dalmatia, where the Austrian government has installed a small herd. As the climate is very dry there, the chances are perhaps more hopeful. The weeds of the steppe are not so poor as they seem to us at first sight for, according to a recent analysis, they contain a higher percentage of vegetable albumen than the juicy grass of an English meadow.

On Dec. 4 we doubled back and bending sharply to our left stuck to the chain of sand dunes which extended towards the north (Fig. 40), their height inducing us to hope that they might lead us far into the flooded district, that is to say, into the confusion of channels, sandbanks and morasses, or, possibly even to a sheet of clear water.

After a few hours' tramping the sandy range came to an abrupt end, forming a promontory which commanded a level expanse in front. This was a delightfully open situation, especially as some of the very highest of the dunes happened to be just here, affording a clear view of the distance and of the third camp which we established at their foot (Figs. 41, 44). The tent stood in a little valley with its back to the sand slope facing the wide plain. Here was the rim of a large peninsula of perfectly dry land, of which the bulge of the dunes formed the culminating portion. Quite near to our camp were the shallow pools and canals of the river water, feeling its way through steppe and reed-beds, while further to the right could be seen the silvery line of the Makhan-darya as it forced its way into the unknown. Behind us was pure sand while the floor of the little valley, where it ran out into the marshes, was composed of a mixture of sand and caked mud.

This was, indeed, a splendid position for a camp, high and dry, giving one that sensation of bracing liberty and free outlook as do downs upon a sea shore. Here we made our home for the rest of the time, for no better head-quarters could we wish wherefrom to pursue our quest. The lake was still invisible, it is true, being too far distant to see its mirror hidden by the reeds, but the sense of hearing convinced us of its existence, for the cackling of thousands of geese was audible for miles.

41. Camp III, Looking North.

This was my first opportunity of trying a device for erecting a tent on sand. It is, of course, impossible to make pegs hold in the loose substance by driving them in. What we did was to bury them two feet deep in the dry sand whereupon they held the ropes as with a grip of iron. This method is called "dateram" and is described in Galton's *Art of Travel.* Its effect is almost miraculous. From my experience there is nothing like dry sand for a camping site, being exceedingly clean and comfortable, especially when, as in our case, fresh water is at hand and no wind mars enjoyment by gritty additions to the cooking pot. The evening was perfect and moreover enhanced by the outlook upon two fat ducks which my cousin Sabina turned in front of the fire on a spit made out of the slender bough of a tamarisk. Carruthers was skinning his manifold collection of small birds the while telling us of his life with the gun in England, Africa and Palestine. The nights had now become very cold, the thermometer sometimes going down to twenty-seven degrees of frost (5° Fahr., – 15° C.) and in the morning we found from one to two inches of ice covering the shallow bights. This melted, however, during the day, it being very hot at noon in the direct rays of the sun, though cool in the shade, as proved by patches of hoar frost remaining all day in sheltered spots under low bushes. The same effect of a dry and transparent atmosphere is observable on glaciers and high mountains. Our melons were frozen because we had not thought of burying them in the sand but otherwise nothing occurred to spoil our pleasure in this bracing air and glorious sunshine.

The yellow enclave near our camp was a constant source of interest. Here we often walked about, up and down the yielding sand, looking back upon our sunken tracks as they receded over the crests. As a whole this is what is properly called sand steppe, boasting as it does, of a fairly good vegetation. Although the plants are not close enough to retain a sufficient layer of dust for forming loess they owe their existence to the fine powder which is constantly being blown across the desert, thus manuring it. But the further the sandy stretches are away from the mountains the less occasional rain they receive, so that finally we have the ever-moving desert, blank and bare. Likewise the tops of the highest dunes of a group are always free from growth of any kind, being exposed to the strongest wind (Fig. 40). It goes

42. Sand Steppe.

without saying that the closest and most varied flora are found near the edge of the sand where it begins to pass into the clay steppe (Figs. 41, 42). In the middle is the most typical sand flora (Figs. 40, 43), while the most elevated ridges are pure sand. It must not be supposed, however, that this vegetation can stop the drift, although the rough surface, presenting many obstacles will retard its progress a hundredfold.

As water can pass in and out between the weeds of a pond, maintaining the same level, so sand can move through the hills, entering at one end and issuing from the other. But the progress of sand as a whole in the shape of barkhans is totally prevented as evinced by the formation of the surface, where all the typical crescent-shaped dunes have been effaced, flowing together into a system of irregular waves. This shape is preserved by the plants and their roots, which, acting as so many fixed points, determine the outline of the sandy mountain landscape. If therefore vegetation succeeds in obtaining a first footing it ties down the shifting barkhans, but still lets through some drift. If near some water or a steppe the plants can secure a still firmer hold and invite more grass—which catches the loess dust so well—to join them, then there is hope of a firm soil originating in this place. Such is the process artificially initiated along the railway line, where mud is thrown on the sand in order to tempt the more exacting weeds and grasses.

Nevertheless the result of centuries is not very hopeful. So far the travelling sand has always been the gainer, slowly but surely, as proved by historical tradition and the reports of old people. The reclaiming of larger tracts of desert is not to be thought of, involving as it would the drenching and mixing with mud of hundreds of square miles. We must not forget that the desert is the dump of the mountains, that it must be increasing like the dump of a mine in full work. Where sand advances in the fiendish shape of the sickle-dunes its onslaught is terrific, as it simply smothers everything before it under heavy loads. Where these border immediately upon cultivation the latter is doomed unless superhuman efforts are made. Fortunately such places are not numerous, the fields being mostly separated from the moving desert by belts of steppe or sand-steppe which only allows a thin dribble to pass. Such drifting sand falling, say into a garden, in small quantities is absorbed by the soil, mixes itself with loess, and

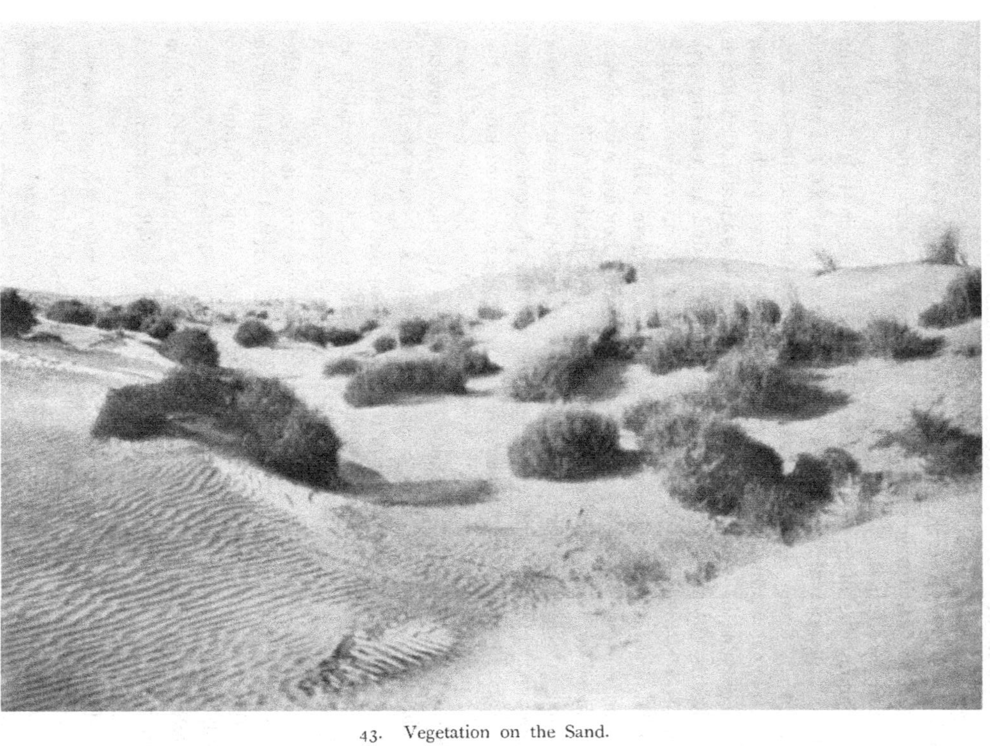

43. Vegetation on the Sand.

will find its way back to Lake Aral by water. By the construction of expensive belts of defence and constant watchfulness it may be possible one day to preserve the present extent of human property, but it will be a hard and tiring struggle, all the more so, for being defensive instead of offensive.

Among the highest dunes we still found some crests of pure sand of the half-moon shape. These have not yet been fixed but march to and fro a dozen feet or so as the winds change with the seasons. Receiving new sand from one direction and losing as much in another, they remain the same size. The zone of their annual step-dance is marked by thinner vegetation (compare the big sand heap in Fig. 40). Dunes are called hills of accumulation by the physiographer, together with moraines and volcanoes, as distinguished from hills of circum-denudation sculptured by water. These sandy heights have always reminded me of the snowy mountains. A mountaineer will at once notice the close similarity between the shapes of sand and those of snow, between the miniature mountains and the phenomena of glacial peaks. There is the long slope of the nevé, the windward side of the barkhan, gently rising to the curved ridge. On the other side is a corrie, a cirque with precipitous sides. Over this hang the cornices built out over the edge of the crest, and these eaves of snow are imitated to perfection by the sand which builds them out over the lee-ward crater. Such cornices, delicately poised, will break off, descending as avalanches, growing in volume as they travel down the steep incline to the bottom of the corrie. The surface of alpine snow is rippled like that of sand (Figs. 40, 43) and whoever has watched the gale-driven feather of ice-crystals standing away from the summit of Mont Blanc will be equally interested in its sister that flutters so lightly and straight withal, over the curved sandhill of the Duab. To complete the illusion only a few rocks are wanted sticking out of the small mountain ranges of the desert.

If a few Latin names interest the reader I may mention that the plants most frequently found in and near the sand belong to such kinds as *Haloxylon*, *Ammodendron*, *Calligonum*, *Pterococcus*, *Calliphysa*, *Halimodendron*, *Anabasis*, *Ephedra*, *Atraphaxis*, *Tamarix* and *Carex*. As shown by the photographs the denizens of the sand favour a globular shape. They are in fact like captive balloons anchored to the ground by a single root, from which radiates the intertwined network of their branches. This constructive principle is the same as that of the Jericho Rose. Many of these big balls, often measuring six feet across, become detached when the winter gales are blowing. Bounding through the air, joining others and gripping them in tight embrace, forming gigantic spheres as they roll over the sand, they career through the howling steppe like ghoulish phantoms shrouded in their tunics of storm-swept dust, leaping fantastically to the shrill music of a thousand demons galloping through space on the manes of the wind.

At night we amused ourselves by collecting as many of these sprawling globes as we could. The task was by no means easy for, owing to the impenetrable maze of prickly thorns the root had to be approached from below, by kicking at it, and generally it was so strong as to resist attack. Then we dragged them into camp—hence the multitude of tracks on the picture—and, piling them as high as one could reach or throw, lighted a bonfire, that primitive delight of man.

With the sole exception, perhaps, of the tamarisk, the whole of the vegetation was now dry and crackling. Much of it is literally ground to dust 'tween wind and sand in the course of the winter. One often notices blackish streaks of pounded and powdered vegetable remains caught against sinuous ripples of the dunes. The roots of the sand bushes are ten and more times longer than the stems above the surface. Here and there one can see their pale strands exposed, creeping along like burrowing snakes, diving up and down.

Of all animals living underground only the susliks (*Spermophilus;* ground squirrel, chipmunk) had kept awake, sunning themselves at their front doors and scurrying off as soon as they heard the slightest noise. Nearly all inhabitants of the steppe roam at night, but their doings are writ upon the sand. We read how little mice have scampered from bush to bush for fear of meeting cat or weasel, whose tracks they cross with a line of pearls, the impress of their tiny feet. Hares have cantered and the wily fox has drawn his single string of dots. The heavy boar will also plough the sands, but here we did not see the striped monarch's mighty paw. In day time it is the birds that leave the signet of the three-pronged fork or the stroke of the flapping wing. For him who hath eyes to see even these outposts of life are full of miracles untold.

Next morning, after the first night in our third camp, my wife and I set out to reconnoitre. Carruthers was obliged to stay behind skinning his bag of small fry of the previous day, for curiously enough, in spite of the cold weather, the birds would not keep long. Sabina had undertaken the duties of the cook and in that role not only pleased herself but our consumptive inclinations as well. Teal and sandgrouse were to be the dinner of the day, so we made haste to start on our errand and be back in time.

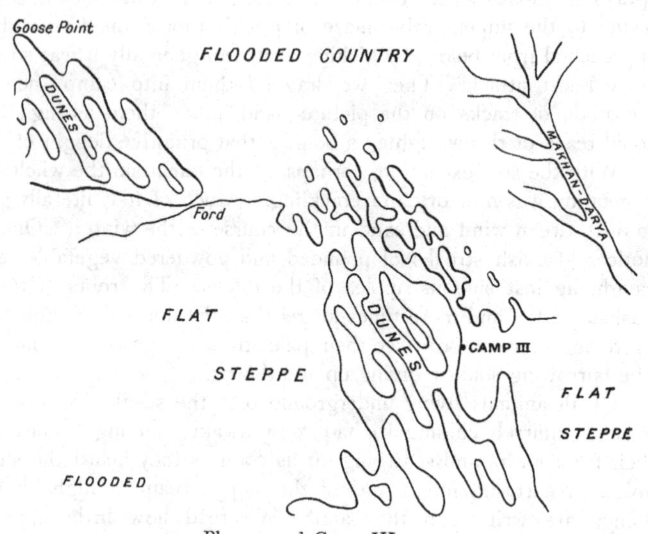

44. Plan around Camp III.

Going north at first we gained the edge of the water lapping up the land (Fig. 44). Our intention was to get round our jutting cape of dunes and to follow the dry shore as far as was possible without wetting one's feet. Thus we hoped to find some point near the, as yet invisible, lake whence on camel back or by dint of shallow wading we might have some prospect of getting within sight of open water. Uncertainty gave zest to our search, for we knew we were in the very midst of a battle ground where moisture was taking temporary possession of the realm of drought, where the wet level was taking advantage of the least depression in the ground, worming its way into every hollow and corner. We can describe this process, which is the same in the Karakul delta, as an attempt of the Zarafshan at finding a road to the Oxus. In summer there is too little water to break through anywhere, while in winter it finds blocked any channel it may have begun to work upon during the preceding high flood, and which in the intervening summer was blown full of sand.

Most curious of all was the meeting between those two contrasting elements, between sand dunes and river water. Following the dry shore we soon became aware of the fact that our route traced the outline of complicated peninsulas projecting into the flooded area (Fig. 44). Often we would come to a narrow channel which was but the mouth of an inland bay spreading itself inside like a branching fiord, obliging us to make elaborate detours around its various prongs. Sometimes we succeeded in cutting off, but this was not always feasible, owing to the difficulty of overlooking the low, wavy ground. I was reminded of Bantry Bay in Ireland, with its bights and islands. The explanation in our case was that we were travelling among the smaller dunes which were being invaded by water at the end of our promontory. Thus half-drowned hills were formed, embayed mountains as Davis calls them. A rising water level hugging a receding and advancing contour line of equal height creates a coast of submersion which is a horizontal section through the hollows of the land surface. But once out of the reach of the sand hills topography became simpler and soon we were able to pursue a fairly straight course along the border of a flat steppe.

Marshalled in line the parched and brittle scrubs were waiting for the deluge. Gradually a bank opposite to our fringe began to materialise out of chaos. The two shores approached until a channel about fifty feet wide took shape, in which a deep westward current was discernible. This narrow passage must be an important landmark, for the ends of a broken dam projected from either side leaving a gap for the free ingress of the water. Undoubtedly this lock was to be shut again after the flood had attained its maximum. Thus in some basin away on our left a store of water would be held back to last through part of the summer, either for irrigating a few fields or for the sake of the cattle. Across the muddy flow we saw a shoal of dunes, their yellow whale-backs rolling away in distant perspective. As they probably formed an island among the swampy broads they promised a route for penetrating further north-west.

We made a note of the spot and returned, this time taking a bee-line towards a thin pillar of smoke which we knew must be the axis of the world with our dinner as the hub. The steppe through which we walked was flat on the whole, only here and there interrupted by a low drift of sand or a dry rill. Occasionally a deep hole had filled with water by seepage from the inundated parts. To him who can adapt his eye and imagination to the lesser stature of the plants which stand for trees, this steppe is a veritable park, with its groves, open spaces, isolated shrubberies and thick coverts. In between were lanes and clearings of hard clay, which made going very easy. The "forest" was on the average breast-high and composed of all sorts of thorny bushes, high weeds, tamarisks and rare clumps of golden reeds. The colours of these copses and jungles had mustered every shade from darkest brown to brightest yellow, giving a fine, rich effect in the sun-light. Frozen dew lay in the shadow of every bush and a breeze, blowing very gently, brought cool, bracing air. These signs of winter impress one curiously in surroundings generally associated with intense heat.

In summer it is difficult to appreciate this landscape with undivided joy, but during the cold weather the bushy steppe and the sand dunes are a source of delight to him who can surrender himself to their varied charms. It is saying a good deal that even I, who nourish an aversion from plains, felt fascinated and loved to ramble through this miniature savannah, to which the presence of small game added but another attraction. Imagine yourself in a very lonely part of England, on a sunny, frosty day, on an immense plain, interrupted by sandy downs and studded with clusters of very high bracken, broom and heather, and you will almost realise the unspoilt beauty of our play-ground at Makhan-kul.

On our way through the mountains of sand we climbed the highest hill, the monarch of the range, rising a hundred feet above the plain. Far away to the south-east we could just distinguish the line of trees which marks the borders of cultivation. On a distant loess hill the pale spot of a white-washed sepulchre stood like a phantom of mirage. Nearer to us, but beyond ear-shot, a lonely shepherd touched the sky-line, holding watch on a sandy rise, looking and wondering; human beings catching a glimpse of each other, meeting across space without a spoken word, then vanishing, who knows whither. Sometimes a wedge of ducks would cleave the horizon to tempt an eagle sitting on a stunted tree. Then we tramped into camp, leaving the saucer-like scoops of man's clumsy feet in the sliding sand.

At night all our wraps and blankets barely sufficed to keep us warm, and the morning toilet had to be performed over a hole broken in the ice. Taking provisions for the day we all went together accompanied by a camel and our servant Juma, the one to act as ferry, the other as retriever. The ford was quickly found and the ladies crossed first, when we men mounted the camel. It rose before we were ready for the spasmodic upheaval. I was thrown into the sand, missing the water by a hair's breadth, while Carruthers slid into the deep arch of the camel's neck, where he remained astride, his legs dangling over the stream. We enjoyed it very much and so did the ladies on the other side, who were certainly placed more conveniently as spectators of this simple but effective performance.

We walked into the dunes and felt that this must be the promised land, for the chatter and gaggle of the geese filled the air. Suddenly we topped a ridge and there we saw the lake. Towards the right were the spreading broads of flooded country (Fig. 44), where land and water, mud and vegetation were equally intermingled. But in front of us, surrounded by a wall of reeds was an open silvery sheet which, in places, came quite near to the sand. On this lagoon there were immense numbers of geese of all kinds, light and dark, as well as other water-fowl. Here and there a couple of swans made dots of brilliant white in the assembled multitude. "A kingdom for a punt," we cried. After watching the magnificent sight for a time, we proceeded along the shore in search of game within range. Many ducks flying past were killed but the geese had so far evaded us.

Finally we came to the furthest point of sand near open water. It so happened that a perfect migration of geese was flying low over this point that day. They came in their flocks of ten and thirty, in almost endless succession and always passing over the same spot.

As we did not have many cartridges with big shot, Carruthers was only able to drop one pink-billed goose, which fell into some reeds but was promptly retrieved by Juma. A pair of swans also sailed by, rowing heavily with their resplendent wings. Although not so satisfactory from the game killer's point of view, that lake with its teeming life made us happy and content.

The valleys between the sand hills were covered with fairly thick scrub in which Carruthers saw a herd of fifteen wild pig. It is astonishing how these animals manage to hide themselves, for in spite of diligent search we could never find any of them again.

We paid another visit to this place next day, but with poor results as far as shooting was concerned. We thoroughly enjoyed it all the same. The sun-rise was glorious beyond description, and on the fine sand there was hoar frost like the tenderest peach bloom. Our ruthless foot-marks destroyed it, spotting the glistening surface with brown blots, where they had burst the skin and wallowed in the sand.

The evening we spent in packing and burrowing for lost things among the loose foundations of the camp. At night the thermometer went down to zero Fahrenheit (–18° C.), and a vigorous morning, the 8th of December, sped us heartily on our march to the station of Yakatut. I shall never forget how some of the most pleasurable days of my life were spent in what many people would call an abominable desert.

CHAPTER V

BOKHARA AND THE ROAD TO KARSHI

WHEN in the opinion of orthodox fanatics the Osmanli had linked Mecca with the West, Bokhara became the spiritual centre and mainstay of the Muhammadan world. From here the mollahs of Turkey, Egypt, Morocco drew fresh impulse to religious fervour.

Bokhara was a defiant stronghold of Islam, the last bulwark of the followers of the Prophet. Here they lived in strict seclusion inimical to all outside influence, displaying and maintaining to our day the pomp, the power and the haughtiness of olden times. This was the tower of Muhammadan learning, the Rome of the Moslem, whither students congregated from all parts of the world. Here, on hallowed sites never trodden by the foot of unbeliever, the suras of the Koran resounded from the lips of venerable men. Wrapt within the heart of a continent, belted with dread mountains and fearful deserts, fringed with the lairs of murderous tribes, Bokhara-al-sherif resisted contact with the hateful Occident. The rulers on her throne were the heirs of a proud history, but they divided into small states, the caricatures of past greatness, where vaingloriousness stood for fame, oppression for might, cruelty for statesmanship. Bokhara became a scourge, the pages of its heroic traditions clotted with gore and the name of the last independent Amir the worst blemish on the mirror of history.

Already in the past the Muscovite giant had extended his arms towards the eastern seas; another stretch of his mighty limbs and the walls of Khiva crumbled and the battlements of Bokhara fell. It was not an easy kick, but it was the kick of a titan against a rotten tree. The greater difficulty then lay in making it bud with new shoots, and in that Russia has done her duty. She deserves to reap where she has sown. Where the thunder of the cannon heralded the dawn of day, the whistle of the locomotive now sounds its challenge to the hoary mosque. The troops of the White Tsar stand in a peaceful land, into which the railway infuses fresh blood and new life.

Where Vambery tramped under cover of a perilous disguise, where no European was safe from death, the Russian schoolboy now takes a holiday trip. Bokhara has no horrors for the tourist attracted by the sight of its wonderful bazar. But whoever wishes to see should make haste. Western ideas and fashions will not remodel the whole, but the number of Europeans with their following of Jewish, Caucasian and Levantine rag-tag and bobtail is increasing, and will one day assuredly strike a jarring note in the delightful originality of the town. Even to-day it is not quite the same as when I first saw it eighteen years ago. Nevertheless we still find the true and unspoilt picture of a magnificent capital of the Middle Ages, genuine in every detail. Wherever we look the tales of Sharazad are realities; the narrow lanes are full of eery phantoms, mysterious doors, beggars, princes and fairy dreams. Only two or three Russian houses have so far been allowed within the precincts of the city wall, and the Sart is not at all likely to adopt trousers and a billycock hat, but the number of Europeans grows apace. Make haste then if you would fain see the glamour of Bagdad, the Khalif and the Grand Vizier, the hunchback and the priest.

The number of inhabitants of Bokhara city has been grossly exaggerated by most travellers, some even speaking of a million, influenced no doubt by the sight of the bustling crowds in the bazar, which are, however, mostly people from outside. Now we know the length of the city wall (visible behind the tombs of Fig. 49) which is seven miles. Hence, assuming it to be a circle, the greatest possible surface within is four square miles. The average size of a house would be about a hundred feet square or ten thousand square feet, which, allowing for streets, gives 2500 houses per square mile or 10,000 in all. A liberal estimate is eight to ten people per household, including children and servants, so that from 80,000 to 100,000 is my number of the population of the town.

We enter by one of the noble gates the massive doors of which are locked at night, nobody being allowed to pass, nor even to walk the streets, after sunset. At first there are nothing but deep alleys between the windowless walls of dwelling houses, until we reach the covered

45. On the Rigistan at Bokhara.

arcades of the central bazar, the seat of business, "the City." We stayed with my old friend Sidikh Bai, who lives in the street leading to Amin Sarai from the western corner of Labihauz. It is the third door from Makhsum's fruit-stall. You bang the heavy ring of the knocker against the knob of a big nail-head and a servant appears. He reports to his master; then there is a scurrying of women's feet and you are welcomed to make free of your humble servant's house this side of the harem's wall. If he is rich he has another house somewhere else; if not he has no home while you are there, but goes to stay with a friend.

"My home is my castle" can be more truly said by the Moslem of the cities than by anyone else. Every house is a fortress defended by a wall high enough to exclude even the glance of a rider on camel-back. The main entrance is defended by a shield which screens the courtyard from peering eyes when the door is opened. A Sart's domicile is rarely without a square tank or pond and small garden, but in the middle of a city these have often to be sacrificed owing to cramped space, unless the person in question is very rich indeed. But there is always a courtyard for the indispensable horse and donkey. The chief apartment is generally high and airy, but devoid of what we are accustomed to call furniture, there being no chairs or tables. One side is provided with niches sunk into the wall and serving as shelves for teapots, cups, books and oddments. These niches are often made of alabaster carved in Persian ornamental style and gaily painted. Near the door is a sink over which to wash, and in the middle of the room we find a square hole intended for the brazier in cold weather, for there are no stoves. On the floor carpets are spread and a pile of bed-quilts and pillows is stacked in the corner. Native bedsteads exist, but not everybody uses them, the ground being considered the best support and common basis of man. Besides many other advantages this simplicity has that of cleanliness. As there are no female servants, at least not for the male and public portion of the establishment, the sweeping and tidying up is left to lusty fellows who would not poke their broom under every chest or draw a finger across the mantle shelf to see if it needed dusting.

In spite of the ready hospitality of the Sarts there is always something lacking, something we miss unconsciously and that leaves us unsatisfied. I ascribe this feeling to the absence of a lady of the house who rules supreme over the internal arrangements, even were she to remain invisible forever. But there is no mistress; there is only a lord and master, and she the upper servant with no authority. Hence we miss those little touches of which we men are hardly ever conscious, but which we appreciate thankfully all the same. There is no soul of the home making every part of the house an organ of expression, welcoming the stranger with flowers on his table and speaking to him softly through everything he uses.

46. The Roofs of Bokhara.

The reduction of furniture to an absolute minimum has given the Sart a most enviable training. He can sit or sleep anywhere; he can write with pen and ink standing up, holding the paper in the palm of his hand; with the exception, perhaps, of liquids he can tie anything into bundles which he stores in a nook or bunches about his person; his fingers he uses as daintily as we do a fork or spoon. The putting of one thing to many uses plays as great a role in Bokharan civilisation as the inventing of special instruments does in Europe. The time which we spend keeping things in order is probably time saved by the Sart which we see him employ in the pursuit of rest or talk. Here furniture is synonymous with disorder, filth, ugliness, for there is nobody to love and tend it, even if she or anyone else knew how to use it, and next to care there is nothing like constant and proper use for the preservation of household articles.

Wherever in Oriental countries our furniture appears it is either laughable or depressing. The "European" rooms of eastern kings look like neglected museums of Western folly; the "modernised" household of Armenians in the Caucasus is a caricature. Cracks in the walls of a mud-built cottage are natural and not an offence to the eye as whitewash becomes when falling off in patches. What have we but scorn for a stuffy hole crowded with ramshackle chairs, a washstand reeking of spilt soap suds, a greasy basin, a jug like a coal scuttle, a blind looking glass, a table which dances a jig on its elegant bow-legs, a spring mattress which secretes the fluff of ages, cheap prints in gilded frames that fascinate the bug in search of lodgings, draperies that smell and artificial flowers that do not, and a sofa with a highly ornamental skin but gaping wounds, revealing the inner jungle inhabited by innumerable tigers which even the boldest hunter dare not face nor turn his back on.

Contrast with this a room, however rude; let there be nothing in it but a fine, large carpet, and it will be comfortable as well as beautiful. A good rug need not cost more than a full suite of ordinary furniture, it may cost ten times more, but where it lies, lounges and curtains, paintings and statuary are superfluous. This is the secret of the carpet—which we do not know in the West—that it is meant to lie free in a long, sunny room or under the trees of the garden, where there is sense in the glory of its lines and colours unhampered by all sorts of wooden legs and boxes. A carpet and a roof make a house.

The scarcity of household articles, their moveability and the impermanence of fitments ensure that general cleanliness which compares favourably with the habits of many nations and classes of Europe. A sweep of the hard floor, a shake of the rugs, a rub of the dishes are sufficient for putting things straight in the morning. True, we must not look too closely at cups and saucers and at the kitchen cloth, but I have seen worse not a mile from Charing Cross. Put an Oriental servant into an English kitchen with its many fixtures and paraphernalia and watch the result, which will be a fine, rich effect to be remembered by at least four out of your five senses for the rest of their existence.

Nearly every Sart can cook and he is able to cook anywhere. All he needs is an iron pot and a support, which is quickly made out of clay or can be bought at the bazar in the shape of a tin tube. Most of the cooking, even in winter, is done outside, in a verandah, backyard or garden. Often have I watched a bachelor on his little balcony in Samarkand preparing the evening meal over a small charcoal stove. After supper everything had vanished, the kitchen into a corner, the food to its orthodox destination. So there is no scullery full of things gradually growing into a neglected, nauseous collection of smells and messes, such as is observable in the poor quarters of a European city or, worse still, in a Russian hotel of the Duab. Altogether the natives live very much in the open air, also admitting it freely to their houses, no matter how cold it is. Not even with the poorest have I ever found that musty fusty, thick and greasy stuffiness of which the European peasant is so enamoured.

47. The Turquoise Cupola; Bokhara.

Although there are accumulations of dead matter and dead men inside and outside of the cities, no refuse stagnates in the house. The Bokhara house-wife or house-man has no stock pots, or flour bins, or anything, in fact, resembling a larder, only a few dry goods such as tea, sugar, spices and sweetmeats being kept in the house, with the addition, perhaps, of some bottles of jam. Even the wealthy live from hand to mouth, so to speak, buying their provisions when needed. What is prepared is eaten up at once, the wives, servants, the servants' friends, the poor and the dogs making a clean sweep of anything left of a meal by the master or his guests.

Food is neither wasted nor allowed to get tainted; once safe inside, nothing detrimental to man can happen to it; all or nothing is their motto. This principle is the more easily observed as there are practically no fixed meal hours, a Sart being ready to eat anywhere, at any time, and as often as you like. But he can fast also or wait patiently for the cooking of his dinner which he has just brought from the bazar. Unpunctuality is no bar to his appetite, nor irregularity a danger to his digestion. This happy go lucky programme accounts for the Eastern custom of always having sweets and tit-bits going, to silence the roaring of the inner lion at a moment's notice. Intermediate feeding does not make a Sart unresponsive, whereas European guests swear inwardly because the dinner to which they are invited is only put in hand when they arrive. Both waiting and nibbling are equally capable of offending their stomach and making it less receptive.

The universal use of goloshes is also a great factor in cleanliness. Nobody enters a house with the soil of the street under his soles, but leaves the slippers at the door. Leather stockings, i.e. soft riding boots are much worn, but even the well-to-do often dispense with them, walking barefoot on their carpets and donning nothing but goloshes when going out. Such at least is a general habit in the provinces, only the citizen of Bokhara, leaning more towards smartness in dress, is fond of exhibiting highly polished top-boots of flexible leather. Socks, stockings and gloves are unknown, nor do I remember having seen knitted garments of any kind, all being woven, sewn or embroidered. Loosely shod feet are also more convenient as regards the frequent devotional ablutions and in this respect I am reminded of the Bible which is illustrated by so many things in the Duab. Constantly we meet the doings and sayings of Holy Writ realised in actual life, whereas mediaeval scenes are recalled chiefly by the style of building in the towns, by social institutions, handicraft and the like. In the Old and New Testaments the ideas about hygiene were neither more nor less than those prevalent in the East at the present day. A saying attributed to Christ expresses an important view on the subject: "He that would be clean hath no need but to wash his feet, but is clean every whit." Amongst the Sarts I have often been struck by the cleanness of their feet when their garments left much to be desired in this respect.

Were we not hypersensitive on the subject of hygiene we could hardly do otherwise than admire the frugality of the Muhammadan in all that concerns the impedimenta of life. Differentiation in clothing, though it may indicate a higher stage of civilisation, immensely complicates our existence. While we are occupied in multiplying wants and heaping up articles for every conceivable use the Sart contents himself with the fewest possible paraphernalia. No man has more time or travels at shorter notice. The only thing which makes the Oriental backward is his want of social organisation for the purpose of division of labour and the saving of time in the aggregate. Individually he saves, and spends on lounging, the time which we waste on dressing, reading newspapers, keeping knick-knacks in order, and so on.

The love of talking is considered a paramount weakness of the Asiatic. It lends itself to jocular comment and I have myself never resisted that temptation. But has anyone ever heard the remarks of a native of the Duab on *his* adventures amid British tribes and the graphic tale of his visit to an exclusive clan or totem. Well, if he is just polite, he says nothing about the gossip there; if he is very polite he says that he was too stupid to discover the bed-rock of their conglomerate of talk. One thing, however, he made out with certainty and this was the unwritten but sacred law that every conversation, be it ever so short, must open with an incantation to the weather-god. By elaborate researches and a close study of literature and life he was also able to prove with a certainty amounting to probability that in literature one had to use "elect" words, that this was as it should be and the same as in good Arabic or Persian. But that there also existed a secret organisation called "society" with undefinable boundaries, to which everybody belonged and from which all the rest were excluded; that in this "society" there existed equally evasive "sets" who used "select" words, syllables and intonations. By these an expert could always know to what set a man or woman belonged, but to which in nine cases out of ten they were not admitted.

We have daily papers which gossip for us, so need not speak unless for exercise or pleasure. All the same, half of the business of the

London city is done by talking coupled with the fine art of "pumping." The market prices of Bokhara are the outcome of bazar gossip. To a people which has no newspapers gossip is a necessity, nay, almost a dictate of self-preservation. The Oriental talks with a purpose, with a full consciousness of what he is saying or wishes to say. His mind has a forethought for every question and an afterthought with every answer. The system of our detectives, cross-examiners and diplomatists is a daily practice with which every native is familiar. How could there be so much verbal tradition unless the same thing was repeated over and over again.

A man with news is a man with something valuable in his possession which will secure him hospitality, friendliness, food wherever he goes. If a Kirghiz has news he gallops off to an aul twenty, thirty or fifty miles away, sure of being received with open arms as an honoured guest. Hardly have the listeners absorbed the gist of the narrative than one of them sets off intent upon capturing a new audience. This explains why in Asia, under equal conditions, native news is quicker than European information; why gossip from Kashgar to Bokhara travels faster than a courier; why bazar talk at Samarkand reports an incident at Tashkent much sooner than the telegram sent to the morning paper.

48. Listening to the Reader.

When we see some merchants of Bokhara sitting together for an hour or more they are simply reading their *Times* after breakfast; when a messenger or servant on his errand catches up snatches of conversation in the bazar, calling out inquiries, he is reading a paper in the train; when a group of listeners sit round a lecturer at the corner of a mosque, they are reading their Bible, or their Shakespeare, or maybe a shilling shocker, which gives more shocks for less money and compared to which our yellow volumes are missionary tracts. What we read they hear, and if count were taken of the hours thus spent I do not think that one side can accuse the other of waste of time. In order to read we must sit; you can walk, ride, eat, see, hear and listen while you talk. I have heard of shortsightedness due to reading, but never of lock-jaw traceable to an overdose of conversation.

What one sees of a man's ordinary dress are his goloshes, the chapan or gown, generally known by the Russian name of khalat (dressing gown), and the chalma or turban. The inner layer consists of a square, sack-like shirt and short pants resembling those worn by athletes, but sufficiently baggy for holding three men in proper training. Over this one usually wears an undress khalat, above which is donned the gaily coloured smart chapan when one goes out. A belt or handkerchief is used to keep the inner khalat together; only officials and officers enjoying the privilege of an outward and visible waist sketched with the help of an ornamental belt. When it is cold one increases the number of top-coats or invests in one lined with fur. Gloves are unnecessary because the sleeves are so long that one can always withdraw one's hands into the inner atmosphere heated by the body. This plan is decidedly preferable to gloves, especially fingered ones.

The Bokhariot's pride is his chalma whose stately white bulge forms a pleasing set-off to the vivid pattern of the khalat. Its size and shape are determined by the station and rank of the individual; mollahs and ministers having the largest and finest. Neatness of clothes and distinction of manners are a peculiarity of the citizen of Bokhara. Here we find high-bred aristocrats, smart clerks, swells and dudes, everybody trying to look as much as possible. This, one would think, cannot be difficult, seeing that nothing but a flaring dressing gown need stand between you and the outer world. But class distinctions not easily noted by our eyes are expressed by certain rules of dress.

There being no uniforms in a European sense except for the common soldier, birth and rank are denoted by the gorgeousness of material, the costliness of the belt, the twist and width of the turban and other niceties of apparel. An extreme case is that of the Jews who are forbidden the chalma and coloured khalats in the street; they must gird their waist with a piece of string and may not ride a horse.

The dress of the women is the same, on fundamental principle as that of the men; but there is more cut in the robe to suit the female figure. Jewelry and make-up are the usual adjuncts of Eve's daughters all the world over. As to the deeper mysteries I am told that they betray more colour and variety of material than those of the men. In the bazar I once bought an inside kind of neck-to-waist garment of red cobweb silk which I can crumple up and hide in the hollow of my fist.

On Russian territory even male explorers have opportunities of studying the appearance of the other sex, but in Bokhara the conspicuous woman is conspicuous by her absence. Nor does the institution of eunuchs exist here. No man ever sets eye upon a lady not his own, for in the street she is nothing but a perambulating sack with a black horse hair screen where the face is likely to be. All architecture and domestic arrangements are influenced by the traditional seclusion of the women, who live in a strictly separate part of the house often having its own courtyard and pond. They never formed part of the various scenes of native life which our intercourse with the people enabled us to witness. Only now and again one meets them at dawn or nightfall, stealing out furtively to fetch water. They shrink at the sight of a stranger and veil themselves in all haste, though their share of the curiosity attributed to the sex generally prompts them to take a good look themselves at the passer-by.

Our ladies sometimes saw the inside of the harems and my wife reports as follows. We visited the widow of Mansur surrounded by friends who for a period of forty days bewail with her the loss of her husband. The women were all dressed in dark khalats, the widow herself in that of her husband, a blue one. There were a considerable number of old hags present, furrowed and toothless. The children, of whom the usual quantity abounded, were suffering from sore eyes, a result of the all-pervading dirt amid which they live and the pestering flies too ready to take advantage of defenceless babies.

Our next visit was to the property of a rich Sart whose house is well built and has a fine garden. The wife attired in velvet and silk had a look of suffering on her face. She is a victim of the greatest tragedy that can befall a woman here, she is childless, but so great is the influence of relatives and such like in this matter that though her husband longs for an heir and wishes to buy himself another wife, he dare not do it on account of his wife's opposition. The poor thing sat there decked with jewels and gawds, but with this sorrow continually gnawing at her heart. Hence it may be taken for granted that the women are not altogether at the mercy of their owners. Muhammadan law, in fact, endows them with certain rights which they can exercise to good purpose.

On the whole the women make the impression of children and in the outlying districts of savage children. This comes of being shut up and confined to a narrow circle without a chance of developing knowledge or independence. They are inexpressibly filthy in the

49. A Burial Ground at Bokhara.

villages and are everywhere on a far lower social grade than the men. One may say that the highest woman in the land is inferior to the lowest man. Children they have in abundance but as more than half of them die in infancy, their labours in this department seem somewhat vain. As regards needle-work they are very skilful in embroidery, while poor wives excel in patching, for the garments of the male members of humble families are marvellous examples of this art, and seem in many cases to have been put together by an indiscriminate use of the contents of the rag bag.

Woman is a cheap article in Bokhara. A man in search of a wife can get one in exchange for several sheep and a little money, or a horse, as the case may be. Ordinary people have one wife only and monogamy in practice appears to be the rule. Those higher in the social scale and better endowed with the world's goods know no restrictions except those imposed by their own conscience or caution. A few years ago a brother of the Amir died leaving no less than twenty-eight to mourn his loss, together with about a hundred children. The ladies were distributed amongst the various provincial governors of the Khanate, who have no choice in the matter should supply exceed demand or quality be open to criticism. The saying of the cat in the sack often ceases being a metaphor to the unfortunate winner of an inferior lot of bliss to be wed. He has to accept the Amir's bounty with a smiling good grace. Mashallah ; methinks the vendor of arsenic will do a trade !

It is not my intention here to furnish a complete guide to the sights of Bokhara as many good books on the subject exist in the English language. A few impressions must suffice for there is still much ground to be covered in our travels.

A view over the flat roofs (see Fig. 46) presents a striking appearance and one could easily ramble over the greater part of the town. But this is strictly forbidden as one might catch glimpses of women in the courtyards. To the left of the photograph the massive tower of the minar Mir-arab is seen. Throwing criminals from its height was one of the pleasant varieties of carrying out the death sentence as a change from hanging, throat-cutting and impalement. The amenity of being forced to sit on a sharply pointed stake has yielded to modern squeamishness, but not so the practice, officially sanctioned, of holding back a man's head as if for a shave, then drawing the razor right through from ear to ear.

Grave-yards are to be found in many places inside and outside the city gates. The body is placed on the ground and a vault of brick raised above it (Fig. 49). Where space is restricted these crypts will be built on top of each other often raised to four or five storeys. In the course of centuries the lowest gradually crumble, being reduced to earth, so that the whole settles down successively heightening its own fundamental platform. That these grim gatherings of the dead are not more of a menace to the living must be due to the climate. Part of its wholesome influence we may ascribe to the germ-killing properties of sunlight, though this only holds good where the microbes are not embedded in their breeding medium or kept moist by water. Moreover there is a respite afforded by the shades of night.

The most important hygienic factor is the general dryness of surfaces which isolates all centres of putrefaction and the refuges of microbes.

50. The Labihauz at Bokhara.

Were it not for religious and moral reasons the dead would probably not be covered at all, instead of being covered as little as possible and exposed to ventilation. At all events they are not "buried." Carcases and refuse left to sun and air vanish, one might almost say evaporate, very quickly in this country, whereas if put under ground they would stew in moisture, heat and darkness, and thus be sure of contaminating soil and water.

We Europeans are simply spoilt in matters olfactory. Eastern cities have proved long ago that frank smells do not represent so deadly a condition as we think. Firstly the bacteria of decomposition are pretty harmless, secondly contamination through the air is infinitesimal compared to other channels; and thirdly the few dangerous diseases which might be possibly communicated at a distance do not, unfortunately, betray themselves by odorization of their surroundings. Atmospheric pollution affects our lungs but nobody has ever caught consumption from the last strong tribute to nature which the horse or camel exhales on the sunbaked loess of the Duab. Nor is the richly tainted garbage on the roofs of the crowded city of Bokhara likely to cause pneumonia. Drainage—quite apart from the practical fact that nobody would keep it in order—would make the smells more evenly universal and the sanitation worse. Just as little as the river drainage of Inner Asia can remove the waste of the mountains from the great depression, as little can a system of pipes serve the health of the population. No one in his senses would run the sewage into the rivers which carry it on to the sweltering stagnation of the rice-fields. The only possible destination would be the desert, but as desiccation can be done at home, there is no need for an elaborate re-arrangement of organic matter before allowing it to crumble into dust.

What we have here is drainage by air, one which cannot get out of order. The climate acts like an oven of heat and radiation while the transport is done by wind. There is nothing like dryness and sun for localising decay and its attendant molestations of eye and nose. I am pretty sensitive myself and am positively sure that the range of action of a dead cat is far smaller in the Duab than in England. In our climate we try to collect and despatch refuse as quickly as possible in order to carry it to the sea before it has started decomposition in earnest. Our main difficulty lies in keeping all this stuff from oozing out somewhere into the houses or the subsoil.

We also paid a visit to Shirbudun the Amir's pleasure house. The palace is built of mud and wood and the interior decorations consist of stucco. It is an interesting example of the best effort of Central Asian architecture on the wane. Although often pleasing in effect, most of the work is poor in quality and execution. The walls and ceilings are ornamented in a style recalling that of the Alhambra, the various and intricate patterns being painted in red, blue, green, white and yellow. In parts the work is very rough and uneven, designed for distant effect rather than close scrutiny. Here and there instead of hand painting the ornaments are produced by means of wall paper of similar design. The rooms, of which there are many, are large and lofty and so far modernised as to have glazed windows. The gardens surrounding the palace are charming, nature having been allowed to run wild. Fruit trees abound and drop lavishly around their burden of apricots, apples and cherries. Vines trained over arcades of wooden trellis work afford a delicious shade from the pitiless glare of the sun and suspend their clusters of grapes rich with the promise of harvest, within easy reach of the passer-by.

A delicious place is Labihauz, the large tank in front of the chief mosque. Groups of gaily dressed men sit under the big trees or on the steps leading down to the water where the faithful perform their ablutions before prayers and where the water bearer fetches his supply in bulging sheep skins. The open tea houses are a-buzz with people handling their cups with elegant grace and listening to the strains of some minstrel whose instrument is of the most primitive kind and whose voice of the highest pitch.

Vendors of eatables go about; water melons in tempting slices beguile the thirsty. A thick creamy liquid made of sugar is much in vogue, and shashlik, bits of meat roasted on small spits over a charcoal fire, allure us by their odour. All is life and colour; colour everywhere; in the trees which flame in tints of green, saffron and orange below the blue sky and against the background of glazed tiles that adorn the temple; in the people with their gowns of every hue and their clean white turbans; in the melons which hang like golden globes in nets of reeds; in the grapes and purple prunes, and in the apples whose crimson vies with that of the pomegranates and the strings of scarlet peppercorns; in the carpets and matting of the tea booths and the loess walls of the houses. The theatrical artist in search of something supremely beautiful and sensational for the stage of Drury Lane need only make a truthful copy of Labihauz and its scenery, full of gorgeous colour and genuine Oriental life.

These places and the bazar are the club of the Bokhariot and as he is always there, he and his fellow citizens are the greatest clubmen in the world. As we are here near the fag end of the Zarafshan, necessitating the most stringent economy of water, the contents of

Labihauz are far from appetising, being renewed very rarely. To describe it as a mess is paying it a compliment. Nothing is done to keep it clean, for as soon as fresh water is let in, the boys of Bokhara are allowed to swim about in it, amusing themselves with all sorts of tricks, drinking tea in the water, shouting, making merry in every way, thus doing their best to rake up the mud of the bottom as energetically as possible. During the rest of the time people content themselves with washing their feet in the green pool and drinking from it, spitting into it, or using it as a slop basin for the tea houses around.

Here, in stalls along one side is also the centre of the barber's and hairdresser's trade. Plenty of instruction and amusement may be derived from watching proceedings. Anthropologists can study the shape of native skulls as in the full light of the sun they are reduced to polished balls by the glistening razor. Or we can, without charge, experience a superb thrill by following the extraction of a tooth. The head of the patient is jammed and held fast between the knees of the operator who then proceeds to fathom the oral mysteries with a formidable crow-bar.

Scientific inclinations are met in watching the sufferers from rishta whom it is the barber's business to relieve of their unwelcome visitor. The rishta—the word signifies, literally, "cotton thread"—is a worm, thin as sewing cotton and sometimes five feet long, coming out under the skin. It is the result of polluted water and a fourth of the population of Bokhara may be said to be infested with it. It also occurs in other places where water does not circulate enough, such as Kattakurgan, Karshi, Jizak, and is thus fortunately localised. The native barbers and doctors show great skill and experience in the removal of the animal, winding a few inches each day on to a match until the whole is extracted. Sometimes they succeed in withdrawing it at one sitting, but generally one has to proceed more carefully as a breakage of the worm leads to inflammation.

Like most parasites of man the rishta is subject to an alternation of generations, that is to say, it passes different stages of its existence in different animals or hosts. When the microscopical larvae are deposited in stagnant water they invade a species of *Cyclops* (i.e. a water-flea, belonging to the *Entomostraca* or *Copepoda*). Anyone drinking of this water may also swallow an infected water-flea. In the human body the

51. Shakhzinde at Samarkand.

young rishta grow up and mate, after which the males die while the females penetrate the intestine and settle under the skin there to develop to full size, maturing meanwhile the thousands of eggs they contain. Thus these female animals are full of a new brood, which, when brought into water repeats the circle if it meets any *Cyclops* there. As the water-fleas do not care for running water and as they are indispensable for the propagation of the species, only towns with very bad water suffer from this scourge. The preventive is simple, for boiling or even heating of the water kills the germs at once. But the natives are too careless and fatalistic to bother about filtering or boiling their water; often they are too poor to drink tea every time they feel thirsty. Even the Amir once had seven of the parasites at the same time.

There are Cairo and Bagdad, Fez, Algiers, Tunis and Timbuctoo; there is Samarkand with its monuments, and as long as the world has stood there is no spectacle to equal a Durbar at Delhi. But for an old-time capital city of Islam there is nothing like Bokhara in the uniformity of its character and the homogeneousness of its massed effects. And the glory of Holy Bokhara is her bazar. Nine-tenths of the people wear clothes of the same cut and the same white turban, thus ensuring that note of underlying similarity which is the secret of so many impressive harmonies of art. For embroidered upon this web is a riot of colour, a rainbow run wild, an animated kaleidoscope which nothing can beat on the face of earth.

The bazar is roofed and there reigns for ever a penumbrous shade. The fact of their being thus low and narrow pervades the crooked alleys with a sensation of intimacy akin to that of a Japanese village in an exhibition building. The whole of the passages of the Bokhara bazar could be reproduced in the Crystal Palace absolutely true to scale, sacrificing nothing in space, size or realism. Imagine yourself in the same crowd as at a show, shoulder to shoulder, pushing in opposite directions like currents that filter through each other, forming whirlpools at the cross-ways. But in addition to that let there be donkeys laden with sacks which pin you helpless to the wall, horsemen on prancing chargers and sometimes even a line of camels ringing their tuneful bells into the rumbling murmur of the voices.

And this din, and hubbub and eager throng flashes with all the daubs and streaks that ever graced palette. There are khalats quite green, or white, or yellow; black with red stripes, green with blue bands, purple with white lightning, or any two stripes or three or four of any tint you like, bright or subdued. There are lines, dots, circles, snakes, flowers, splashes shouting their shapes of crimson, scarlet, orange, emerald, or brown from a background of ultramarine, grey, sapphire, gold or silver, yet are they never gaudy, such is the secret of their beauty in this place. And all this gay phantasmagoria ebbs and flows in the dusky halls, quivering with sudden flare where struck by a sheaf of sunbeams slanting through a loophole.

In the stalls are silks and stuffs of flaming hue, jewelry and carpets, and countless sparkling things of tin or glass. And ever on the man-high surface there floats the multitude of snowy chalmas, like innumerable, glistening doves, white and pure, that have alighted amid a streaming galaxy of flowers. To this the music of a throbbing noise marking the pulses of the human tide; to this the smell of Asia's earth, the scent of mutton grease, the pungency of camels and of men. But over all is wafted the fragrance of nutmeg and cinnamon and a hundred spices. O how sweet to my nostrils is the memory thereof! Where are ye now, O incense of the Magi, O perfume of the East!

The highway from the capital to the southern and eastern provinces of Bokhara is the ancient trade-route to Karshi. This I followed in 1896 and 1898, using the hill tracks from Samarkand on subsequent occasions.

Ten days is the best record I have ever been able to make for the journey from London to the gates of Bokhara. After that comes another considerable reduction of mileage. With one's own horses four miles an hour is a very good average on a long journey.

The first essential for a visit to these lands is the permission of the Russian Government transmitted to the Imperial Political Agent at Bokhara. This resident official controls all foreign affairs of the Khanate and its relations with Russia. The Amir is confined to the interior administration of his dominions, where he can disport himself more or less to his heart's content, only the most draconic measures and drastic persuasions having been toned down. Thus Russia has practically no expenses connected with Bokhara while the customs' union gives her every commercial benefit.

In the summer of 1898 I made the journey to Karshi accompanied by my wife the late Dr von Krafft (who afterwards died in the employ of the Geological Survey of India) and my old, faithful servant and interpreter Grigor Makandaroff of Batum, generally called "Mac." A Karaul-begi, an officer of His Royal Highness the Amir, was given us as guide, his duty being to see that fitting accommodation for ourselves and our horses was provided at the various halting-places on the route. The presence with us of this official was tantamount to a letter of safe conduct, and assured for us a friendly and courteous reception at the hands of the native officials, with whom we came in contact. As he always sent a messenger in advance to announce our arrival at any station where we intended putting up, we found ample preparations made for our reception.

As we approached our destination we were usually met by a crowd of gaily dressed officials, whose white turbans, brilliant garments, and richly caparisoned steeds recalled some scene out of the Arabian Nights. These, saluting us with true Oriental dignity, escorted us to their houses, where we found provision made for our comfort according to the Bokhariot's idea of a European standard. Thus, though dispensing with tables and chairs himself, he always provides these articles for his guests at the "meiman khana," i.e. the house for strangers. Being of native manufacture they present some truly remarkable features. In very few cases do tables and chairs correspond, and the legs of both are of most uncertain height and stability. If the table is raised some two feet above the ground, the chair generally towers some two feet above that, while if the table is an ordinary size, the chairs are frequently several feet below it.

As regards hospitality, the Bokhariot upholds the traditions of Oriental lavishness. The tables literally groaned under the weight of fruits and sweets of every variety. Soup, fowl, and mutton with rice (pillau), the national fare, formed the more substantial part of the repast, which never varied. In spite of this monotony, the diet did not pall upon us, for fresh air and constant exercise probably whetted our appetite. If put before the choice I would rather eat a good pillau every day of the year than the dishes provided by an ordinary restaurant.

As regards our quarters they were uniformly the best the place offered. The native bedstead, with its net of woven rope, can be made a most comfortable resting place if covered with wadded quilts and rugs. In the deliciously cool nights of early summer, the traveller can with safety sleep outside, the dryness of the atmosphere doing away with all danger of chill, while mosquitos scarcely exist, save in certain marshy districts. Dangerous snakes or insects are almost unknown. In late autumn one has often to make the best of the native dwellings, owing to rain or cold, but air is to be had in abundance, as the doors seldom fit, and cracks and crannies in the wall provide sporadic ventilation in unexpected quarters. In the better houses fleas are far less numerous than in Italy or France, only poor people keeping larger flocks as well as the creepy-crawly. Bugs are foreigners to the Duab and where discovered a sure sign of European visitors.

Beyond minor discomforts, there is nothing to affect a traveller with sound health and normal nerves. Of attacks from the natives there is no danger. The people are quiet, too much in awe of their rulers to make themselves aggressively disagreeable; and are effusively polite when it is to their interest to be so. That it is possible not merely to travel, but to enjoy the private hospitality of the natives in this interesting country, speaks volumes for the mighty change which has taken place in it since the days when no European dared venture in safety within its borders.

Our little caravan, consisting of some twelve horses, left for Karshi on June 27. The first part of our journey was through a region of sand dunes alternating with steppe as far as Khoja-mubarak. There it gives place to partly cultivated steppe until Kazan is reached, where begins the luxuriant belt of gardens surrounding Karshi. Up to that point it is three days' steady ride through endless aridity where nothing breaks the monotony of the scene save the half-devoured carcases of animals fallen by the wayside; and a blazing sun turns earth, air, and sky into a huge oven for the greater part of the day. It was already fairly late in the afternoon when we passed through the Russian colony of New Bokhara, for getting under way is a task of some magnitude.

The horses are not yet accustomed to each other and to daily routine, but are continually fighting and running away. So for the first five miles we had to turn ourselves into cowboys capturing fugitives or separating stallions who were having a little discussion in spite of their heavy loads. Soon however everything went quite smoothly and the horses rarely took to fighting. Though inclined towards a bout of teeth and forelegs with one of their sib, their behaviour to man is on the whole exemplary. I have never seen them kick at a stable boy, nor need the rider fear many surprises in the bucking and rearing line.

Gradually settling into an even jog trot we advanced slowly and steadily. There are several stretches of genuine shifting sands, dry and bare, with none of that vegetation we found at Makhan-kul. Hence many barkhans of characteristic shape could be seen. The thermometer stood at ninety in the shade and not a breath of air was astir. In spite of this desolation bevies of the fleet-winged sandgrouse whirred from under our feet, while here and there a company of vultures hovered over a dead horse.

From time to time there appeared over the horizon a pole with a tattered rag indicating a mazar, the grave of a prominent saintly man, a mollah or a chief. Things smouldered in the silent blaze that went streaming forth from the round fire-hole of the heavens. Then evening descended; the stars began to twinkle in a metallic sky, and in place of the burning glare there came a sultriness, which drove perspiration out of every pore, but later yielded to a gentle coolness. On and on through the lonely plain; Venus casts a faint light on the path which guides us through the darkness like a pale sheen. Sometimes a tinkling of sonorous bells strikes the ear; swaying, rocking shadows spring from nowhere and the long string of camels, treading with silent foot, passes, disappears, taking with it the dying tinkling of the bells. Nobody speaks; in single file we meet the silent hours of the night that come and go; patiently the horses pursue their ambling gait, their hoofs brushing the blades of grass that wither on the ground; onward, onward, trot, trot, trot. Finally I fell asleep in the saddle as if it were a cradle, until a barking of dogs, announcing the nearness of human dwellings, aroused me from my dreams. The traveller in these desolate regions is not without provision for his safety by the way. At Karaul, the first rest-house after Bokhara, there is a fine well or reservoir covered with a large cupola of brickwork. Close by, the ruins of an imposing karavansarai recall the efforts of former rulers to mitigate the dangers of desert travel. These and similar remains elsewhere along the road are ascribed to Abdullah Khan, an Amir of Bokhara during the sixteenth century, and one of the few historic names still lingering in the native memory as associated with the past greatness of the country. It was these buildings we saw looming as dark shadows when after midnight we rode into our first hostelry. In these dreary surroundings the bill of fare is scanty and the water brackish. We managed however to hunt up a juvenile and therefore tender fowl.

Nearly all karavansarais throughout the Duab belong to a mosque or some other pious and charitable foundation endowed by the rich. The traveller can use them free of charge for himself and his caravan, only paying for such food as he obtains from the caretaker. The simplest of these refuges are nothing but a wall enclosing a courtyard with heavy stakes for tying up the horses. On the inside are a few sheds with a raised platform representing the rooms, more or less public, of the guests. In the daytime the stable yard swarms with scarabs busily turning the balls which serve as nurseries for their young. A constant unrest proceeds from the neighing and shuffling horses, occasionally accentuated by the powerful and penetrating clamour of a donkey. When heard from a distance I have often been startled by the likeness which this sound bears to the hoot of a motor car. But on coming near one is pleasantly relieved by the appearance of a harmless ass signalling to us as brothers of the road.

At nightfall on the second day we were overtaken by a fierce tempest. It became so dark as to obscure the path completely. We tied the horses to the trunks and bales forming our luggage, crept into our blankets and waited for the dawn. In spite of the rain and the wind, which overwhelmed us with clouds of driving sand, we could have slept very well, but were kept awake by our anxiety for the animals. Tugging at the packs and dragging them over the ground they erred about in the darkness, and if it so happened that two stallions met, a fearful row of whinnying, stamping, squealing was the inevitable result. We had then to get up to restore peace, otherwise they might have broken their ropes and vanished into the steppe.

At Khoja-mubarak we were installed in a real room in which various cats disported themselves with impunity. As our shelters since leaving Bokhara had been mere wayside shanties, we looked forward with some curiosity to what Karshi had to offer in the way of quarters. Very welcome after the miles of weary sand and clay was the sight of the gardens of Karshi with their mulberry, poplar and apricot trees. One often reads in books about the joy of the pilgrim when he spies the green oasis on the horizon, but here only have I been able to realise this sensation to the full. The thick luxuriance of the vegetation, the cool shade and the moist exhalation of the water, however muddy, give one a feeling of reverential thankfulness that easily links the dreams of paradise with religion on earth.

We found our quarters nice and comfortable and typical of those we were to receive generally throughout our journey. The house consisted of a series of one-storeyed mud structures built round a courtyard and made in the roughest manner possible, but simple and clean to look at. The flat roofs were covered with a mixture of mud and straw, on which grass often grows. The interior walls were quite plain and in the raftered ceilings swallows build their nests, unmolested by the owners. Carved wooden doors of very small dimensions, but great number, serve at the same time as windows, being closed during the heat of the day. The unwary European, forgetful of their height, frequently gets skull cracks from violent contact with the tops of them. Carpets, wadded quilts and pillows are strewn about the floor in abundance.

The table and chairs were fearfully and wonderfully made. The table especially suffered from weak legs, rickets and knock-knees, and was warranted to stand in one place only, any attempt at moving it being followed by disastrous results, such as the simultaneous collapse of all four legs or the temporary disablement of at least two of these members. Knives, forks, spoons, and even table-napkins, though not on the inventory of a Sart's establishment, were also supplied. The napkins were a curiosity, consisting generally of strips of new and unwashed calico, on which, in some cases, the English trade mark was still visible. We had a suspicion that some of these had done duty before and would probably be put into requisition again before being considered fit for the wash-tub.

Tea was our only beverage. Water, save in the mountains, the traveller dare not drink, as its source is usually the house tank, inefficiently supplied with fresh water, in which the different members of a family wash themselves, their clothes, and their cooking utensils. During these sojourns we were waited upon entirely by men.

At Karshi I went to see an Afghan banker by the name of Mirvaksh, to whom I had a letter of credit from the Prime Minister of Bokhara. Mirvaksh was the agent of an influential Afghan banker of the capital, a man enjoying the confidence of the Amir. Through his agents I was enabled to obtain cash in several places instead of being obliged to drag about the whole of my money in coin. Nevertheless I had to stock a fair amount of silver and copper, for in the villages change is hardly ever forthcoming, nor are the peasants likely to betray the secret, if they had any. The counting of the money occupied nearly the whole of the afternoon, and my fifty pounds' worth of tengas and puls came almost to a horseload. The silver tenga, formerly subject to erratic fluctuations, has been fixed at 15 kopeks, or about fourpence by the Russian government. It is divided into sixty-four puls of brass.

We had also to pay our respects to the local beg. His palace, a collection of houses, stables and gardens surrounded by a crenelated wall, lay on the rigistan or market-place. At the gate we found a loitering swarm of courtiers and flunkeys, among whom gossip and barter were in full swing, outsiders being admitted. Everybody haggles and drives bargains, from the head of the state down to the beggar. The minions were holding a sort of stock exchange, though perhaps horses and old clo' were all in the day's work, forming legitimate objects in every sphere of business. It required some strong words to cause these people to climb down from those regions of condescending impertinence where this fry usually abides. The Beg received us on a carpet in his square, and we could just catch a glimpse how with sour face he slipped a gold embroidered khalat over his cool undress pyjamas. To this resplendent surface his naked feet stood in curious contrast, although they were so clean that one could have shaken hands with them.

After a ceremonious greeting we sat down. I inquired after the latest news of the Amir's health, asked how his grace's own health prospered, how that of his father and grandfather, of his children and grandchildren; expressed satisfaction at the state in which I had found his district and the way I had been received; invited him to call on me at my home; and generally used him as an object for practising Oriental flattery, although I fear that our interpreter was not up to the mark in rendering the tropical flora and dazzling imagery of my speech specially acquired for these occasions by a laborious study of the poets. The imposing dignitary in his turn failed not to invoke blessings upon the sacred head of my sovereign, nor to mention my humble physical condition as well as that of my male relatives into the third degree upwards and downwards; my invitation he gratefully accepted, taking care to find out how many days it was there on horseback, and the names of the karavansarais on the way. To my intense regret his language did not betray that flowery ornamentation which I expected. It must be granted however that he was unprepared and at a disadvantage, not having had time to learn quotations from the poets in honour of his guest. Coupled with this exchange of polite words was the absorption, by us, of as many sweets and cups of green tea as we could conveniently stow away, his duty meanwhile being to watch and try to look exceedingly pleased. Having acquitted ourselves manfully with much honey which had flowed from our lips and that which had re-entered them, we took our leave with stately bows, commending the venerable lieutenant-governor to the grace of God. Through the door ajar we caught a last glimpse of him, his turban was already by his side on the floor, and the scintillating gown on the way thither and in the course of unveiling the white pyjamas that enshrouded the portliness of his imperious figure.

Next day we proceeded to Guzar, a small town nearer the mountains and situated on a route which we shall traverse later on.

CHAPTER VI

SAMARKAND

An architecture of stone dominates the high mountains of the Duab. Nowhere is there so much rock-hewn art overruling the vaults of ice and basins of snow; nowhere has so much been saved on the white lace and shining embroidery which drapes the Alps. The angular nakedness of stone lifts itself proudly to heights undreamt of. With the weight of Babylonian heaviness is mated the reckless daring of the Gothic style. Here we think again of the old problem of all masters, that ever stands before them as they build. How do I raise an unshakable fastness, and then, how do I lighten it by beauty, but finally, how can I endow the seeming flimsiness with a strength that convinces the eye? Necessity and joy, pressure and ease, how do I unite them? How can I combine the power of props and braces with the cowardice of vision? Men ask me to weld into beautiful symphonies of reality the two eternal elements of heaven and earth! Where is the golden mean that ever satisfies? Such may be the musings of one who lets his view sweep from the summits of Hazrat-sultan to Timur's avenues.

Where the mountains, whence issues the all-generating Zarafshan, overshadow the plain, there lies Samarkand, the queen of the world, like a lovely woman reclining on her couch; she who is mother and child, in whom are conception and birth. Beside her verdant bower the mountains stand with a paternity protecting and austere, while near her lies the untouched steppe. To them she is fulfilment and a promise, she the ever-youthful, beatific, and crowned with the glory of Tamerlane. Seeing her we feel that the towering giants of the south are a symbol of virility and the ardent plain, of hopeful desire. Otherwise they would mean the never-to-be of stern frigidity and unquenchable barrenness. Out of the passionate longing the creative power uplifts the miracle of growth and blossom, calling forth the young down of corn, the swelling bosom of the trees and the ripening fruit. Thus she of the thousand breasts reclines resplendent in the sun, and with tender fingers draws new virginity from the waiting steppe.

Bokhara is the commercial, Samarkand the residential capital. Thus one might oppose their characters, drawing a parallel with Glasgow and Edinburgh. One has the money, the other the charm. Bokhara is a yellow labyrinth of hamsters, where during business hours an incredible coil of humanity squeezes itself through the winding passages. From above one sees a desert of clay roofs, below which run the endless twisting mole-tracks in semi-obscurity. At Samarkand there is but a small city nucleus, while the rest of the straggling town must be looked for in the shape of peninsulas and islands lost among the trees. Undoubtedly Bokhara owes to trade its recent position as the chief place of Sogdiana, for only a great warrior like Timur can rule and build where he wills. Muhammadan fanaticism must needs centre upon Bokhara, where a native prince holds sway, where haughty officials pace the streets, where ascetic mollahs proclaim the unadulterated truth, where the people still keep up their traditions, manner and dress with almost demonstrative obstinacy, and, one must say, fortunately, to the unbounded delight of the lover of genuine folk life. The Samarkandi often dresses shabbily, drinks wine in public and frequents without shame localities of ill repute. But then he is also frolicsome and hearty, for he drinks the water of the Zarafshan, fresh from the mountains, while Bokhara only gets the dregs, with which it likewise seems to have imbued the internal decay of its outward elegance and superiority of manner. As my friend Yakub used to say, "the Bokhariot must wear a fine chapan, even if he has no money and nothing underneath, or else the Amir will cause him to be flogged." This statement which need not, perhaps, be taken too literally, nevertheless illuminates with fine sarcasm the irony of brilliant respectability.

But Samarkand is not dependent upon the originality of its inhabitants, for nature and the command of Tamerlane have richly endowed her. Curiously enough the seat of the government is still at Tashkent surrounded by muddy rice-fields. Thus Samarkand, the healthiest and most beautiful of all the cities of the Duab, remains a simple provincial town. But the few decades which have passed since the conquest are as nothing in the life of one who looks back upon a thousand years. When the mountains will have opened out to a flourishing industry of mining, she may yet awake to a magnificent future. Of the Russian settlement which stands about a mile away from the native town, there is not much to say, beyond that it makes a pleasant impression, owing to the foresight of General Kauffmann, the best administrator Turkestan ever had. Every street is a vista of poplar, acacia, willow and elm, hiding the low and unattractive houses on each side of the broad avenues where four carriages can drive abreast. The most imposing of the boulevards has six rows of lofty poplars. Unfortunately there is a talk that they will probably have to come down as many of them are rotten at the core, thirty years being a green old age for this fast-growing tree.

The position of Samarkand, 2358 feet, can best be compared to that of Munich (about 1600 feet above sealevel). Both stand on a high plain ventilated by bracing winds while to the south rises the serrated wall of snowy mountains. The distance from this is about the same, being forty miles (70 km.) in round figures, but with the difference that the Hissar range is ten thousand feet (3000 m.) higher. The view on a clear November day is very grand. The air is crisp, still and free from dust. Before us a maze of yellow houses nestling among the bulging wealth of trees now turning to bronze and gold; in the middle distance the turquoise cupola of Bibi-khanum flashing and glistening in the sun and behind it the immense peaks sharply outlined against the steel-blue sky. We know that they are distant because a mile-devouring stretch of flat country lies between us and them, yet they are near, rising upwards like the threatening wall of a thunder cloud, not merely forming a rough inequality on the horizon. In the retreating middle of the background tower the rugged, massive blocks of the Fan Group, now robed in white; on the left, the broad-backed end of the Turkestan Range stands like a rounded pier-head, while on each side these dominating points of the perspective are drawn out into lowering ridges which lose themselves in the western plains. A sight, indeed, to quicken the pulses of the mountaineer, for in two or three days one can reach the glacier snouts which propel their frosty cascades into the amphitheatre of Archa-maidan, the "Place of

Junipers." How little Russians must care for mountains, when one thinks that up there no summer resorts as yet exist. How would the English have jumped at this opportunity, for little is needed to make the road to Panjikent fit for speedy motor cars. Thence to a spot called Kuli-kulan would be a ride of half a day. Here, at a height of six thousand feet, lies a group of morainic lakes, fringed with juniper trees that are dark and stately like cypresses, and overhung by the rocky terraces and ice-falls of Chapdara. The tarns are fed by rivulets which meander over green pasture or lose themselves among thickets of roses and honeysuckle. But the country is still new and who knows if some day a Russian Darjeeling may not be founded in this lovely spot.

The native town with the mosques and medresses lies amid some broken ground cut out of a loess shelf by gullies and irrigation canals. To the west it is dominated by the old fortress whence a fine view may be obtained of the monuments around the Rigistan. A sharp eastern boundary is formed by the plateau of Afrosiab, where on the edge of a cliff over the reed bazar we can judge of the erstwhile magnitude of Bibi-khanum, erected by the great emperor to the memory of his favourite wife (Fig. 53). The buildings must have been magnificent beyond description, judging merely from the dilapidated portions of them which remain. Five hundred years of exposure have as yet not dimmed the lustre of the colouring on the glazed tile work. Turquoise and ultramarine still glow in their pristine splendour in every variety of pattern on the fast decaying domes. The arabesques adorning the walls with their graceful curves and the originality of their designs are a constant pleasure to the eye. In looking at the edifices of Samarkand one comes to the conclusion that, intentionally or unwittingly, the greater care or skill was expended on the external decoration. The glassy enamel of these wonderful tile mosaics is incorruptible, still showing a hard and fresh surface, while the masonry has come to the

52. A Boulevard at Samarkand.

53. View from the Edge of Afrosiab. The Reed and Timber Bazar with the Ruins of Bibi-khanum.

term of its life. Even assuming that frequent earthquakes are responsible for some of the havoc done, there can be no doubt that the artistic ideas and workmanship of the ornament are on a higher level than those of the structure. The brick material of the masonry is relatively bad and one can almost feel how people struggled with the constructive problems of large bulk. An observer of the mosques was meant, it would seem, to stand on the Rigistan only, for from the inner court the back of the frontal archway betrays itself as nothing but a screen, a veritable piece of stage scenery. Towards the square this fraudulent wall is covered with a lavish profusion of enamel, but behind it is a mere prop of accumulated bricks without the least effort at palliation or colouring. Evidently the architects were no match for the keramic masters.

This disproportion between building and ornamentation, this lack of durability accentuates the tragic aspect of the ruins. Here we see a coloured carpet knit in stone still shining with the glow of olden days, but dragged to perdition by the crumbling bricks; the vitality of the jewels is greater than that of the wearer. Five centuries are quite a decent age even for a temple, and contemplating these wrecks one may ask if we should not console ourselves with the inevitable. True, the Heidelberger Schloss dissolves with convincing necessity, but in Samarkand the spirit and will of beauty make a desperate appeal to sunny life while a decaying body pulls them to the grave. The question of preservation has often received serious attention, but all is vain, for the glazed tiles fall off, because the wall and cement cannot hold them any longer. Perhaps it is better to be saved from the hand of the "restorer," who here is powerless. Let Samarkand battle with death, heroically and regally, and let her die in beauty.

The authorities have caused large photographs to be taken of all ornamental detail to be found in Samarkand. Besides their artistic and historical value these pictures will afford us some clue to the rate of weathering, since the process of artificial disintegration due to curio hunters has been stopped by draconic punishments. As far as I could make out the north-western sides of the cupolas are most exposed to architectural "denudation." One asks if, after all, the climate was suitable for this style of outside decoration. Rain followed by hard frost and strong sunshine is a not infrequent combination and its effect

54. Ornaments of Glazed Tile in Shakhzinde.

upon the mosques can easily be imagined. The glaze is well nigh indestructible, but the joints between the bricks and tiles are lines of minor resistance. Earthquakes do not count for much, I should say, save as affording a geological proof. Catastrophes like those of Andizhan and Vierny would have levelled these buildings to the ground, so Samarkand cannot have been visited by such strong shocks for the last five hundred years.

A favourite belvedere is the roof of the Shirdar medresse or the top of one of its leaning minars (Fig. 55). Below is the Rigistan, a square of crowded life, filled with the hum of voices and the murmur of shuffling feet. Amid the white splashes of the awnings and framed with the wondrous embroidery of glistening tiles tremble the many-hued streaks of raiment and quivers the ever-moving scintillation of countless turban specks. The rising air, with its dust-breath of the land, carries up whiffs of ripe melon and of the nourishing softness of the fat-tailed sheep broiling in the booths. Once Samarkand also boasted a covered bazar which was, however, destroyed during the conquest. A small Russian garrison had been left behind as a guard after the surrender, but was fiercely attacked by a rebellious crowd when the main force had marched off to Kattakurgan. The brave soldiers held out against overwhelming odds until the regiments came back post-haste to their relief. As a punishment for this treacherous act the city was given over to plunder in the course of which the bazar was burnt down.

To the east is Afrosiab, the site of ancient Marakanda where Alexander rested from his battles and killed his friend Klitus. This place is an irregular plateau, about a hundred feet above the level of the town; a square mile of surface full of humps, hollows and ravines. On most sides it breaks off in steep loess cliffs (Fig. 22). There are also some flat-bottomed cauldrons or bowls without visible drainage, the two largest of which serve peculiar purposes, the one nearest the town being the knackery, the destination of numerous carts with their wobbly load of intestines from the slaughter houses, the other the baiga course, where congregate the horsemen fond of sport.

Afrosiab is oldest Samarkand and at the same time the second youngest, for on the topmost strata of time the generation before last is buried. The mazars, where pole, yak-tail and flag indicate a deposit of distinguished bones, are surrounded by the collapsing vaults of minor mortals. Many layers have here succeeded each other, some of the dead with tombstones and offerings, some of the living with houses or workshops. The whole of Afrosiab is full of remains from the days of Bactria to recent times, and with every step one kicks a fragment of the past. In some of the ravines potsherds, glass and coins can be found in great numbers carried thither by rains and spring floods, thus representing what might be called anthropological sedimentation. In

55. The Samarkand Rigistan with the Ullugbeg Medresse.

other places I have noticed ruined walls partly buried under loess in historical times. Take all the crockery and china of London, Paris and Berlin, smash it to pieces, mix it with two or three million cubic yards of clay and bricks and bones, such is Afrosiab. Everywhere one comes across holes where men have dug after coins and unbroken vessels or the hordes of pariah dogs have burrowed in search of well-preserved bones. We seem to sniff the odour of centuries, but on pursuing it come to the vultures' club, where brownish-reddish lakes of blood uplift their stench towards the fiery firmament and purple streamlets issue forth from piles of bowels that spread like sluggish glaciers.

The deepest and most varied deposit from the cross-currents of history has taken place nearest the town, where the road, after passing Bibi-khanum (the round dome of Fig. 53) climbs up, just in front of the modern mosque (Figs. 56, 59). Between here and the mazar in the picture of Fig. 57 the ground is still rough with the crumbled tombs of recent generations, and one frequently encounters fine marble slabs with inscriptions. Unfortunately for the archaeologist these too evident witnesses of modern burial check excavation in what is perhaps the most curious necropolis in the world. Further away the foundations of old palaces have been traced and partly laid bare by explorers from St Petersburg. At the extreme edge of the southern rim of this deserted plateau, where without transition it borders on the lower level of green luxuriance, stands Shakhzinde, a group of small mausoleums connected by a passage and containing, on the inside walls, the best preserved examples of tile work. Outside the buildings are almost white, the brightly coloured skin having nearly all peeled off, only patches remaining here and there. Yet there are some walls which plainly show that they never were adorned with those marvellous ornaments, and again we are struck by the apparent incongruities of Samarkand architecture. They seem so impossible to our minds that any other explanation is more welcome than that of difference in artistic standards. History was such in these countries that perhaps nothing was ever quite finished before a new conquest came and with it destruction or at least neglect of all things belonging to the previous rule. Timur has left the greatest monuments, but not even he seems to have found time for putting on the finishing touches which nearly all his buildings appear to have needed, at least to our taste.

Wherever one steps down from the loess heights of Afrosiab, one does so through a cutting resembling those which lead from the cliff tops of Margate to the beach. Framed in the opening of this passage one sees below the joyous green of the watered plain surrounding the dry island like a billowy sea of tree-tops. When the gloaming blackens the clumps and clusters they become a sombre wall, while the upper twigs and leaves still faintly glitter with a reflex from the silvery sky. North-east of the town, and separating it from the Zarafshan, is a low hill called Chapan-ata (see Map II and Fig. 57), crowned by a small sanctuary. To the left the distant Nurata hills are visible.

I have already mentioned that the baiga-ground is at Afrosiab. The baiga is the national game, taking the place of polo. A slaughtered sheep is flung on the course and the rider who succeeds in carrying it to the goal wins the game. Helter skelter a hundred horsemen rush for the prize. Oftener than not it is dropped at once by the man who first lifted it, owing to the pressure of the rushing crowd, or it is torn out of his hands by sheer force. Clouds of dust are raised and the furious scuffle lurches to and fro in a dense yellow fog. It is wonderful to see how a rider in full gallop will pick up the fleecy football, often weighing fifty pounds or more, and try to escape from the whirling knot of excited horsemen. Compared to this massive onslaught, swaying and swelling to the music of clattering hoofs, an American football scrimmage is a very mild affair. Sometimes a strong man on a fast charger will hold on to the carcase, the field after him in full cry, drawn out into a long, dark streak flecked with the foam of darting turbans. Obstacles are unheeded and a man frequently whips his steed up the precipitous loess slope of the natural amphitheatre, until the animal, well nigh spent, nearly falls backwards. Then he turns round in headlong flight hoping to break through the line of his pursuers who come clambering up the steep incline in hot haste.

The hapless sheep is ball as well as reward at a small affair or practice game, but on public occasions or at ambitious entertainments prizes of money or silken khalats are given. When the game goes well the sheep will lose its legs or may even be torn to pieces, the bearer of the most important fragment of its anatomy being proclaimed the winner. The goal is scored by throwing the sheep, or what remains of it, at the feet of him who gave the prize, or in whose honour the function was arranged. A grand baiga is given every year by the various Russian governors of the provinces. A single game will often last for hours, being won by the horse with the greatest stamina and toughest lungs, but serious accidents are very rare. After seeing such a performance one cannot wonder any longer why most horses are not faultless from our point of view and that the majority are crippled, however slightly. Every promising foal is tried at the game, in order that it may earn a reputation. What it is pretty sure to gain is some internal or external trouble. These people do not treat any sport with reason, scientifically, so to speak, as we do. Our idea is to know the exact breaking strain of man, beast or instrument, to extend it by training and construction, and to get as near as possible to it during the game or race. Here one simply "goes for it," regardless of consequences. Otherwise the people are just as keen as Europeans, and though not everyone is a sportsman, in the cities, the proportion of players to spectators is far greater at Afrosiab than at Kennington Oval. Psychologically one cannot help feeling a little puzzled watching the fast and furious pace, the enormous bodily exertion, the strength and skill of the men. How does this agree with our notions of their laziness and general worthlessness?

I think our judgment in this respect far too general and superficial. We know that these nations are behind us in technical civilisation, at our mercy in economic and technical wars. Searching for an explanation we jump to the conclusion that they must be loafers because we see them resting more frequently than we do, that they must be cowards because they generally run away before the battle. But the peasants work very hard and they can fight splendidly when properly roused and the battle has got into full swing. Probably they have their own wise reasons for not working more than necessary and for not fighting the frivolous wars of their princes. On the whole the races of the Duab are neither degenerate nor effeminate. More children die, but more healthy men remain. The sallow Sart of the bazar is every whit as emaciated or obese, as dyspeptic or livery as our merchant, shopkeeper or scholar. But their peasants, artisans and servants look quite fit; there is not that enormous gulf between wealth and poverty, between capital and labour. The standard of life of the population is practically uniform from the top to the bottom rung of the social ladder. The rich have unpatched garments, sleep on silk, keep several wives and eat more than the poor, but that is the only difference. Who has ever heard of the problem of the unemployed in these parts?

Such reflections, constantly recurring, have made me most chary at venturing general statements or passing judgments of merit on the traits and qualities of foreign peoples. Ethnography and folklore in a scientific sense are more and more becoming exact reports of facts

56. Bazar Day at Samarkand; the Carriage Depot.

divested of the words "bad," "beautiful," "brave," "lazy," etc. The subjective impressions of the writer reflect his own personal views and even in these he will apparently contradict himself. He is liable to calling a bad tribe good again before a dozen pages are past, having, of course, thought of a quality which shocked him in the first instance and of a, to him, laudable habit in the next. Likewise the opinions of an English traveller will not be interpreted in the same way by German and French readers. Consequently I believe that human estimations of foreign races—apart from purely scientific measurement and description—must be read subjectively, like a novel, with not too close a criticism of truth, for in popular and practical psychology there is no objective standard of truth, unless it be that of the reader himself or the public opinion of his countrymen. Whenever one tries to gain a just standpoint for the comparison of social progress, happiness, morality, health and wealth in different races, one finds oneself led into an endless sequence of equations for the reduction of relativities, imponderables and incommensurate values.

After all, there are two kinds of truth, the one we feel and the one we measure, the real and the ideal. Knowledge is a means to an end, is the organiser of the supreme driving power, which is will. The opinions individuals or nations have of each other, and their actions towards each other are ruled by their angles of view, their needs, desires and impulses. Science ignores the terms "pleasant" and "unpleasant"; even "hot" and "cold" are, strictly speaking, not admissible. Hence ethnology should not have opinions of any kind, but devote itself to the description of the shape and colour of dress, of marriage ceremonies, agriculture, and so forth. Now ethnology is a science of man and therefore also concerned with the feelings which other tribes have for us and we have for them. In their practical bearing these feelings are hard facts of greater importance than all other facts, scientific and unscientific, taken together. Diplomacy and politics would therefore be branches of ethnology. Like literary criticism, history, or the study of comparative art, ethnology strikes too near the human heart. Having to do with feelings it must often find itself tripping, scientifically and theoretically, over the subtle transition from the objective to the subjective. A poet writes a romance with the inventive help of his imagination; but from the point

57. Afrosiab; Looking towards Chapan-ata.

of view of exact science an observer can write an ethnological romance by accurately describing a people as he saw and felt it, without the least recourse to embellishment or exaggeration. Exact science stops dead when it cannot express a fact in the standard units of a scale or instrument, and the larger part of the domain of psychology will never be a science.

As long as we cannot say how many square inches of cruelty a Chinese has more than a Hottentot, or how many centigrades of love are the average in Eskimo maidens, so long the strictly scientific section of ethnology will be very limited. Yet we need not despair, for as often as statistics have destroyed superstitions, so often novels and impressions have proved more true or more productive of results than learned disquisitions. Let us be fair equally to the logical or systematic and to the impulsive or empirical side of human nature. Science is real but not realistic; it is a common bond and symbol, but one that must be retranslated into practice by the hands, energies, loves and hates of men.

I must ask the reader to gain a general impression of the natives from what I tell of them, but he should not expect a concise abstract, which is impossible. The sun can be very fully described on a page of exact figures, but a race, even the smallest, can only be painted in a thick volume by a scholar and artist of the first rank; or one can roughly sketch it by giving a few intimate details, by throwing occasional sidelights upon life and character. While this is all I can do, it would be futile in any case to attempt the elimination of all seemingly contradictory valuations of quality. It may even be said that such contrary judgments are strong proof of the truth of one's impressions, that one has not evolved them at the writing desk.

Differences are the greatest reality of life. I am sure of having often called a man good and bad almost in the same breath, but that is the daily experience of everybody. I have called the Sart clean, whereas I might just as well have described him as dirty, for compared to the average Englishman he leaves much to be desired. On the other hand, I am not quite sure if the worthy Briton always knows what he carries into his house with his boots, or is acquainted with the kitchen mysteries of his favourite restaurant. The three main standards—for there is no officially recognised academic, international metrical unit of dirtiness—

which are always unconsciously present to me and which I apply indiscriminately on the spur of the moment are: the well-to-do Anglo-Saxon, especially when he is also with-nothing-to-do, who is the maximum of cleanliness; the Polish Jew, who is absolute unsavouriness; and thirdly, an estimation of a kind of just, general average of international sweetness. The Sart represents this inter-racial more or less clean and dirty condition, being dirtier than the Englishman and very much cleaner than the Polish Jew. My wife in reporting

58. Gur-amir, the Tomb of Tamerlane.

the women and their quarters dirty, naturally had the nearest comparison in her mind, that is to say the men and their rooms, thus characterising the women as a lower grade within their surroundings. Should I ever call a Sart dirty it will be on account of certain individuals or special habits such as the drinking of tainted water.

While at Samarkand we lived in a native house for some time. A complete property of the inner districts generally consists of the dwelling house, a pond shaded by big planes or karagach (elms), a grove of larger trees, a young poplar plantation, a small vegetable garden and a vineyard, the whole surrounded by the inevitable wall. Rich people have larger estates outside, and no fields are found on the more expensive land near the city. The vines are planted in rows between irrigation furrows, their long tendrils being arranged over a frame-work of bent branches so as to form arbours which are however very low. When such a plot is in full leaf one looks across a thick, green tangle about breast high. Under this roof the grapes are hidden dangling into the shady tunnels below. Everywhere one stumbles across the water channels, for the running moisture must be brought to every tree and to every blade of grass.

Wherever anything grows there a ditch is sure to be, and owing to this necessity the cultivated vegetation is nearly always arranged in parallel lines or some symmetrical pattern of endless meander. The absence of lawn in the untrodden parts of a garden seems curious to the new-comer, but the channels are deep, so that only the penetrating roots of trees and larger plants derive a benefit. A thick crop of weeds or grass is therefore restricted to the banks of the conduits, unless we have to do with a patch or field systematically soaked. All vegetation is made use of. Often, when on a walk, have we tried to find a convenient resting place, but it was impossible to espy a piece of unused grass under a tree. The best seats in the shade are always the worse for wear, having long ago proved attractive to man and beast. The water here is very dark from the glacier-ground particles of slate. It seemed to me that at Bokhara the colour was more yellow, and this suggests a problem. Is it not thinkable that the Zarafshan takes a good deal of loess from the neighbourhood of Samarkand, depositing it on the fields of Bokhara? Is the lower oasis perhaps more fertile or more densely settled?

We often paid a visit to the river near Chapan-ata where its many branches stray widely over a shingly plain. Here there is much swampy ground with high reeds which represent a certain value, playing an important part in building and the manufacture of many household articles. We saw herons, egrets, large and small cormorants, black and white storks, plovers, an occasional flamingo, bitterns, ducks, ospreys, vultures, pheasants and quail, but all these birds were extremely shy owing to frequent molestation by gun-dischargers from Samarkand. Carruthers collected many small birds, most of them being varieties of species familiar to English people. Starlings, gaily coloured wood-peckers, and that jewel on the wing, the bee-eater, were among the most conspicuous of the feathery tribe. During spring the clumps of trees were swarming with cuckoos evidently on their way north. Near our house was a rotten tree in which our sharp-eared naturalist discovered a company of bats, betraying themselves by their chirping. Having watched the hour of their

59. A Modern Mosque at Samarkand (Hazrat-khaizar).

nightly appearance we held a butterfly-net over the hole and managed to catch six while another twelve made good their escape.

Knowing Samarkand at all seasons of the year I can only speak favourably of its climate, which strikes me as being the best of any large city in the Duab. The short spells of rainy season, chiefly in March and April, are disagreeable merely on account of the unfathomable slush to which they reduce the roads and courtyards. July, August and September are almost without a drop of rain. The dryness of the atmosphere makes the greatest heat bearable, to which must be added the sharp contrast between sun and shade. I remember once, how after a long ride I felt chilly in the shadow of a tree, but upon consulting the thermometer it was found to register 70° F. in the shade (21° C.). Late in autumn this condition almost approaches that of the Alps in winter, where one is baked and blinded in the sun and trembles with cold round the corner. There is much less malaria than anywhere else. In July the night temperature often went down to 60° F., while sometimes the thermometer would be as low as 80° F. (26° C.) in the middle of the day. In November 1907 I observed a very curious drop in the temperature in a short space of time. At 2 p.m. on the 8th the thermometer stood at 86° F. (29·5° C.) with a southerly wind; next morning the air was crisp (8.30 a.m. 6·5° C.), a beautiful autumnal day, the wind veering round to the north during the afternoon, whereupon followed clouds; on the 9th it was almost freezing in the morning and at 7 p.m. snow began to fall. Thus within less than 48 hours a change of temperatures which we are accustomed to identify with summer and winter. I certainly prefer the climate of Samarkand to that of any place I know in Europe, and the summer there seems to me far more agreeable than anywhere I ever lived. To this must be added the vicinity of the mountains as a factor of scenery and with the opportunities they offer for excursions. When in winter one looks across the ruts and ridges of frozen mud on the grand boulevard even the long line of the low Nurata hills is impressive, for the snow cap makes them appear ten thousand feet high.

How strong the light is in this region the photographer has the best means of judging. A view taken of a group of houses, say at ten or two, will give a pattern of flat, irregular geometrical figures, some chalky white, others inky black, produced by the sharp boundary lines of walls in the light, and walls in the shadow. At the same time the plastic detail of the trees will prove that the general exposure was perfectly correct. How the climate acts upon man, the traveller has the best means of judging, being out at all times of the day under all sorts of conditions. The sun cuts silhouettes into the land and fetches a wealth of colour out of it; the faintest tint is distinct from its nearest gradation, everything which can shine, glimmers and glitters. Coupled with this sharp light is the penetrating dryness which desiccates the body and stimulates the brain. Will can only be paralysed when threatened by that most dangerous and indirect result of drought, the collapse of the body through thirst. Of course this dry air is not and need not be an obstacle to that deliberate or instinctive laziness towards which man inclines in all circumstances and under all latitudes. But it is a glad idleness with which one withdraws to the cool shade from the glare and heat of midday, in order to appreciate the joys of existence, lolling under the spreading elm.

The day of the steppe knows not that state of weariness and lassitude, that sultry oppression apt to undermine the energy of the strongest character, that tropical moisture ready to destroy the sensations of pleasure and hope. The vertical ray of white heat cast from the zenith is not a caress and loess dust not a refreshment, but they are not creative of relaxation, for one remains attentive, active and rather snappy. The body is not being stewed in soft steam, but contracts without visible and almost without noticeable perspiration, having even to be protected against excessive loss of humidity by thick clothes. A European can keep his alertness, fitness and appetite, and is not subject to brooding. Ninety in the shade, unbearable on the Rhine or Congo are nothing in the Duab, while alcohol, so invidious in the tropics, can be consumed in normal quantities. The physiological effect increases with height, so that for instance at Tupchek (11,500 feet) one feels as if loaded with energy, a condition however which would turn into nervous surexcitation during a long stay at still greater heights. For us the dry air of the Duab has something stimulating about it, and perhaps that may to some degree account for Russian successes in Turkestan. Siberia was slowly colonised, the Caucasus vanquished after bloody wars with the natives, but Turkestan was subjugated with comparative ease. The Transcaspian railway was a piece of work almost American in the smartness of its execution, apart from the fact that Russian life and methods show many striking analogies with the ways of Uncle Sam.

I append a short report of Carruthers' journey to the Nurata hills in quest of wild sheep. These animals are ubiquitous in Inner Asia, occurring wherever there is a rise of the ground and forming many varieties of species. It is interesting to note that some of these mountain sheep are almost animals of the plains, coming down to the level of the steppe just as the last outliers of the mountain ranges lose themselves in the lowlands. My friend wrote to me as follows in January 1908, "I have done the Nurata trip, being away 15 days. I had to go four days N.W. of Bogdan (near Jizak) before I found the arkhal. At the extreme end of the range I killed four, one a magnificent old buck. I have prepared the whole skin complete. As far as Bogdan I was able to take a cart, after that walking with donkeys and coming home over the range to Kattakurgan. I tried for gazelle in the desert to the north of the hills, but found none. Of birds there were very few, but all I got were new, including a fine golden eagle. I shall be interested to see what the sheep turns out to be; it may be the same as the Koppet-dagh one or it may be another variety. It was fearfully cold and I got back just in time, for on the following day we had a very heavy fall of snow."

CHAPTER VII

THE ASCENT OF KEMKUTAN

STANDING on the edge of Afrosiab and looking towards the south one sees between the two blocks of the ruined Bibi-khanum a grey mountain outlined against the sky (Fig. 53). According to the Russian maps the highest point visible from Samarkand is called Kemkutan (7268 feet; 2216 m.). It is the culminating summit of a group of hills forming part of the large mass between us and the valley of Shakhrisiabs, which are sometimes known under the oft-recurring name of Aktau. Rising, as it does, nearly five thousand feet above the city, and by reason of its conspicuous shape, Kemkutan must needs tempt the mountaineer. On its slopes the members of the Samarkand Alpine Club received their first lessons in climbing and hill-craft, but I believe that by now they have found out that it is much wiser to admire mountains from below. In company of my wife, Carruthers and others, I have visited the district about ten times, including three ascents of the highest peak. As some of these excursions were made at different seasons, I shall give a separate account of four of them.

To the village of Agalik at the foot of Kemkutan is a distance of ten miles which we generally traversed in a carriage and pair. This sounds very distinguished especially when no less a personage than Phaeton has stood sponsor to the vehicle in question. But the Russian "phaeton" is a four-wheeled hackney carriage of the victoria kind, which can be found at every street corner, unless one happens to be in a hurry. As a Russian does not walk when he can help it, and as the tariff is very low, one meets this conveyance in every colony, be it ever so small, of the Caucasus and the Duab. There are always one or two at the meanest railway station; there are a hundred waiting for the train at

Samarkand; at Northampton there would be five hundred, at Charing Cross at least five thousand. Although not over clean and often held together by bits of rope, they have springs that seem to laugh at any bump under three feet high and a load under a ton. As long as the road is just wide enough, the crests and hollows of its waves not more than six feet apart, and the river ford not deeper than to the horses' necks, the driver will take you anywhere and without notice, if the

60. Agalik with Kemkutan.

distance is under ten miles. Unmindful of obstacles the two lean horses keep up a gallop most of the time and the native Jehu has a very quick eye for the estimation of depths and angles. In spite of violent lurches among the breakers or a heavy list from shifting cargo I have so far not been capsized, although it once was my happy lot to get stuck in the mud owing to a giving out of motoric force.

In October 1907 we made our first attempt. As far as the Dargom Bridge (see Map II) the road passes through the garden country, but beyond the river there is a broad belt of dry steppe. The Dargom, flowing between vertical loess walls about 120 feet high, is a branch of the Zarafshan, and began life as an arik or irrigation canal, for such it is called. The top of the loess shelf at the bridge is 200 feet below Rakhmetabad, the point where the Zarafshan was tapped. No difficulty lay, therefore, in bringing water along this line by a channel of moderate or normal depth and of very gentle grade. But from Dargom Bridge down to the main river there is a drop of 600 feet, and over a shorter distance. The water descending this slope with a steeper and therefore more rapid current began to cut down into the soft material, gradually deepening its bed backwards, i.e. up-stream, in order to produce an even fall from end to end. It is the tendency of every river to attain uniform grade from source to mouth, and the same law caused the overfall of the Dargom arik to dig a trench through the loess table, so that now it has nearly accomplished an incline of sixteen feet per mile, instead of running almost horizontally for a long distance, followed by a sharp drop. The length of the Dargom is fifty miles, that of the Zarafshan, between the same points, forty miles, the difference of level 800 feet, so that the Dargom by reason of its greater length will finally have a gentler grade than the natural river. So far the task is not quite done, for at first it is still horizontal while there are some mild rapids in the chasm which begins several miles to the west of Juma-bazar.

The Agalik steppe is deeply gashed with rifts and fissures, which one does not see until close to the edge. Sometimes the water from the mountains has wormed its way underground, collecting into subterranean streams which have burrowed tunnels and crevices, the roofs of which cave in from time to time, forming funnels and pot-holes. Large cubes of clay are strewn about the sole of the canyons; at the sides where faint trickles have occasionally escaped they have modelled pillars and cones out of the riparian bluff. These, expectant of their ultimate collapse, form a salient feature of the loess together with the crevasses, sink-holes, and nullahs.

Owing to its natural hollows and the facilities it offers to cave diggers the yellow earth harbours a vast population of animals. It is late in the year now and nothing is visible beyond some blue pigeons, a few hungry birds of prey and an occasional fox or hare. But under our feet sleep myriads of creatures, innumerable species of insects with their eggs or larvae, many reptiles, and several kinds of amphibians and rodents, all awaiting the call of Baldur's wand. When the steppe has donned its garb of spring there is much creeping and crawling, buzzing and humming by day and by night, and the steep banks are alive with bee-eaters, swallows, rock-pigeons, falcons flying to and fro from their nests, to be relieved, after sundown, by owls and bats.

On our way we see small hills from forty to sixty feet high, rising out of the flat surface and conspicuous from afar. Known as "kurgans," a name occurring frequently in place names, these hillocks are too big and shapeless to be ruins of buildings, too soft to be remains of denudation by water. Hence they must be artificial mounds such as were made in the olden days around the important towns. Declarations of war were not customary in those times and every city had to be on the guard against surprise. But as the plains offered no view points these kurgans were heaped up to serve as watch-towers, and they were most numerous in those directions whence an attack was most likely to come. Thus a long line of such look-out knolls extends from Tashkent for fifty miles on the road towards its old enemy Kokan. At present they are generally occupied by a buzzard or hawk surveying his hunting ground.

61. The Top of Kemkutan.

The first sign of dense vegetation is a small garden in the steppe watered by what remains of the Agalik river after having passed the village. A Russian is trying to cultivate this plot of land which has become his own on the strength of the old law, which awards all land to him who first reclaims it. It does not seem to thrive very well, however. And hereby hangs a conundrum. Why do the inhabitants of Agalik live inside the somewhat stony valley where scant space and an uneven surface must make tillage and irrigation difficult? I have asked some of them, to which, of course, they replied that there was not enough water, which goes without saying, considering the fact that it is used up by the village fields. What I want to know, and what the native cannot explain, is why they went into the valley at all instead of tapping and distributing the river as it emerged upon the fertile steppe beyond the foot of the mountains, as is done in the case of the Sokh (Fig. 34). As Samarkand thrives in the open it cannot be a question of valley shelter, wind and temperature. Does the high loess absorb and evaporate too much water—owing to its porosity and depth—and did the village for this reason grow up-stream when the people saw that they could not extend it outwards? Thus the first settler may have been further out, but in the course of centuries succeeding generations sought to capture the water above the fields of their fathers, so that gradually the whole of the village withdrew into the mountains, the lower peasants not being able to go outwards or remain outside, because the upper ones left less and less water for them. So now Agalik is a broad triangle of gardens (Figs. 60, 64) flush with the rim of the mountains and extending upwards in a green strip to the last houses of Upper Agalik.

At the entrance to the valley there are some outcrops of limestone strata giving employment to native lime kilns in which nothing but the dried scrubs of the steppe are used for fuel. The process is most elaborate, each kiln being built afresh every time in the shape of a cupola of round stones bound together with clay. The mountains consist almost exclusively of granite, quarried on a large scale by the Government which uses the stones for the erection of irrigation dams on the Imperial Estate at Bairam-ali near Merv. The numbers of heavy carts have made the road no better than it was, much to the disgust of the Russians who have bought gardens and built summer houses at Agalik. There are several dozens of these "dachas," as the Russians call them, and some of them have been made very beautiful with flowers and ramblers in the European style. There is a complete absence of malaria, the nights are very cool and the river affords excellent bathing, so the well-to-do send their families here during the summer, visiting them for week-ends.

Having stood still at one time, looking at the view, I distinctly felt some slight shocks of earthquake which the others did not notice as they were walking. It was one of the many vibrations which followed upon the destruction of Karatagh on the 21st of October 1907. We felt that same shock at Bokhara, but it was stronger at Samarkand, causing, however, no injury, save a panic.

On the way we met many peasants taking their produce to town. Live fowls are carried on the pommel of the saddle, head downwards, struggling frantically and often arriving with broken legs in consequence. Little donkeys almost disappear under huge bundles of straw and chaff, the latter being transported in large nets. One often sees stones wedged in between the ropes on one side to make the load balance properly. At Agalik we found Carruthers, who had preceded us some days before, and who had already amassed a good collection of birds, rats, mice, shrews, and other rodents. Not being acquainted as yet with the topography of the neighbourhood we decided to try one of the nearer summits. We chose the one visible in Fig. 60, just over the big trees on the left; Kemkutan, which is much further away, being the high peak nearer the middle of the picture. These hills are arranged on the usual system of branching spurs, not on the parallel pattern so frequently observed in the Duab and developed to perfection in the Nurata Range and the ridges of the Turpak Pass (Fig. 196). Thus their ups and downs are similar to the general formation of the Welsh and Cumberland mountains. But the divides are never deeply notched so that no individual shapes are singled out from the mass which rises as a whole, however deeply riven and scarred its flanks may be. All culminating points are merely accidents of surface, like rocks on a crinkled mound and from a distance we see a broad, curved saw with fine teeth against the sky line. The ridges rise as grass-grown whalebacks from the plain, but small blocks soon begin to pierce the skin from the underlying granite core. The higher we go the oftener we meet sudden outcrops of heaped up and cloven boulders until we reach the summit plateaus where grassy plots are intermingled with abrupt cox-combs of isolated cliffs. In the narrow valleys, of course, the granite heart is cleft to the ground and their sides represent a wonderful spectacle of cyclopean buttresses and stairs.

The biological character is entirely that of the steppe of which these hills form part in a climatic sense. I was therefore greatly surprised at the size of the stream, whose water must come out of the rock, seeing that there is no permanent snow, and there had been practically no rain since the end of May. But granite is a good water-holder owing to its many fine clefts which retain a considerable store allowing it to percolate slowly. In order to get some idea of this capacity we might calculate as follows. From Kemkutan to the plains is two miles in a straight line, so that this mountain dominates a plane surface of sixteen square miles. Allowing for valleys, etc., this block contains about three cubic miles of granite, a third of which drains towards the Agalik river.

62. Among the Rocks of Kemkutan.

This cubic mile has a total of internal surfaces of 7000 square miles, if we assume that the granite is not cleft into blocks smaller than half a cubic yard. Then we obtain, in round figures, 20,000,000,000 (twenty thousand million) fissures of one square yard between two walls. Each of these clefts can hold, let us say, one cubic inch of water, allowing it to flow together slowly. This gives seventy million gallons or a steady flow of eight gallons per second for three months. In this manner one can easily understand how it is possible that absolutely rocky and barren hills can still supply us with abundant water after many months of drought.

We penetrated into one of the side gullies fringed with fantastic rocks and traversed by a tiny creek of crystal water. Here we found a charming idyll hidden among the arid ridges, a hut with a few trees around, a clump of golden reeds surmounted by its silvery feathers, and a miniature waterfall. We saw dippers and snipe along the edge of the streamlet, wall-creepers (*Tichodroma*) and a steel-blue thrush on the rocky slabs and, curiously enough, a great bustard, an animal which inhabits the plain, as a rule, thus furnishing another instance of the gradual transition from lowland steppe to mountain steppe. Rock partridges (*Caccabis*; a relative of the chikor) were plentiful, but already somewhat wild. We safely accomplished our climb, judging the summit to be about five hundred feet lower than Kemkutan, and then returned home after a cold drive in the chilly wind of the steppe.

We resumed our attack on the 15th of December, the party including my wife, Miss Sabina Rickmers and Carruthers. Going up the main valley as far as possible we spent the night at Upper Agalik (3400 feet; Lower Agalik 3000 feet) in the house of the aksakal or village elder. After a frosty night we started at six o'clock next morning in order to have plenty of time for getting over the lower portions of our route. Although our goal is not really far, a number of gorges intervenes and one has to tramp along the main valley until one can strike a convenient side gully. The path twists itself round many spurs with the contortions of an epileptic snake, making the distance three times longer than it is found by the flying crow. Emerging from the labyrinth of rocks we ascended the open, northern slope of our mountain, but found the going very disagreeable as the ground was frozen hard as stone and the thin grass covered with a light hoar frost. Still higher up we trod on snow which lay in drifts and patches among the boulders, forming treacherous pitfalls. The track of a hare and two lämmergeiers circling in the air was all we saw of life. Until 11.30 we were continually in the shade which only yielded to bright and warm sunshine when we reached the broad, plateau-like ridge, from which the final summit rises in the shape of a solid block of carved granite. On smooth slabs exposed to the genial south we lunched and rested, then roping together made the last assault, to which Kemkutan capitulated on the 16th of December 1907, at noon, this being the first time man had set foot upon its crowning crest.

63. The Summit Rock of Kemkutan.
(A flaw down the middle of the photographic negative has been repaired.)

The clear view was magnificent. First we turned towards the east, where the great peaks of Hazrat-sultan stood fiercely above the smaller fry. There was something almost uncanny about their grandiose bulk with its ice-cold shadows and glistening diadems of ice. In the basin of Shakhrisiabs wide stretches of water, run astray, were blinking in the sun. Letting one's eye sweep over the northern expanse one saw the horizon bounded by the low, even wall of the Nurata hills. In front of these lies the oasis of Samarkand stretching towards us like a dark shadow, which does not however reach us, being bordered by a strip of arid yellow. At our feet the furrowed slopes descend to meet the belt of steppe. Into it debouch the funnel-shaped valleys like estuaries where plain and mountain intermingle. These funnels are marked out by the garden villages which fill them from a narrow beginning in the river gulch to the broad base of the alluvial triangle, almost cut flush with the outermost rim of the hills. It is as if the groves of Agalik wanted to expand and merge into the bounty of Samarkand, but that the steppe was meant to intervene as a gulf of cruel separation.

The rock formation of the summit of Kemkutan is peculiar and repeated by all the other summits, although to a lesser degree. The knob sticking out of the plateau is a solid mass with a bulging, flowing smoothness that endows walking and scrambling upon it with a special charm. It is neither regular nor polished, but looks like the solid piece it is, bossed with swellings, lumps and globes betraying a firm connection with the main mass. There is no part which suggests that it has been stuck on, or broken off, or that it is about to fall. There are big basins sunk into clean rock (Fig. 62), fissures and clefts, but these hollows, grooves and flutings are scooped out and all edges are rounded off, so that the surfaces of the stone sections run into each other by longer or shorter curves, not being bounded by broken lines and acute angles (see Figs. 61, 63, 66). The divisions of the rock are due to cleavage and are seen everywhere, the rounding off of the edges is due to weathering and may also be observed lower down; but the general, sweeping lines of the solid summit block must be ascribed to glacial agencies, especially to what is called "glacial protection." This term expresses a theory which, stated in a few words, maintains that in mountains the portions covered by ice have been less affected and destroyed than those open to the attacks of air and water. Ice only scraped off layers of rock, leaving soft, broad surfaces and

64. View towards the Plains from Kemkutan. For locating the Moraine compare with Fig. 65.

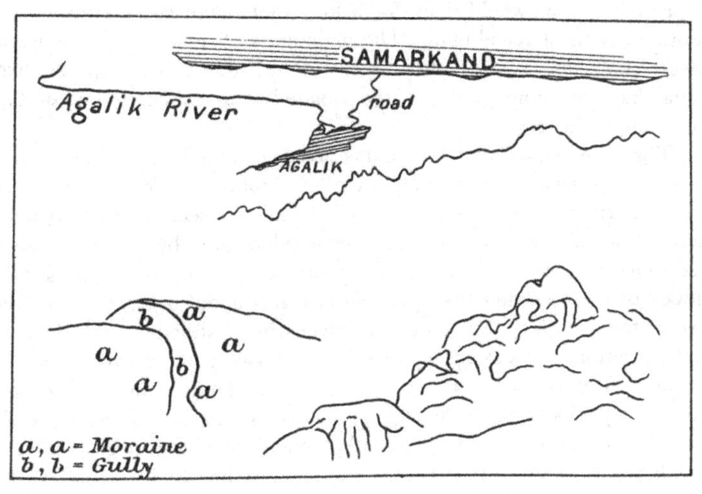

65. Diagram explaining the View of Fig. 64.

trough-like valleys or basins, while water eroded chasms, cut fissures, and washed away all friable material, thus creating gashed flanks, serrated crests and deep gorges. The question is a very complicated one which cannot be answered one way or another in a summary fashion, for much depends upon local conditions. But on the whole one may state with reasonable safety that ice works with a lesser waste than weathering and water, that it does not weaken the bulk of the mountain so much by sawing and undercutting. "Protection" is therefore to be taken in a relative sense, meaning "lesser destruction." That glacial protection was active on Kemkutan seems to me proved by the fact that extremely few loose blocks or other débris are found at the base of the top-knot, while piles of tumbled stones occur in the lower valleys. Hence weathering cannot have made much progress since the summit was cleared of its ice cap at the termination of the last period of the great ice age. At the same time this will afford us some clue to the enormous length of time required for the disintegration of granite in this climate. Later the culminating blocks on the broad ridges and small plateaus were probably not entirely sheathed in nevé but only covered with broken patches of snow or tongues reaching up a gully; a snowfield may, for a long time, have lain on the gentle southern slope of Kemkutan. The main mass of the ice rested between these pinnacles, so that they stood out of the white expanse like nunataks, the name given to the isolated crags sticking out of the Greenland glaciers.

That the Samarkand mountains were glaciated to a considerable degree is proved beyond doubt by a fine moraine which I discovered on the northern slope of Kemkutan. It can be seen on a photograph taken from above (Fig. 64) and rendered clearer by an explanatory diagram (Fig. 65). Forming the gentle outrun of the long, grassy slopes of the mountain its upper surface lies about 1500 to 1700 feet below the summit. Its outer or lower rim distinctly marked by a curved outline has a steep drop to the main valley of the great basin or cirque. A gully has cut the moraine to its foundations laying bare huge, grey blocks protruding from the sides and still retaining their sharp edges, as distinct from all other boulders of the neighbourhood, which are rounded. The branch of the moraine to the right of the gully abuts on the side wall of a small canyon cut through solid rock, by a stream, so that here the end of the glacial deposit is safeguarded from erosion. This phenomenon is the third example of the kind I have found in the Duab and will be more fully described in connection with the glaciers of Tupchek (Chap. XV) and the barriers of the Dandushka (Chap. XVI).

At the mouth of the gully the morainic accumulation is about 150 to 180 feet deep, but this thickness decreases rapidly towards the top where the banked-up talus thins out into a veneer of pasture land which still further up is soon pierced by solid rock. The surface of this moraine, which forms one of the best grazing plots of the district is strewn with loose boulders of great size. Such a powerful witness to glacial times always makes a strong impression in a mountain landscape now dry steppe far below the snow-line, which, in the Duab, lies from six to seven thousand feet higher than the summit of Kemkutan. As this latter is a little above 7000 feet and as the moraine ends at about 5000 feet (1500 m.), the snow-line in the Duab during an important period of the ice age must have been somewhere near 6000 feet (1800 m.). One feels tempted to put it a good deal below 6000 feet in order to account for the great bulk of the moraine compared to the moderate size of the Kemkutan basin or corrie, but it will be safest to say that the snow-line was not above 6000 feet. It is not necessary to assume a vast glaciation in such a case as this where the slopes leading to the "dump" are very steep. Moreover favourable conditions of preservation often make the moraines left by small glaciers look more important than those of larger ones, but which were partly destroyed by subsequent erosion. A morainic heap under an insignificant but steep glacier is, so to speak, an "indirect" shoot-cone or scree slope, because such a short and broad ice patch has hardly any pronounced medial or lateral moraines, but ends on a declivity before reaching the valley. There may have been something akin to the Alaskan rock glaciers described by Capps.

Strong testimony of the individuality of the glacial phenomenon is, however, borne by the characteristic stamp it sets upon all results of its activity. Even the talus which has, as one may express it, merely fallen through a snowfield betrays by its internal structure the moisture and muddiness of its short intermediate condition. A dry cone of débris does not easily cover itself with sward, especially in a dry climate, whereas a moraine does so very readily, turning into grass-steppe in the Duab. The larger and longer a glacier, the more conspicuous its great forms become in the landscape; the more distant and complicated the road from source to end the more independently the glacier works, the less its role as an intermediary, the grander its own creations out of the raw material. Even an expert might overlook the Kemkutan moraine unless he was on the watch.

As to when this moraine was built, whether at the Günz, Mindel, Riss or Würm period, I cannot say with certainty. While it is easy to trace morainic or fluvio-glacial deposits on the higher mountains of the Duab this becomes despairingly difficult low down or in the plains on account of the thick covering of loess which hides the different features of the underlying ground. Nor have the known deposits of higher levels been classified, owing to the small number of observations. Thus we have no base in the plains like that afforded by the Bavarian layers of morainic material discovered and explained by Penck. Unless one has the whole scale, identification is very speculative. But assuming that the Alpine and Duabic periods ran parallel, against which there is no serious indication to the contrary, so far, we may venture a careful guess. According to Hess the last glacial period in the Alps (Würm) was the coldest with the lowest snow-line. For the Oglio Glacier in the Eastern Alps he puts it at about 4000 feet (1200 m.), while to-day it lies at 8200 feet (2500 m.) on an average, so that on this basis we must ascribe the Kemkutan Moraine to the Würm period. Simple as this may sound I record this opinion under protest, for there are many complications to be considered. To-day the Duab

66. A Nunatak on the Kemkutan Plateau.

snow-line can roughly be estimated at 14,500 feet (4400 m.), that of the Alps at 8800 feet (2700 m.), to take a high figure, the difference therefore being 5700 feet between the modern Duabic and Alpine snow-lines. But there are 8500 feet of vertical distance between the ancient Kemkutan and the present Duabic snow-lines, while in the Alps this difference only amounts to 4800 feet in round numbers. In that case the rate of desiccation or deglaciation during the post-glacial period must have been more rapid in the Alai-Pamirs than in the Alps. Assuming our figures and suppositions on both sides to be correct there was only a difference of 2000 feet (600 m.) between the Würm snow-lines of Alps and Duab, whereas now it is nearly 6000 feet. The ratio of desiccation was quite different in the two mountain groups, which seems to corroborate a foregone conclusion as to the extremes and rapid changes of a continental climate. The insignificant quantity of weathering and erosion which has taken place on Kemkutan since the moraine was formed also warns us against dating it too far back.

Leaving the summit at 3 p.m. we turned round its eastern side and crossed part of the plateau before descending into the cirque. With their ledges and chimneys heavily cloaked in white, some of the bold rock flanks paraded effectively as challenges of alpine magnitude to the mettle of a climber, while their general appearance reminded me strongly of Caledonian fells in winter. The flattened ridge over which we walked bore some resemblance to the irregular plateaus of Dolomitic regions and more especially to the rugged surface of the Karst. Moreover, the sculpture of the granite discloses traits of kinship with the grooves and flutings of that peculiar limestone erosion known as "Karren." In places we crossed a space smoothly covered with a foot or more of snow, where, looking in certain directions one saw nothing but the dark granite knobs thrust from the even, white sheet, like tors from a down.

Surveying the long range of the Samarkand mountains as it receded into the distance, one was forcibly struck by its solidity. Water has scooped out valleys and etched the flanks, but it has not succeeded in cutting through the ridges and dissolving the main mass into separate peaks. As far as large features of the landscape are concerned river erosion has not yet encroached upon the higher level which still shows the typical imprint of snow pressure. The central spine of the system is one uninterrupted hog-back, sometimes laid out in fairly regular zig-zag, with a branching spur at each corner of the rampart and a cairn, like Kemkutan, to mark the spot. One had only to imagine this winding monster covered with a thick back-cloth of snow with white flaps, shreds and tassels hanging down into the folds of its wrinkled withers. I cannot for a moment believe that considerable changes have taken place in the relief of these mountains since the last glaciation. The stream which has furrowed a trench through the old moraine has not yet scoured the bed-rock to

67. New Snow on a Ridge of Kemkutan; looking East; Kughi-surkh in the Distance.

any appreciable extent, while the gorge which I have mentioned may owe much of its depth to the glacier stream, the erosive activity of which went concurrently with the heaping up of the glacial detritus. Some of the larger currents of water may have gnawed vertical gorges into the granite, but the denudation of the hillsides, of broad surfaces must have been practically nil.

Descending towards Agalik we soon entered the region of boulder cascades. Where the granite was not polished by ice to resist weathering, and where, lower down, it is not covered by turf, many steep tiers and ravines have been reduced to gigantic piles of stone.

Here the fastness of the mountain has slowly broken asunder under the influence of air and rain, in the course of thousands of years, but there being no means of transport the wreckage of the breaches has remained where it fell, or simply sunk together. The fragments of the granite wall are never far from their original position, for on the whole they have merely settled down owing to the removal of substance from the joints between the ashlers. These embankments of broad-stones suggest the disintegration of the cyclopean walls of Mycenae or the collapse of Egyptian temples and pyramids. Heated by the sun, moistened by clouds and scrubbed by winds through centuries untold the surfaces of the blocks have been corroded, their angles and edges rounded off, so that quaint shapes have issued from nature's workshop. There are high piles of sacks as in a store house, or the contents of a ship's hold turned out into a tumbling jumble of goods in boxes, bags, squashed bales, distorted casks and water-skins; or a heap of broken toys and oddities a child loves to collect, curious shapes and figures, rumps, limbs, animals, tree trunks, cannon balls, kettles, or anything fancy cares to restore from the torso of a form. Sometimes we find the swollen legs of a giant suffering from severe elephantiasis or the strangulated columns of a cactus. Here the mountain turns towards us the lamellae of an enormous toadstool; here queer fish are huddled together or fossil corals have joined to raise a mound; there overhangs the bulging side of a sponge, that must measure a hundred feet, full of round holes and concave cells. Elsewhere stunted turrets, contorted banisters are sketched out of the rock-wall, and rude pilasters run up the sides of a staircase of titanic blocks. Innumerable is the furniture of fairy land we thus can call to life from childhood's gladdest dreams. Even science obeys the spell and calls these masses "woolsacks" owing to their soft and pliant bulk. They are cubical, bench-like or elongated segregations of the rock, separated along their joint planes by weathering and corroded, honey-combed or marked with small-pox by the chemistry of air and the sandblast of wind.

Owing to the strains and pressure to which a heavy mountain is subject, even the "healthy" granite far below the surface is traversed by fine cracks dividing it into angular pieces of different shape such as obtained by breaking a sugar loaf. The mountain is such a loaf, cracked but still held together, except on the outside where the uppermost blocks are laid bare. Weather works along the lines of breakage and cleavage, rotting the surface of the blocks, while running water or wind rinses or blows out the fine grit and powder. Thus the exposed stones become gradually detached from each other and their edges are rounded off. As these boulders are internally sound they do not split and crumble readily, nor can the weak streams pound them, or, still less, carry them away. On the whole these blocks remain near the same place or slide imperceptibly towards some convenient gathering point, there forming loose piles. Only from very steep places they will crash down to the foot of a cliff. Sandstone is equally ready to form woolsack shapes as can be seen in Saxon Switzerland or on the picture of the Baissun sandstones (Fig. 200). The property is shared by all homogeneous masses cleft into large sections solid within. Thus at the lower end of an ice-fall we see the seracs and ridges between the crevasses robbed of their sharp corners by the sun and forming a labyrinth of tapering cones and rounded backs.

Driving home we met excited people who had just witnessed an attack by masked men at Dargom bridge on the cashier of the quarries. Loading our guns with the biggest shot we proceeded to search for the robbers, but they had already vanished in the labyrinth of loess canyons, so we got no chance of a nice shot.

Our next expedition to these mountains took place on the 19th of April 1908. The poplars and karagach were just budding; the fruit trees stood in the full glory of their snow-white bloom, and peeping over some low garden wall one could see the upright twigs of the peach studded with pink blossoms rise from a knee-deep field of juicy lucerne, or admire the round white clouds of cherries in flower lining the brink of water rills. The vineyards were being uncovered and put in order. The road was mostly dry, but here and there careless people had allowed ariks to overflow into lakelets. Then the driver would whip up his 2 h.p., make a rush across the slope of the pool with one wheel on the dry incline and the carriage dangerously tilted, while the inside wheels swished through the yellow water like the keel of a racing yacht. Carruthers dropped a kite from a high mulberry tree near the mosque. We met many Sarts on their way to the grand baiga for to-day is a first-class tamasha.

The steppe is just sprouting forth a tender down, but the spring birds have not arrived yet, flowers and insect life seem to be behind time. Our goal was the lower summit, but we found it hard work, for the sun was broiling owing to moisture in the air, and I came very near a heat-stroke. Hovering over Afrosiab we saw a towering column of dust that gave tidings of the baiga in full swing. On the very top thousands of lady-birds had congregated, crowded together, hundreds at a time, among the lichens and crevices of the highest rocks.

Again on May 3rd. The steppe is brighter and full of fighting tortoises clanking their cuirasses as they pound against each other with a rush. A fair one is the prize and great are the deeds of valour. Among the green trees of Agalik the gay birds were holding converse, discussing marriage, house-hunting or a continuation of the journey, for who had ever seen a place infested with so many shrikes, half a dozen sometimes sitting on a single thistle bush. Sparkling bee-eaters rocked themselves on slender branches, the roller was blue in a blue sky, the oriole whistled its familiar song, red finches, buntings, larks, and all that bright company filled every hedge with life, while the cuckoo with officious voice announced the opening of the season. Two miles above the village and beyond the radius of action of the week-end-shooter we put up a couple of breeding rock-partridges about every hundred yards. Investigating a new side valley we stepped through a sombre gorge and up a flight of boulder stairs, suddenly to emerge into a shallow basin carpeted with green herbage, and dotted with buttercups. Around was a circle of battlements and cleft rocks, making a haven of retreat of this hollow in the mountain. A yurt (nomad's tent) was standing there and a few cattle were browsing on the meadow. Near by was a high, bulging precipice where in a narrow slit a large colony of swifts were already feeding their young, busily flying to and fro.

Our longest day was the 10th of May, when we tried to explore the upper reaches of the Agalik river with a view to reaching a pass leading over to Shakhrisiabs. We had taken horses in order to save ourselves the weary tramp along an endless path bent in every direction of space. In one lateral ravine the government has made an experimental plantation of ailanthus trees. They cover a slope several hundred feet high from top to bottom. Those living low down are already full-grown, but half way up they are getting smaller and smaller, until at the edge of the crest, we see crippled individuals doubled up with an incurable stoop acquired from the wind.

After many hours we came to a confluence of two streams, and choosing the left or larger one we gradually got to high and open pasture ground alive with natives and their herds. It was too late to climb the final ridge, but we obtained a good idea of the country. The surroundings were quite alpine at this time of the year, patches of snow still lying on the divides and undulating folds of emerald coloured grass sweeping down to the noisy torrent. Here we made a fire of pine cones which we had brought from Samarkand and boiled our afternoon tea.

68. The Plateau of Kemkutan; looking East, towards Panjikent.

Looking at some of the summits it seemed to me that one or the other might easily be higher than Kemkutan.

When we had returned over the worst part of the road someone suggested that we should make a night of it, for the sickle of the moon was already fairly fat in the pale sky and would shed sufficient light on the homeward track. At a nullah which seemed lonely and deserted we alighted and left our horses in charge of our servant. Here the grass was already parched or eaten and much to our disgust we found the trodden circles indicating the former position of a hut. Pools in which water collected had been covered up very carefully with sticks and stones in order to keep earth or small animals from falling into them.

Near the summit of the ridge we discovered three groups of huts, hidden in basin-shaped recesses among the reefs and bars. These were inhabited by a motley crowd of ragged men, women and children, but they were very friendly, calling off their dogs and inviting us to step near. A few thin tricklets of water were flowing here from springs among the rocks; in one corner there was even a willow tree and the camp looked very picturesque indeed. On the wet soil I found many naked snails. In the air were a few vultures, and kites descended quite close to wrangle over a bone with the dogs. A few hundred yards further was the top of the broad back of the range, where we enjoyed a wonderful sunset. Blood red the orb of day sank down between yellow streaks, casting his last lurid rays upon an ascending spur beset with teeth and looking like a saurian monster with dorsal spines climbing up from the steppe at the approach of night.

69. Camping.

CHAPTER VIII

A TRIP TO THE MOUNTAINS OF URGUT

BETWEEN Samarkand and Panjikent and about seven miles due south of Juma-bazar (which is marked on Map II), lies the important village of Urgut, famed for its vineyards. As from a cornucopia it flows out of a recess of those mountains which the Russian map calls Kirtau and which form a branch of the Hissar range (Map I). The Takhta-karacha pass, which leads from Samarkand to Shakhrisiabs is the deep gap separating them from their further continuation to the west, the Aktau (Kemkutan).

As seen from the minar of Ullugbeg the snowfields of Urgut look very near, and even ridiculously near when one is told how long it takes to get there, namely eight hours. The distance to their base is 26 miles by the road and 22 miles in a bee-line. A not too ambitious City express with four stoppages on the way might traverse the intervening space in an hour. This would enable the business man of Samarkand to leave his office at one, take his lunch and board the train at three o'clock. Reaching Urgut at four he could have some tea, and then walk or ride to some high camp the same evening. On Sunday he has all day for climbing peaks or traversing ridges, and may catch either the 7 p.m. excursion train or the late slowcoach at 10 p.m., so that in any case he will be home and in bed by midnight.

I have not yet decided whether to christen Samarkand the Vienna or the Munich of the Duab, but so much is sure, that from the steeples of neither of these Alpine cities can one see such a bold wall, squarely set upon the flat, and having only very short outworks or supports. It is as if this pier invited us to embark here for an uninterrupted mountain walk far into China or even to the Pacific coast, always in touch with rock and ridge, no day without at least one patch of summer snow.

70. Loess Architecture.

The profile of Kemkutan is set with fine teeth, but the detail of Kirtau is engulfed in the shadow of its flanks and the crest seen at right angles from the north shows a moulding of curved notches on the top of a solid embankment. From the Takhta-karacha this escarpment maintains a uniform average height of 8000 to 8500 feet for 25 miles. It is this stretch which like a true rampart is suddenly reared from the level, and whatever spurs it may send down are foreshortened to our view. After this come big, single peaks and severing cleavages, and then the forceful uprising of Chapdara, the titan who with mighty spring has wrested himself free from the bondage of earth. Around and beyond it is a gathering of summits transforming into a vista what so far had been the dark and unbroken outline of a battlement.

Thus within a hundred miles from its beginning in the steppe the mountain chain has overtopped Mont Blanc by two thousand feet (Chapdara 18,000 ft., 5400 m.; Mont Blanc 16,000 ft., 4800 m.). Here is the angle of the great fork of converging lines whence issues the Zarafshan from the gate of the mountains. The green garden-island is everywhere separated from the foot of the wall by a belt of steppe from two to four miles wide, owing to the gentle rise of the land towards the sides which prevents irrigation from below. The torrents breaking out of the hills are too small to water much soil, wherefore the green sea in the middle never touches the cliffs, but laps a broad band of grey and yellow foreshore. Into this arid zone are thrown emerald triangles from out of gloomy bights, and the size of these cultivated alluvial cones gives a clue to the importance of the river together with the quantities of snow above. The greatest of these bosky fans within easy view of Samarkand is that which enfolds the rich village of Urgut, and letting the eye soar above it, we notice that here the ridge favours a display of white slopes.

In April 1908, my wife, Carruthers and I paid a visit to Kirtau. It was raining and the Zarafshan rose. Pale and dim and dull the showery curtain trailed over garden and steppe. Spring had arrived on the wings of the western wind and thrown out its veil, like a magician, to cover a surprise. The grey sheet of the clouds was ever pressing forward into the mountains who gathering the fleeting sponges of the mist with open arms, let the oozing moisture run down their rocky limbs, to pour it forth into the valley below. Again the waters of life welled from the bosom of the mother who had rested from her labours of the year before. Old guardian winter had left and now she awaited the coming of her lord and shivered as he shook the spray from his sombre travel cloak. The river heaved its breast and the great ariks that stood ajar engulfed its swell with gurgling greed, and the tide, drawn in with oily swiftness filled them to the brim.

The spate had leaped from the mountains a dark and slaty colour, but now it swept along the loess banks which had been dry for many months. Cracks sucked in the water with bubbling sound; clods and lumps slid helpless into the current; the earth, crumbling and dusty, melted in the sluices of wealth to be; and pelting rain washed the dikes with muddy rivulets, so that the grand canal became a turmoil of thick, ochre liquid which urged itself forward in turbid swirls. On the surface, spinning with the eddies or dancing to the waves sped a ragged fleet of straws and leaves which, scattered by November winds and caught in drifts in empty gullies, were now rising to go back to the land which bore them. The middle-sized dividers now took in the outburst from the main feeders and their walls, already coated with greasy mire from the drizzle, added another share of slime to the troubled waters. Here the straight flush found its swiftness broken by many turns; it now lifted its bulk to swamp the crooked byways which received the flood with a steady and silent indraught. Then all the smaller branches, the ditches and furrows gorged themselves, but they did it more slowly, and the water thus checked in its eager search for an outlet began to overflow. Leaking through the dams of the channels it first spread over the reed-beds, the safety vessels alike of the bounty of nature and the negligence of man, and the hard shafts of autumnal stubble were drowned in a lake. Next came the terraced rice-fields, which stood like square dishes one above the other. When the uppermost had gulped in as much as it could hold at once, it spouted out the overtake into a basin below, so that ponds of regular shape succeeded each other in tiers, all connected by little waterfalls. To their gush and babble was joined the dripping of the rain as it pock-marked the sheets of water on the fields. Meanwhile the soil underneath drank and drank, that it might turn into a knee-deep squashy sludge, into a soft cradle for the seed.

But still the Zarafshan was rising. The water found many places which gave a level of rest or offered short cuts that avoided the twistings of narrow gutters, while some rills which could not keep pace with the stress of supply made use of the nearest hollow as of a store. The flood was pouring on and the moving water sought to rush the large ariks or wriggle and glide with frantic haste through the network of innumerable arteries, for there was ever coming more. Where it found the winter's refuse or forgotten sods in its bed, or where a damaged dike suggested relief, it eased itself and spread, making marshes, bogs or shallow lakes in which wavy currents betrayed the line of submerged canals. It was not however an angry deluge that boded ill, but merely a congestion of abundance, for much of the land was only drenched with rain, letting the streams pass by in their orderly path.

71. A Karagach Tree or Turkestan Elm.

Man now believed the message and Sarts stood in the damp haze with wringing khalats tucked around their hips and drops running from their beards. With ketmen and spade they endeavoured to coax the water or ward it off, made inlets or dams, widened breaches or stopped them, as their case demanded. Everybody saw after his own, so what was public ground depended on the mercy of the gods and to the road they were not merciful. On the whole it was a course of quagmires and bilges of stagnating peasoup, but sometimes it sloped into pools of unfathomable treachery or was crossed by a leisurely stream. All horses had bespattered bellies, saddle-felts were pulpy and riders were sorely bedraggled; camels floundered, their legs went apart, their splay feet slid aside and with the suction of a squeegee drew broad smudges in the mire; people stalking to the other side lost their slippers in the middle of the sticky morass.

The heavens sprinkled and squirted and in alliance with the moisture rising from below steeped the world in wet. Streaks ran down the outer walls and the turtle dove on the trellis of Rustam Bai's vine-bower had to shake off beads of water while sitting on her eggs. The gargoyles spurted and the budding trees shed tears to every fitful gust. The rain sank into the flat roofs, and trickled through the ceilings into dank and chilly rooms. Whoever was not called out, lay low, for around was sloppiness and slush, seeping humidity and vapour; a dry seat was worth a king's ransom, a good roof was heaven, and stepping-stones were stumbling-blocks no more. In the tea-houses and sarais besmeared travellers talked, listened to the patter of the rain and looked out upon the universal saturation.

Everything was wet and dirty, yellow or brown, flooded and soaked. The first touch of the master hand was not altogether lovely to behold, but the kneading of the clay was a necessary beginning. Earth had lain stiff and stark, and as one aroused from her slumbers she was clumsy and slobbery in her uncouth haste to drink from the pail of plenty; she was not yet the beauty clad in radiant green and bedecked with flowers. Spring was come. The mountains had known the tidings first and had been the first to help. On the next sunny day they will don their chalmas of white cloud to feast the vernal tamasha and like naughty boys messing in a puddle they will stand laughingly around the dripping chaos to watch what will arise. For yet a while nature wallowed in a bath of mud which hid the charming miracle to be. The Zarafshan had risen; the oasis was swelling with a thousand turgid veins.

This time we stowed ourselves and all our belongings into an araba or native cart (Fig. 4), with wheels eight or nine feet in diameter. Some of them specially intended for passenger transport have a semicircular hood. The araba is in so far like a hansom cab that it has two wheels, one horse, a curious seat for the driver and that the better ones are upholstered and provided with a looking-glass inside. The arbakesh (driver) rides on the horse guiding it with short reins and a short whip. The araba fears no obstacles and is undoubtedly the best vehicle for the roads of the country in their natural state, whereby

I mean that condition which the Sarts are in the habit of regarding as normal. Deep mud, hills and dales, six feet above or below the general level, ditches, pools, nothing daunts the immense wheels set so wide apart that a tilt of forty-five degrees leaves a wide margin of safety Should a felloe miss the edge of one of the many so-called bridges which span the irrigation canals and descend into the water it matters not: a bump, a heave, and out we are again in the smooth waters of the lesser ruts. One follows the line which the cart traffic of the time has decreed as the best, the horse's sailing directions being given by a deep narrow groove just wide enough for its moving feet, and which hundreds and thousands of them have trodden before it. But it may also, in its turn become the inaugurator of a new variation rendered necessary by obstacles.

72. The Sacred Grove at Urgut.

The araba is one of the numerous instances of the best instrument adapted to surroundings. For transport over the roads nothing can surpass it for the ease with which it overcomes rough ground. Anyone travelling in the plains or steppes should use it for his camp outfit, baggage and servants, riding on horseback himself, if he prefers it. If I were a resident obliged to take long journeys frequently, I would have an araba made with strong springs, which would enable me to travel by night while comfortably asleep. We built a thick layer of camp bedding and pillows in order to deaden the jolts, and covered with mackintoshes such of our limbs as went beyond the shelter of the hood. It was raining cats and dogs, and at one place we drove through a lake over half a mile wide, but only three feet deep. Here I thought of the mud-flats which connect some of the North Sea islands with the mainland and over which people used to pass in carriages at low water, but fortunately we had no rising tide to fear. On the edge of a dismal pool we surprised a native lady probing the abyss with a stick, intending to cross. At our approach she was evidently undecided whether to drop her veil or her skirt, but she saved her face, stepped bravely forward and made the red silk of her undergarments swish through the yellow flood.

Schwarz who lived at Tashkent and whose veracity cannot be doubted tells a story how once he saved the life of a boy who had fallen head foremost into a slimy hole in the bazar. At another time eleven women were drowned in the main street of Tashkent. Their cart happened to break asunder in the very middle of a large puddle, and they were unable to extricate themselves from the sticky morass. Undoubtedly their efforts at keeping on their veils had much to do with their helplessness, as also the reluctance of passing strangers to touch them. One will realize this when one hears that during the earthquake at Andizhan the fanatical Moslems drove their women back into the houses when they came rushing out frantic with terror and the walls collapsing over their heads.

When about halfway to Urgut we began looking for a nice roadside inn where to refix our bones for a bit, but before we had made a choice our arbakesh was already unyoking the horse at his favourite hostelry, as is the wont of waggoners all over the world. As equally little is obtainable at all chaikhanas (tea-houses), we agreed and settled on the stoop, where trays were placed before us on carpets of black and grey felt. A man was sent to the baker for bread, another to the raisin seller, and the innkeeper provided green tea ; coffee, by the way, is not drunk by the natives.

The highway to Panjikent is a very busy one and as the rain had ceased for a time the life and movement of the road beat again with quicker pulse. Nearly everybody was riding either on horses or donkeys, and sometimes a whole family, father, mother and one or two children are seen on the back of one patient ass. Nobody ever walks if he can help it, a habit sufficiently explained by the state of the roads. For this reason the natives of the plains are useless on their feet and the formation of an army of serviceable infantry is a hopeless undertaking. The Bokhariot foot-soldiers are always transported in carts. Short stirrups are the rule and a regular trot is never seen, because one either goes at a walk or at a gallop; on journeys a quickened walk is used. The highest prices are paid for amblers, while the trapata (a Russian word) is another favourite gait. It is said that to produce the trapata all four feet are moved one after the other in circular order, not alternately, but although I tried to investigate this point I always found the horse's legs quicker than my eye. One rarely sees the best horses which are those of the Turkoman

73. The Kughi-surkh and the Pass from above Urgut.

breed, said to be equal to the Arabs in everything save beauty. The Kirghiz and mountain varieties are also hardy and useful, but the Sarts generally give preference to the exterior of their chargers, paying fancy prices for an isabella with red eyes or a piebald in the manner of a chess-board.

I have already mentioned the fact how seldom one comes across a restless animal and it also seems curious that tales of horse-breaking and horse-breakers never crop up in conversation with Europeans who know the country, although feats of endurance of horses and riders are frequently mentioned, such as that of a courier who, of course with relays, galloped from Bokhara to Karatagh in forty-eight hours, a distance which we took a week to cover with our caravan. The baiga also proves that the natives are not to be despised as horsemen. In a large measure this tameness and obedience may be due to the way in which the horse passes its youth, and especially with the nomads, it is a very domestic animal, following the caravan as readily as sheep. The greater spirit of our European horses may possibly also be accounted for by richer food, while the cruel bit employed by the Sarts is probably to be reckoned with as an element of discipline. There is no reason to suppose that the Arab horse is more fiery because it is "nobler." It would be interesting to read the opinion of a good horseman who understands the subject thoroughly and who has studied the behaviour of his friends in all parts of the world and under various conditions. Nobody who wishes to keep up appearances in the Duab ever rides a mare, and geldings are unknown, but apart from their inclination to fight I have found the stallions as quiet as lambs. Peasants, who do not mind riding a mare, always take the fillies with them, and I have often noticed a small package, perhaps only a few pounds in weight, on the backs of very young foals, so that in this manner they are gradually made familiar with the incubus which is to be their fate. The Sarts and especially the Kirghiz are also lovers of horseflesh after it is dead, and a joint from the young and fatted steed counts among the greatest delicacies imaginable. Very many of the passing travellers hum a low and monotonous tune as they absorb the weary miles and this added to the apathetic and indolent bearing of the rider strengthens the impression of a jogging automaton.

The camels with their wool coming off in patches look very ragged and moth-eaten, their beauty apparently standing in inverse ratio to their usefulness. Fortunately it is not necessary here to ride them and their duties are confined to the carriage of goods. Even where the camel is not absolutely necessary it is employed as much as possible and as far as the softness of the road will allow, for it is easily fed on dry sticks and carries a big load. Long rows of them, linked nose to tail, are a frequent source of delay on the road as alternate animals have opposite ideas as to the side on which they had better pass you. In front, the man in charge rides on a donkey holding the nose-string of the leading camel. One often sees him asleep, for through long practice he has learnt to sit far back on the long pack-saddle resting his head on the folded arms in front. In the Duab meet the domains of the dromedary or one-humped camel and the Bactrian or two-humped camel. The former is more frequently met with in the desert, while the Bactrian is found in the cultivated parts and among the Kirghiz, who take it up to the highest mountain steppes. As long as the path is not too rocky or narrow the camel can climb anywhere.

74. The Kughi-surkh from the Pass; looking South.

In very hot districts, especially in Chinese Turkestan the animal is only made to work in winter. Crosses between the two kinds remain procreative and are specially bred in several districts.

In a garden near our sarai we saw some very fine karagach. The karagach or Turkestan elm (Figs. 17, 32, 70, 71, 128) is one of the typical trees of the Duab landscape, chiefly owing to its tendency of forming leafy balloons of great regularity. As the main limbs shoot upwards like a bunch of flowers gradually unfolding near the top, and as the secondary boughs follow this example, the inner space of the cupola of foliage is a tangle of close-set ramifications which retains dead branches as well as the flotsam and jetsam of the air. As moreover short spikes and twigs grow out everywhere the whole forms a disorderly nest of bark, leaves, sticks, and straws affording protection to many small birds.

After having rested, we again proceeded, reaching Urgut nine hours after leaving Samarkand. Allowing an hour and a half for stoppages this works out at three and a half miles per hour. We found the belt of steppe to be very gravelly; it was strewn with rounded pebbles of fluvial origin. Urgut itself is built among cliffs of high loess and consists of three or four different villages united by an uninterrupted succession of houses, rising one above the other in the steep valley. In order that they should not be washed away the streets are paved, if mountains thrown into soft earth can be called a pavement, and we had a mile of this rack before arriving at the uppermost suburb of Sufian. With that cool indifference which is the courage of the Oriental our arbakesh and his horse attacked the labyrinth of steep and narrow lanes. Sometimes the hubs grazed the mud walls on either side and a helping hand applied to the spokes was necessary in several places. Fortunately we never met another cart. What would have happened in such a case is a puzzle unsolved to this day. I had determined not to give way, so that we should have stuck there for ever, unless the ill-fated other man, yielding to persuasion, had taken his araba to pieces and carried the parts back to the nearest turning.

The situation of Urgut is extremely picturesque. From the flat, green delta of the plain the town is built up in a series of shelves, receding into the valley and drawing more closely together as they are wedged into the spout of the funnel. The trees grow thinner, but the dwellings seem to increase in number. Bluffs and platforms of yellow loess are the larger steps of this perspective, divided into smaller tiers by equally yellow houses. The ledges of the flat roofs were grown with a soft new down of grass and many scenes of child-life were played to us on these carpeted stages as we passed. It was a Friday, and we saw many groups of women and girls wading through the slush on their way to the baths, not in the river, be it understood, but in the "Turkish Bath." It must be a mean hole, indeed, which cannot afford these sounding vaults filled with steam.

Just beyond the last houses, at a height of about 3500 feet, there is a place of pilgrimage. Sheltered by the canopy of a sacred grove of giant sycamores lies a small mosque with pillars of wood and painted friezes (Fig. 72). A few lonely guardians sat in the outhouses and altogether the place looked somewhat forlorn and dreary in the rain. The great domes of the trees were still a network of brown twigs and sprigs in which the young leaves hung like a golden haze. Storks, jackdaws, starlings, doves and sparrows were busily and noisily occupied around their nests. On the sodden ground, in between rare patches of grass, one discovered numerous heaps of charred wood and ashes, the remains of the camp fires, where the pilgrims had held their picnics. At the upper end of the grove a spring of fine, clear water wells from a fissure and the large pool is inhabited by fish enjoying the protection of taboo.

75. Our Big Tent.

Such places of pilgrimage are nearly always associated with water and big trees, while judging from reports which I have read the "tame fish" are a favourite institution. If this had not been a somewhat sanctified locality I should have asked to be accommodated in the mosque, which would certainly have offered the most comfortable quarters. The reader may be astonished, but in the villages we were very often taken to the meshed as the best house in the place. In the large towns, of course, conditions are somewhat different. The mosque is not really an abode of God, but a meeting-hall. Owing to this the ordinary temple is not defiled by the foot of the unbeliever. Only those containing sacred tombs or shrines are looked upon as consecrated ground. It is altogether the presence of the Kafir which is unwelcome anywhere, in the open as well as inside any house, private or public. Wherever this objection is not or cannot be enforced, wherever the unbeliever is allowed to show his face, the soil of the prayer-house forms no absolute restriction. This may possibly be different in other Muhammadan countries, such as Turkey. But we must remember that the western Moslems, whom we generally have before our mind, have always been in close touch with Christian nations, whose respect for the church building they may have observed, exacting a similar respect for the mosque on the tit-for-tat principle. The eastern native takes his shoes off at the threshold of every house and therefore also at the mosque. Whoever suffers a European to enter private rooms with his boots on cannot possibly make a fuss in the case of the meshed, even less so. One looks upon the Western boot manners as upon a piggish habit which must be borne with well-bred equanimity followed by a cleaning up. In the medresses of Samarkand nobody ever asks you to take off your boots. The only difference between the private house and the house of prayer is that the latter is a public place, and that in public bad manners of any kind or breaches of national custom may furnish a ready pretext to excitable and fanatical crowds. From all I have seen in the Duab, in Bokhara as well as in the mountains, I have come to the conclusion that the views of the people in connection with their places of worship do not seem to be based on the idea of Divine presence. The logical inference is, therefore, that one can offend the Moslems by intruding upon their religious privacy or behaving indecently in a mosque, but that one cannot desecrate it.

We found a dark little room in one of the neighbouring huts and went to bed after a hearty meal of pillau and a scrutiny of the clouded sky. During the night a cat tried to steal our biscuits, and broke a glass in the attempt. Awakened by the noise we took a peep at the stars and to our immense satisfaction saw a few of them twinkling through moving gaps. The weather was still more hopeful at dawn, and when the mollah greeted sunrise by calling the faithful to prayer we were already under way, loaded with rucksacks and guns. At first a level path led over open slopes and into a ravine where from shadowy depths we had our first view of the mountains which were our goal (Fig. 73). To the right of a serrated limestone ridge which rose in a clear sky we saw a col or pass, whence descended broad masses of morainic material. The gentler ground and the bulging sides of the range were covered with square patches of field up to the snowline, while in between lay grassy pasturage. The geometrical patterns of the fields showed many colours, the freshly ploughed ones being of a dark brown while the older untilled ones wore yellow of many tints according to the crop or the time they had lain fallow. Here and there a single furrow is drawn as a mark of taking possession of this season's ground, for the mountains belong to everybody and the first comer is first served. The reason is that these fields are not properly irrigated but left to the chance of rains. Here, as on the northern slopes of the Alai there is always the reasonable probability of a little rain, because these snowfields are the most advanced towards

76. A Bazar.

the plains and the first to meet whatever vapour may be drifted against the highlands from the North or West. Where earth had been turned up by the plough we saw many birds hunting for seeds and insects. At 6000 feet snow began to appear and the path crossed several very steep gullies which demanded care in case of a slip. It was a great joy to be among alpine surroundings again, to step on white nevé and to hear the waters rushing. The pass was won at 10.45, but we had dawdled a good deal on the way. We were now standing in a hollow running parallel to the main ridge and separated from the plains by a low summit forming one of the short spurs which the range sends out to the north. This flat basin of the pass, at a height of 7200 feet (2200 m.), is caused by an accumulation of morainic débris which have choked the older and deeper valley sunk into the solid rock below. This is a condition typical of the Duab mountains and represented on the grandest scale by the Pamirs. After the comparatively sudden retreat of the glaciers the streams lost so much power as to be unable to cope with the enormous quantities of detritus. We found ourselves at the foot of an almost uninterrupted snow slope a mile wide coming down from the Kughi-surkh, a line of peaks from 8500 to 8700 feet (2600 m.) high. At its lower end this nevé split up into streaks and patches among hills of old moraine (Fig. 74).

Clouds were again collecting and as there was no prospect of a view into Bokhara we contented ourselves with a lower eminence on the northern side, devoting the rest of our time to the collection of birds. There were many interesting species about, notably mountain finches of white and pink plumage, which had to be stalked with great care, being extremely shy. Their crops were filled with seeds from the dead plants of last year laid bare by the melting of the snow. The desk-like peak which we climbed in half an hour, rose gently to 500 feet above the pass and broke off in precipices towards the valley of Samarkand. Its rock was a limestone reminding me of that of the Karwendel, having that gritty surface, which offers such agreeable purchase to the nailed boot. In the wide basin between us and the Nurata hills flat clouds were swimming like ice-floes in a limpid sea and through the open spaces we caught glimpses of water-streaks shimmering on a drab expanse. The Turkestan range was lost in a distant haze, while immediately east of us everything was wrapped in thick mist. Then we descended.

Next day we returned home to Samarkand and found the road quite dusty in several parts; so quickly does the sun dry the yellow earth. Examining the mountains we had just left, we espied in a shallow valley, to the west of Urgut, what looked like awesome cliffs at least 3000 feet high, composed of limestone with threatening slabs and long chimneys. Something for next time, we thought. The return journey was accomplished at somewhat higher speed and our team made a record of three and three-quarter miles an hour.

CHAPTER IX

FROM SAMARKAND TO VARZIMINAR

On the 19th of July, 1906, we left Samarkand for a long journey through the mountains of the Duab. My companions were my wife, Cenci von Ficker (now Frau Dr Sild), the Tyrolese guide Albert Lorenz of Galtür and Makandaroff, the Caucasian interpreter, who had already accompanied me on seven journeys. Lorenz is one of the best mountaineers of the Alps, while "Mac" may be described as a famous expert in all the broken idioms of Europe and Asia, as well as in cooking and laundry. To the ladies I owe a debt of gratitude for the assistance which they gave me in my photographic and other work. Six horses carried the photographic outfit consisting of a large and a small camera and a thousand glass plates. Exploring with a big apparatus is beset with difficulties tempting a man to waste his hidden store of emergency language. To get this heavy artillery into position from six to ten times during a day is a good test of nerves, but the ladies did all the drudgery, setting up the camera and handing me various objects while I made notes or calculations. Thus the firing off of exposures became a silent and smokeless affair. In the beginning it took over half an hour to unload the photo-horse, to unpack and prepare the camera, to pack and load up again, but later a record of nine minutes and a half was obtained. Some time ago I read a letter written to the press by a famous traveller, wherein he speaks of certain Indian officers as "fortunately unmarried." Well, it only proves that the great explorer must have sadly failed in one branch of human experience, the one which makes me feel thankful for the companionship and assistance of those whom the natives called the two wives of the honourable tura.

77. The Zarafshan Valley near Sasun; looking down stream.

I had originally intended buying my horses, but with the help of some friends I succeeded in hiring a caravan from a dealer. The price for twenty horses and five men for three months was £100, but I had to feed them. According to my calculation the cost in money would have been the same if I had bought and sold the animals, but I gained an enormous advantage on the counts of time, worry and risk. It would have taken me at least a fortnight to haggle for twenty horses, and as long to sell them at a reasonable price after my return, for it would be autumn then, when fodder is expensive and cattle cheap. As it was, my agent provided the caravan within a week, bearing all breakdowns and losses himself, while at the end of the journey I was saved all delay.

Rising at four o'clock on the morning of our departure I already found my men busily engaged, but loading took a long time, and it was only after three hours weighing, balancing and roping that we were able to start. The first day always marks the climax of that battle between theory and practice which precedes every undertaking. This initial struggle represents the birth pangs of either success or failure. The ideal of perfect organisation is never fully achieved, having to contend against an inexhaustible variety of material facts, and although the labour never ceases it can be lessened by a gradual approximation of things as they ought to be, to things as they are. This war between the absolute viciousness of objects and the absolute will of man is a play of forces keeping up life and movement by trying to oust each other. Everything has been beautifully thought out; every button has its place, every animal its office, every man his duty. Such is infallible theory. But a button can burst, even if it should not; sardine tins will find their way into the cartridge case, even if they are not allowed to; and horse as well as stable-boy are but imperfect materialisations of the leading idea. Such is the irony of practice.

At first one is the victim of despair, when one realises mistakes in the outfit and the cussedness of things. Innumerable are the delays on the road caused by slipping packs, by the drawbacks, now apparent, of new devices, by forgotten precautions and the endless variety of small irritations, all of which hang like leaden weights on the feet of impatience. But this fidgety state does not last long if the traveller has some experience and if his kit is at all useful, for soon begins a life of tacit agreement between men, things and events. One discovers that the same end can be served by other means, that certain articles can be put to better use, that many can be replaced in the country and still more can be thrown away. Not only servants and horses have their peculiarities which we must consider when wishing to guide them, but every instrument and every new day have an individuality of their own. The wise man knows that the rule by which matters arrange themselves does not wholly obey his own line of preconceived planning, but that it is the diagonal of a parallelogram of forces whereof he can choose to be the stronger side. Thus all trouble will not have been in vain and outfit really begins to fit, as the name implies, adjusting itself to circumstances. Every campaign, every mechanism, is subject to the complaints of infancy, until organisation settles down to a regular and rhythmic beat, which nothing short of downright misfortune can stop. Of course there are children of sorrow in every camp. Ours were the photographic plates and the ice-axes. Both were carefully watched, in spite of which the servants managed to crack two ice-axes on a difficult snow pass by throwing them after the horses while we were out of sight. I was luckier with the glass plates, of which only two were broken on the journey, whereas I smashed half a dozen in the dark-room at home. At first I felt nervous about my precious freight, especially after exposure, but gradually I became as callous as the captain of a vessel loaded with dynamite, for knowing that I had packed them like babies I must needs leave the rest to the kindness of fate.

There is often something clammy about an early summer morning in the Duab. The dawn is livid and sultry in the steppe and among the gardens, for at sunrise the air contains its maximum of moisture which, though little in itself, produces a depressing influence upon body and mind by contrast to the usual dryness. Add to this the feeling that the task has begun in earnest and our foot is now uplifted for a serious step into the future. Although we have been looking forward to this day with growing impatience, cursing the slowness of the Sarts, there is an attack of the blues as we suddenly realise that the caravan is now launched for better or for worse. But then we shout "aida, aida" to the men, and raising our whips, start the little jumps in the saddle which are to beat the time to a thousand miles of road.

We watched the fingers of the sun toying with the turquoise dome of Gur-Amir where lies the dead Timur, a silent greatness among the vagaries of life; and paying a last homage to the crumbling vastness of Bibi-khanum, which stands a gaunt memento outlined against the morning sky, our jogging cavalcade is engulfed by the tunnels of poplars on the road to Panjikent. For several hours an eastbound current of the traffic bears us along between those garden walls, behind which intense cultivation is practised. In crowded districts it may happen that the traveller is cooped up within the yellow quays of his channel for the better part of a day, always within hailing distance of a house. Sometimes a surface of fifty square miles is nothing but one garden village wherein denser agglomerations of cottages, shops, chaikhanas and sarais are like so many knots of trade and administration. Poplar, willow, elm (karagach), mulberry and cherry are the typical trees which we see everywhere. In between are vineyards, melon beds and tobacco plantations. Above it all hovers the yellow dust, from which there is no escape. The horses' feet stir it up in little puffs and wisps, the lightest lisp of the wind lifts it in floating whirls, while a drove of sheep appears swaddled in sulphureous gloom, so that one wonders how the animals can breathe. But water also makes its presence felt and by strange contrast wet puddles caused by overflowing ariks alternate with deep beds of dust.

At intervals we seek the shade of a tea-house in order to await the baggage-train of horses and servants struggling with provoking loads. There is nothing like the platform of a chaikhana whence to survey Oriental life on the road, for we are not merely spectators of a moving panorama without, but co-actors among the real surroundings of the foreground. In company with squatting figures we look from under the shady elm upon the hot and powdery road; from the stable yard comes the neighing of stallions and the braying of donkeys; the turtle-dove coos and the fighting-quail clucks in a cage suspended from the roof, while the chilim passed from hand to hand pervades the air with whiffs of acrid scent.

After about fifteen miles the compounds begin to thin out and across open fields one is sometimes able to catch a faint glimpse of

78. The Darkh Gorge.

mountains trying to reveal their outline and blinking snows through the tantalising dust-haze which, like the gauze veil cast over a Venus, shows just enough to make us wish for more. At Juma-bazar we made a halt for lunch. Here, behind a row of houses facing the street, we discovered the Dargom in a green flood plain several hundred yards wide between vertical loess shelves. The backs of the houses stood so close to the brink as to form an almost direct continuation of the drop, which shows great confidence in the solidity of the loess. Once a stream has cleared a space for its highest watermark erosion of the steep outer banks comes almost to a standstill as they are removed from the flush of currents, while rain does not readily dissolve loess which is porous as well as fairly firm. From under the foot of the ledge came a spring of clear water in which some fish could be seen flitting about. The country around is to a large extent given over to the cultivation of rice, and in convenient places we observed as many as six terraced fields one above the other, the difference of level between them being about one foot.

At Rakhmetabad, where we passed the night, dry fallows and the scent of wormwood announced the neighbourhood of steppe. Our sarai was very clean and we enjoyed a good bath in an arik, the swift current of which was still young with the vigour of the Zarafshan. Afterwards we climbed a loess hillock which on the other side overhung one of those empty rows of stalls used for weekly markets. To-day we were the only occupants of the view-point which must, however, be very popular with visitors to the bazar, for we found the top of the kurgan worn smooth by generations of patient sitters.

We examined the mountains through Zeiss glasses and drank in the beauty of the sunset. Against the red fire-hole in the west there stood the silent silhouettes of poplars. The atmosphere was filled with liquid gold which transfigured every leaf, every tree, every curve of the hills. One was reminded involuntarily of Turner's great picture "Childe Harold's Pilgrimage." For the vague mystery suggested by his distances you have the same effect produced by the setting sun which glorifies common objects and shrouds all in a halo of romance. Towards the east there are still a few gardens and some patches of young rice gleaming like emerald amid the duller colours of corn and other produce. Beyond is a yellow steppe hemmed in by the lines of converging mountain chains, and in the distance a misty blur hides the gardens of Panjikent. Pointing towards them is the grey ribbon of the road with here and there low dust-feathers trailing away from it and showing the whereabouts of travellers. To the left is the faint silver streak of the Zarafshan with pebble islands and shoals of quicksand, losing itself on one side among the gardens, on the other into the shadow of the mountain forming the last bastion of the Turkestan range. These hills send steppe into the gardens and the fertile plain sends dashes of green into the lower slopes, a picture of the battle of life and death.

Next day, at Panjikent, I went to the office of the Russian chief for a letter to the head-men of the various villages through which we were to pass. The document was brought to me at my inn by an elderly Tatar, interpreter to the Russian administrator. These Tatars from Kazan are found everywhere as dragomans and secretaries to both sides, for they write and speak Russian as well as the native languages and may be said to correspond to the Indian Babu, in a way. The learned personage who called upon me began a long talk to which I listened politely, but when after an hour he still continued telling me things which I had known before, or which I knew to be untrue, I succumbed to his hypnotism and sank deeply into the stream of oblivion until the voluble flow sounded like a far-away murmur. Our friend being very short-sighted did not notice that I had fallen asleep, but kept up his eager address, accepting my nods as signs of approval. When finally he took his departure, my wife derived much amusement from his efforts at shaking hands with a void.

Towards evening we walked past a gaily painted mosque to the banks of the Zarafshan, where we found a low terrace of alluvial loess mixed with gravel. A clear source formed by seepage from the river was strong enough to drive the heavy stamps of a rice shelling mill. Divided into many fibres the stream was rustling among shingly beaches; its water when scooped up left a film of slaty silt in the hollow of the hand.

Opposite us rose a broad flank of the Turkestan range, its lower edge bounded by a narrow frill of trees skirting a steep bank. Above them follows a tier of horizontal strata, surmounted by the main mass of the hills, round and soft on the whole, but knobby and lacerated in detail. In parts the slopes are coloured with outcrops of red, yellow, brown and white marls. The higher flanks show blackish rocks, where gaunt gorges with mysterious depths caught the splendour of the setting sun and were bathed in purple fading into mauve. Near the top we could see a sprinkling of gnarled trees which, where they came into the sky line, made the ridge look frayed. On the other side two limestone crags shone white in the last light of day, while above them was a murky battlement of cloud, bluish-grey and lined with pink,

79. A Terrace Oasis.

which cut off the crowning portion of a mighty snow peak. Among the hushed groves, where the gloaming began to filter up from below, one solitary figure in a white robe performed his evening prayer bending and kneeling towards Mecca.

Between Rakhmetabad and Urmitan many of the right and a few of the left affluents of the Zarafshan are intermittent, ceasing to furnish water in summer and autumn. This condition is, of course, due to the fact that these streams come from low ridges without permanent snow or incapable of storing water like the granites of Kemkutan. Besides that the evenly spread fertility of the plains is now buckled by the upthrust of the land which makes irrigation sporadic, confining it to localities of convenient flatness. Water to be of any use to the soil must be splashed over it, which, on steep slopes can only be done by nature in the shape of rain. Otherwise the water must be distributed artificially from a main feeder and that is only possible on level surfaces of very gentle inclination, such as flood plains or fan deltas. Where this cannot be, the streams flow in deep trenches leaving the banks a desert. Above Panjikent the landscape is very arid and the few settlements, almost devoid of trees, lie in a desolate and thirsty steppe. Before entering the mountain valley of the Zarafshan we spent a night at Guzar, a hamlet about three hours from Panjikent. On the way thither we saw a fine row of organ pipes of red rock on the opposite bank.

We found quarters with a peasant owning the title of yelikbashi and a field of lucerne which latter he sold for six shillings. Into this we drove our horses. Such is the nearest approach to an English meadow in these parts, where grass is never cultivated for fodder. A real meadow only pays in moist climates, as a rule, and on the valuable irrigated land one must grow plants giving the highest possible quantity per acre. Lucerne, a clover-like perennial full of nitrogenous nourishment forms luscious thickets three feet deep. Like grass it can be cut many times, yielding several crops until exhaustion of the surface demands reploughing and rotation. For England lucerne is not sufficiently hardy, but in Argentina where climatic conditions similar to those of the Duab prevail it is sown on a large scale (alfalfa). Nearly all the seed used in South America comes from Russian Turkestan where it forms an important and highly speculative article of trade. Henceforth, while in the mountains, our horses had to be content with barley and dry grass. The question of fodder is the greatest difficulty of mountain travel, forming the largest item on the bill of expenses and a constant source of worry and delay. Next time I have decided to take donkeys only, reckoning about three of them for every two horses. They climb better and can be fed on almost anything.

Guzar is overlooked by a rampart of conglomerate rocks the eastern promontory of which is eroded into pillars. By their tapering form which stands midway between the shapes of earth pyramids and those of the Yakhsu conglomerates one can see that they are fairly soft. We climbed up to this grey wall which on its base of rounded slopes looks like a fortress on a hill. The ground was hard and hot; here and there some dry grasses and herbs struggled for an existence furthermore endangered by flocks of goats. Besides thyme, sage, and worm-

80. A Lane in Rars.

wood we saw a creeping plant resembling a coarse and stoutly built mimosa and which at this time managed to produce large, white flowers. From above we surveyed a characteristic hill steppe which in spite of its barrenness gloried in a fine effect of artistic distemper. The whole was painted in subdued tints of reddish brown, chocolate and sepia, with bold streaks of orange and delicate shades of grey thrown in between. Near the group of houses lay an irregular chess-board pattern of square fields ranging from mahogany and terra-cotta to amber and daffodil. Through the midst of this design in pastel chalks ran a snake line of vivid green where willow and spiraea hugged the banks of an arik.

When we were about to start next morning the yelikbashi arrived with two small bits of wood, saying that they once had been a bridge now broken by the horses and for which he claimed compensation on a magnificent scale. After a long palaver the matter was amicably settled for sixpence. The horses must have gone a good deal out of their way in order to destroy private property seeing that it was much easier to cross the ditch without stepping on the plank. As far as Vardagan travelling is easy, but soon the causeways of Alai on either side come near together and from Yabon onwards the valley is very dark and narrow, the mountains bracing their rocky limbs against the roads and works of man. The section from Yabon to Varziminar is the narrowest and most difficult of the whole Zarafshan valley, and here also the river has its steepest fall, owing to the influx of the Fan-darya. It may even be said that, as far as hydrography is concerned, this portion of the Zarafshan is really a continuation of the characteristics of the Fan valley.

The terraces which accompany the Zarafshan up to the glacier are relatively small below Varziminar, because scant space is left by the towering, dark mountain flanks which compress the gloomy gorge. The path is at times even dangerous. The horses, still unaccustomed to this sort of country, stumble on the angular rubble and often find themselves in precarious positions. In order to skirt projecting noses or to avoid a flight of slabs the path crawls and plods about in the most aggravating fashion. Six times between two villages it goes down to the river's edge, as if to drink from the waters which refresh the mountaineer, and then frantically climbs up again several hundred or even a thousand feet to some gap or ledge it has discovered above.

The slopes of the Zarafshan valley form a veritable mountain desert, for the grey waters below are out of reach and the heavens do not shower gentle blessings. Owing to this nakedness the long trough is a grandiose collection illustrative of physiography and geological history. Every day and at every step we see something new in the shape of simple models teaching us the elementary laws of earth-life. Here is shown the activity of water as accumulator and destroyer, as architect and sculptor. The most prominent feature is supplied by the thick beds of glacial gravels hardened into concrete which once served as a floor built by the glacier for its river by filling the old rock-hewn valley with a basement of rubble. Afterwards the main stream and its tributaries cut down again to the bed-rock, making valleys within valleys and sawing the block or inset into separate terraces (compare Figs. 78, 95).

Everywhere we behold curious pillars in semi-relief and flutings which the trickle of ages has carved out of the hard conglomerate; we see all sorts of ravines with manifold examples of grading and watershed; furthermore many kinds of shoot-cones and alluvial fans presenting the clearness of outline of plaster models; and besides landslides, large and small, we observe the various degrees of the mudspate from the welted furrow among fine grit to the enormous vomitings of the mountain composed of caked mud interspersed with boulders. All the rubble heaps, pebble-beds, mortars, cements and concretes of nature have here been well preserved by climate, whereas in the Alps and Northern Europe they are generally coated with vegetation or have been washed out of shape by the wide surface action of rain.

Dryness and the linear confinement of water explain why we find all friable deposits in such an excellent state of preservation. A diluvial rubbish dump may have been sliced and partly carried away by torrents, but what remains of it, however little, is enough to appeal to the dullest eye, for there it stands, dry, bare, and clear. It is now beyond the reach of the stream, rain cannot fritter it away, no green grass mars with frivolous gaiety the austere correctness of the historical record. Forest and meadows can preserve the large, soft forms of a hilly landscape, but they prevent and hide all finer etching. In the Duab mountains no verdant drapery enshrouds the flanks, and persistent rain would make a clear sweep, but coming with interruptions and not being stored by a vegetable sponge it quickly descends by the shortest line, soon leaving behind the gutters and nullahs as dry as before. Hence the linear character of water-work, the parallel or branching grooves chiselled out of the valley sides and hardened by the mortar of rain drops wed to sunburnt lime. The mudspates caused by melting snow and the floods of cloudbursts already have their accustomed openings of sufficient width so that they hardly ever encroach upon the walls of their chasms; all water is quickly collected and run off in deep trenches.

Thus in comparison to English climate the water here is confined in time and space, coming down at long intervals and imprisoned within well-defined tracks. Hence the clear-cut, if stern and sombre, features. Destruction has acted in the manner of sharp slashes—wounds which the eye easily fills out after short practice—and designs are not blurred by sodden growth. Wherever we look on earth we find moisture associated with transition and softness in anatomy, life and light, while drought favours a harsh and regular formation of rigid limits, for it works with edged tools. Of course this comparison ceases above the timber-line and beyond the Arctic circle, for beyond the belts of animal and vegetable life, things become monotonous and very similar owing to the lesser number of objects of comparison in the shape of forests, fields, villages, rivers, forms of weathering, etc.

Dryness loves contrasts, producing them in great variety, when, as is the case here, the dynamic force of atmospheric circulation is, for the greater part of the year, so to speak split into the two contending lines of sun heat and running water. Thus geometrical and architectural patterns stand out from gigantic fronts with here and there a bold patch of green thrown on the glaring wall; throughout the seasons the Zarafshan carries volumes of water past unquenched soil; inky shadows cleave surfaces of white light; the roots of the Ferghana elm drink the arik, but its branches overshadow the clay of a hungry steppe; nations come as conquerors and end as conquered. Man and stone remain lying where they are awaiting a new flood from the East or a fresh burst from above to tear them from sunny inertia, until at last both crumble to dust and salt in the great depression. Down there solid matter goes through a term of infinite waiting until disturbed by some new epoch. Where torrid air and frozen water rule supreme, as in the sea of sand and in the Polar desert, there we find the immense contrast of gigantic wholes to their distant surroundings, in the absence of variety within. Here individual shapes are forced into strict repetitions under the ban of a general and uniform pressure, which crushes all independence out of the waves of the dunes and ice-packs. And, finally, we have the ocean and the atmosphere, where there is one huge mass and where only movement is form.

The Duab represents one of the many modes in which water and air combine with each other and with the soil; it is one of the stages of imperfect interpenetration. The less thoroughly elements

81. A Double Bridge over the Zarafshan.

are mixed the more each of them preserves its own texture or colour, the cruder is the contrast between component parts. In an oceanic climate this intermingling of atmospheric circulations and their close contact with the earth's surface attain the highest degree; the elements are shaded off into each other. Of this condition the forest is a leading symbol, being result as well as agent of an energetic interpenetration between the kingdom of the clouds and the realms of stone. If so willed and if we only identify perfect blending with the master-hand, we might call the Duab an elementary landscape piled up, carved and daubed by a child. But if this simile were to be taken as an argument of æsthetic value I cannot entirely agree with it. Freshfield has called these mountains hideous from a picturesque point of view and this may possibly be true if, like an artist, one looks at them merely from outside and with the eye only. But how rarely do we judge beauty without interest, seeing without feeling or thinking? Most people love a landscape of wheat, vegetables and fruit-trees, probably because it yields things good to eat; the inside of a wood can, after a time, become as monotonous as any steppe, but we love it on account of the cool shade on a summer's day.

I have given up travelling in search of scenery in Europe because I feel unconsciously as if all of it had already been fingered too much. It is a curious obsession and quite unreasonable from the purely artistic standpoint, but explicable in the light of "civilisation-nausea" produced by tourist literature, tourist traffic and the exploitation of natural wealth. A hundred miles of desert I prefer any day to ten miles of cornfield, knowing that, since the advent of blotting paper, the desert cannot be put to any utilitarian purpose and will remain in its wild state for ever. The magnificent forests of the Caucasus, for which I cherish an intense love, always fill me with sad thoughts of lumbermen and saw-mills, whereas in the Duab what little there is may be considered fairly immune. Add to this the powerful influence of contrast which makes a garden in the desert seem a hundred times more beautiful than a garden among gardens. Quite similar is the fascination exercised by isolated mountains such as the Matterhorn, Ararat, Popocatepetl.

Scientific interest can also enter into the involuntary definition of beauty by the contemplating mind and I feel sure that the character of classical art—Greece, by the way, is a dry country—shows a strong tendency to reasoning on simple lines. I must confess that I have contracted a final and permanent love for Duab scenery because fine weather is written large over its every trait. In spite of the invention of waterproof materials there is nothing like rain to damp scientific ardour or dull æsthetic perception.

Before moving on I should like to describe the mudspate which is an important form of rock-transport, especially among barren mountains, and which is frequently met with in the Alai-Pamirs. The inhabitants of the Eastern Alps call it "Mure," while "mud-avalanche" is the word generally adopted by Englishmen, but I prefer coining the shorter name "mudspate," because in scientific terminology "avalanche" ought to be reserved as much as possible for falls of snow and ice. There are, roughly, three ways, allowing of infinite transitions, by which fragments of the heights can reach the bottom of a slope. Stones detached by weathering and frost-blast can fall by their own weight forming scree-talus and shoot-cones, the results of dry transportation. Then there is the driving force of the torrents where water acts dynamically, tossing the debris downhill and depositing them in alluvial cones and fans. In between these we have the descent of a mixture in a semi-liquid state, where water constitutes, say, one third of the entire volume of the mass.

The typical mudspate consists of mire charged with a great number of rock-splinters and blocks, but sometimes it may be composed almost entirely of "clean" stones ranging in size from a pepper-corn to large boulders. Thus "mud" signifies that the downpour is tough and plastic like a lava-stream or snow-avalanche, while "spate" insists upon the sudden and spasmodic character of the phenomenon. Nor can it be otherwise considering that a perfect suspension of much firm matter and coarse particles in little water cannot be kept up long, because it "sets" immediately when at rest, as everyone knows who has to do with mortar or concrete. Owing to this physical difficulty the mudspate is rare in comparison to the familiar modes of denudation, although in favourable localities it may become very important. As to the limits of definition they cannot, of course, be drawn with mathematical precision, being comprised within a wet landslip and a flooded torrent overcharged with rubble. On the whole we should take into account the beginning, progress and end of the outburst.

When a gentle slope of grit and shingle has been soaked like a sponge by rain or melting snows there may come a time when it bulges out and slides off in the manner of a bog-burst on Irish moors. Slipping into channels and gullies this mass is mixed with more water, attains a higher speed and carries away soft material as well as rocks which it finds on its way. It is during this descent that the mudspate generally acquires its characteristic composition, for only by movement can an even mixture of liquid and solids be maintained. It is neither dry nor is there much free water, but the whole mass appears like a rapid flush of mud, although frequently the rock waste is so rough as not to suggest what is popularly called mud. Enormous boulders will float in this thick porridge like cork on water or iron on quicksilver. A mudspate may also be caused by the sudden bursting of a reservoir of water in the bed of a torrent (or the glacier above) which thus may be enabled to charge itself, for a short time, with an inordinate amount of loose material from the higher banks beyond the reach of normal floods.

The typical mudspate-track does not however readily associate itself with the ravine of a permanent or powerful mountain stream, for the simple reason that the catchment area and bed of a torrent at work throughout the year are already deprived of the bulk of easily shifted material. Operating with a minimum of water the mudspate liquefies itself automatically when, during its descent, it has become too thick. Stopping for a while it dams up the water runlet in the gully and then proceeds again, repeating, if needs be, the process several times. It is as if the mountain were suffering from some internal complaint easing itself in fits and starts.

In the Alps the "Mure" is not so much a regular phenomenon as a catastrophe, because most surfaces of soft soil are covered with grass or forest, and because a steady transport of quantities distributed over the seasons is kept up by the streams, continually flowing. Cloudbursts, small landslips damming a water-course, a sodden piece of pasturage losing its hold on the bed-rock and snow-avalanches are the usual causes of a mudspate in the Alps. On slopes covered with vegetation the violent rush roots up weeds and trees, thus leading to fossil enclosures otherwise rare in alluvial deposits. The frequent occurrence of such remains in the Hottinger Breccia leads me to think that its origin may have a good deal to do with mudspates, especially as the interglacial periods are supposed to have had a climate of the steppe.

Intermittent water supply owing to a dry climate, absence of strong vegetation and barren mountain flanks reaching up to the snowline are the conditions which favour the mudspate as a habitual and periodical phenomenon. Slopes of soft grit (Fig. 150) covered with snow during winter are the best starting-ground. During spring the snow melts evenly over a large surface thus soaking a top layer of the friable stuff up to bursting point. In this manner large quantities of half liquid rubbish are suddenly set free, initiating the process. Conway has given a very good description of mud avalanches in Nagar and Hunza (*Geogr. Journal*, 1907, pp. 501–2), where they occur on a far grander scale than anywhere in the world and with the regularity of clockwork, forming a typical period of the mountains' seasonal life. Conway saw as many as 150 falling over a single broad hillside during one day. On the top of this flank, 10,000 feet high, were slopes cloaked with rapidly melting snow, the water trickling down innumerable little rills which gradually united into large and steep gullies at the bottom. Mudspates were continually being discharged and re-formed, and in one trench five were counted in an hour. This happened late in June and early in July, a time which coincides with my observations in the Duab, for when I passed through the Zarafshan valley the activity of the mud-shoots had already ceased, most of the winter snow having disappeared. Thus I only saw the fresh deposits, unmistakable by their shape and texture.

82. Ox-Sleigh.

When not too liquid the discharge forms a snout or tongue such as seen in Fig. 84, where an overflow from the gully has been thrown over the brim. This is the lobate shape assumed by all viscous matter such as snow-avalanches, glaciers, lava, honey, peat-bog, and the like. But this only happens when the mixture is fairly thick and allowed to rest on gentle inclines. The other extreme is represented by narrow gorges ending in a river which prevents accumulation. Usually the mudspates build up an irregular cone or delta (Figs. 86, 167) furrowed by one or more characteristic gullies. These are deep and narrow trenches with very steep and smooth sides. The sudden gushes loaded with angular fragments act like a rapid, liquid file which rakes and rasps the channel, at the same time plastering it with mud pressed against the walls (Figs. 84, 86). Most of the smaller dejections regularly use this gully, which also serves as bed to an insignificant or intermittent stream. But many downpours miss this chute and large ones overflow it, so that in this way a talus is raised.

As in the case of snow-avalanches the flowing mass has the tendency of forming a gutter within itself, not needing one determined beforehand. When left to itself on an even slope the middle of the mud runs faster because there is less friction, while at the sides, retarded by friction, deposition takes place giving rise to an embankment, so that the crawling leviathan builds its own track. One can observe this process anywhere on slopes of finely ground limestone in the Dolomites, of scaly slate in the Zarafshan valley, and on the ashes of volcanic cones where what might be called the "welted furrow" is frequently seen (Figs. 85, 86, 134). It consists of a shallow rill with a welt on either side. Started somewhere a bulge of soft stuff begins to hiss down like a hurrying snake, never relaxing its speed. It looks as if it ploughed up the surface leaving little bordering ridges. In fact the central current continuously discharges solid particles at the sides, the water thus set free recharges itself with fresh material in front, and so, by an uninterrupted sequence of setting and liquefaction two parallel strips are drawn with a gutter in between, until an obstacle or a hard surface without loose grit, a large stone or a patch of grass causes a final discharge, a splaying out of the water and its load.

An analogy is offered by lava streams which, cooling into a hard crust on the outside, continue flowing within, leaving a tunnel in their rear. In some respects the lateral moraines of a glacier can also be compared to the lateral welts of the mudspate, and whoever studies the various movements and deposits of suspended, viscous or granular matter will note a great number of common points with interesting variations and transitions.

Great importance must be attached to the mudspate as a geological factor. Firstly as to volume, on which point I quote Conway: "Where I was, all through Nagar and Hunza, the mud-avalanche is by far the most important agent in forming the miscellaneous deposits you meet with in the valleys. The moraines, even though enormous, are secondary, and are not so big as the enormous deposits produced by the mud-avalanches...and the volume of debris that they must have brought down in the fortnight was enormous, far greater in proportion than anything carried by the glacier in a similar length of time."

Secondly we must remember that the stones of the mudspate will scratch each other, as well as the rock slabs over which they happen to glide, so that the track could easily be mistaken for that of a glacier and the deposit for a moraine. This is not likely in a broad valley, but a clayey bank richly garnished with blocks and suspended at the end of a steep ravine may often pass, and perhaps, for all we know, has sometimes passed, for boulder clay from a hanging glacier. To this must be added the hypothesis of modern glaciology which assumes a very dry climate during interglacial periods, not only comparatively, but according to present conditions, namely, those reigning in the steppes and deserts of to-day.

83. A Village Mosque.

On the whole I am inclined to believe that the high mountains of the Duab are very much like what the Alps must have looked after a maximum, when the ice streams were in retreat. What with screes, mud breccias, alluvial fans, conglomerates, terraces, cones, loess banks, loess mixed with stones, moraines, wind-drifts and residual soil the Zarafshan valley presents material of endless abundance and bewildering variety to the student of post-tertiary deposits. I can strongly recommend this morphological museum for the investigation of all the possible forms of rock-waste disposal.

Up to Urmitan, which we reached on the second day out from Panjikent, we crossed the river several times by frail bridges. At first we followed the left bank as far as Yabon, where a tributary throws out a large fan forcing the Zarafshan to describe a loop. This cone delta owed its origin chiefly to successive mudspates, the superimposed layers of which could be easily distinguished and which were traversed by a rift from 16 to 20 feet deep with loamy walls. At the mouth of a smaller gulch we found the most perfect specimen of a mudspate snout I have ever seen. It had the shape of an avalanche tongue, having come down as one mass of thick clay of an ochre colour, probably owing to a liberal admixture of loess. Like the snow of some avalanches this mud, which must have been very sticky, had rolled itself out into numberless balls and nodules, now caked together and as hard as stone.

84. Mudspate-track opposite Veshab in the Zarafshan Valley.

The sides of the little valley, whence this disgorgement had belched forth, were splashed with dried mud as if a Sart house-builder had flung loess mortar against the rocks with his trowel. Here also was a deep channel with vertical sides and at the bottom there ran a miserable thread of warm water.

As the Hissar and Turkestan ranges come very close together here, not leaving room for terraces, the mountains slope straight into the Zarafshan. Some of the inclines are covered with a gritty soil baked so hard by the sun that the path is merely sketched out, and walking outside of it becomes very difficult or even risky. Climbers are familiar with this kind of cement into which they have often tried to dig their boot-nails when struggling up an Alpine moraine by lantern light. I always kept my valley-foot out of the stirrup, ready to fling myself on the path should the horse be determined on racing back to Samarkand with the waves of the Zarafshan. Near Yabon we had to negotiate a ledge barely five feet wide drawn across a giddy face of conglomerate, and then we dived down again into the Tartarean gloom of the canyon, which in parts was an almost vertical cleft through strata of reddish slate.

Most of the tiny tributaries issuing from lateral ravines were no better than tepid soup, and only one of them seemed to come from invisible snowfields. So, at least, we inferred from the gay babble of its cool waters with their freight of fine silt. At its mouth was a miniature oasis of barberry and wild rose peopled by yellow buntings with bright plumage. Just before Urmitan we had to squeeze along the foot of a sheer pebble-pudding towering three hundred feet above our heads even straighter than "a beggar can spit," using such foothold as was graciously allowed us by the wild and roaring cataracts of the Strewer of Gold. Although the conglomerate is very solid the sensation that something might drop from above proved most uncomfortable. A man lying at the top of the cliff could have killed the whole of our caravan with a pocketful of small stones.

On arriving at our hard-won goal we were put up at a sarai where we found several of the comforts of civilisation, to wit chairs with legs that stood, a table and a samovar (hot water urn). In the open verandah or gallery which runs round the room the population of the village assembled to watch our proceedings. Closely packed together they sat motionless but keenly attentive so that I feel sure they would have been able to pass an examination on the inventory of our table and belongings. Down below was the courtyard with the stables in which our horses were tethered, and in the starlit night these surroundings vividly recalled the scene of the manger in Bethlehem. Things change but little in the immemorial East, and very likely the sarai where Mary laid her babe was very much like this one.

Here we met two of the Russian forest-guards stationed at various points to prevent speculative depredations on the juniper trees sparsely

85. A Study in Relief. Welted Furrows and Lions' Paws in the Zarafshan Valley.

dotted about the mountains. With the influx of a European population to the Duab the price of all fuel has risen enormously and charcoal-burning has become a profitable trade. These guardians of the scrub were simple people supporting their wives and children on thirty shillings a month and what taxes in kind they might levy from the villagers. One of them next day accompanied us to Iskadar.

We engaged six men to help at bad places, for one of the sills overhanging the grey abyss of conglomerate was so narrow that the loads had to be taken off and tugged across by the porters, while the animals were carefully led over by two men at the head and two at the tail. Below was a fearful drop and above one nearly knocked one's head against a bulge consisting of pieces of slate embedded in yellow clay. I take this to be the desperately clinging deposits of mudspates habitually shot over the brink. Similar mud-breccias overlaying the Zarafshan conglomerate are now frequently met with.

On the mountain side of the left bank opposite Vishken we observed the straight line of a water conduit two miles long, of the kind that is called "bisse" in the Valaisan Alps. Where the slope is not too steep and soft enough the channel is dug out of the ground, being continued along rocky cliffs in flumes supported on props. At one spot a leakage from the aqueduct had caused a landslip of several thousand cubic feet, leaving a deeply scooped-out scar in the conglomerate. The broad upper end of this hollow was marked off by the straight, green fringe of small trees rooted in the artery of life.

At Shakhisia one of the many possibilities by which springs may be formed was clearly demonstrated to us. The small lateral valley ended abruptly in a vertical cliff showing in section a conglomerate wall overlaid by a soft, brownish deposit. All water oozing through the upper layer was stopped by the denser mass, flowing out at the line of contact. The valley as a whole is still narrow, the great flat terraces only just beginning to sketch out their vast bulk. One hardly dares look aside, the break-neck path demanding all one's attention. For many hours we ride through chasms of forbidding severity where the cruel sun beats on the naked rock during the day, where a silent dread creeps through the shades of night. The difficulty and slowness of advance, coupled with the anxiety for the safety of the caravan, heighten the fearsome impression made by the lower part of the Zarafshan valley.

86. Alluvial Cone with Furrows.

After dim gorges between walls of black and brown slate come cauldrons where the sun smites the screes with merciless force. No bush or blade braves this oven, no twittering of birds is heard; the ice-grey river rumbles far below through the masonry of the mountain steppe. It is still the hot air from the sand seas which surrounds us here. This glowing current battles against the snowy world rushing up to the highest valleys so that even on the summits it is not quite exhausted. To meet it the snowfields send down their boisterous waters, and so fierce is the stormy onslaught of the two that they overrun and underslip each other. Above is the air charring the mountain's flanks, below the water thundering through its sluices as if wildly indignant, because the white-hot enemy draws in above, scorching and burning all before it. With us, in Europe, it is otherwise, both being intermingled at the time of summerly vegetation when damp clouds coax forth the Alpine verdure from the wetted earth. Here the powers calling and challenging each other graze past in their quick rage. In truth it must be the madness of love searching its object through eternity. Here the elements are not destined to meet in the sultry embrace of tropical vapours. High up in the blue the desert flame caresses the face of snowy giants; low down, in the plains, the torrent stills the fiery longing of the steppe. A budding and flowering bounty wells from the blessed land. The blast from the desert is a call to the mountains and the answer is revealed in the gardens of Samarkand. Therefore the middle belt of the mountains is poor and dry, but it has its oases where on rare occasions a rivulet is captured on its way. For many hours the valley of the Zarafshan seems a lifeless world with no consolation save the murmur of the stream carrying its message from the glacier, whither our thoughts precede us.

But suddenly, at some bend of the walled-in vista, there is unfolded the vision beautiful. From among the gaunt, grey pillars of water-chiselled concrete bursts a thick luxuriance of refreshing green. A great and glorious gleam of green on ashy cliffs, a shining emerald set in grey steel. Sheer the cemented precipice rises from the canyon, supporting a flat shelf on which clustered groves sway to the breeze. Along the rim the verdure descends into the sculptured sides of the terrace, and in the fissures the green shaft of the poplar stands by the

87. Talus Cones.

bleak pilasters of conglomerate. Coming near we find a village hidden under the leafy roof. Riding through crooked lanes arched over by vaulted branches we seek the carpet of a smooth lawn spread around a square tank filled with clear water. About us are the stems of gigantic apricot trees holding aloft a cupola of foliage dotted with the golden fruit. At night the owlet coos and a tender, lisping wind plays with the cool tree-tops under which we sleep. Such are the villages along the Zarafshan, like pearls strung widely apart, like smiles on the wrinkled and determined face of nature. Dardar, Zaravat and Iskadar are perhaps the finest of these hanging gardens of Semiramis. Zaravat, as we saw it from the opposite bank, is like a grove planted on a mighty quay the walls of which rise clean and bare tinted with neither moss nor lichen. No greater contrast could be imagined between aridity and vegetation, between grey and green. Beyond and above is the soft outline of mountains about 9000 feet high, their long slopes partly covered with red marl, partly with the chess-board design of yellow fields serving as a painted background to the plastic scene. These villages form a very Eden in the wilderness. And truly life, if measured merely from the material side, offers the inhabitants all they desire. Their wants are few and easily satisfied. Their horizon is limited, happily for them, and they know nothing of what we in the West call Divine despair. Their lot is certainly no worse if not better than that of the teeming millions of our great cities who are the victims of the Moloch of industrialism. Their life passes away in a steady, even current, and it is a great event in it when a cavalcade of strangers descends upon them at sunset clamouring for shelter and food. Then the village dignitaries assemble and sitting in a circle at a respectful distance observe minutely every action, nay every change of expression of the visitors.

Before Iskadar we had to pass a succession of interminable and very steep scree slopes consisting of a thin layer of loose rubble resting but gingerly on the hard-baked stuff underneath. Here portions of the track, barely a foot and a half wide, often broke away under the strain of our long and heavy file of pack animals so that an ice-axe had to be degraded into a tool for road making. Seen from the swaying height of the saddle and with the knowledge that foothold was precarious, even in boots, the bold swoop of the even gradient descending to the angry river seemed as giddy as the sheerest precipice.

There is hardly any loess hereabouts, and the thin crust of fertile soil is too valuable for building purposes. Hence the cavernous houses and the garden walls are made of stone, namely of waterworn boulders and pebbles taken from the conglomerate or from the beds of lateral streams. What strikes one at once is the extreme neatness with which the natives here have handled this round and slippery material, the elusive rollers being cleverly and solidly cemented and often arranged in very regular rows. Probably the mortar used is as good as the mason's sense of balance, while to European minds the task seems more difficult than the quarrying of slate or lime. But we must remember that these people have imperfect tools, that powder is too expensive for blasting, and the rocks some distance away. They have therefore made the best of the material nearest at hand. The top of a wall is frequently finished off with a kind of concrete shaped in the mass or formed into bricks (Fig. 80). Nowhere in the Duab have I seen the architecture of rough stone brought to such perfection, a fact which seems to agree with the general condition of the locality, for of all the more inaccessible mountain valleys that of the Zarafshan is the most prosperous. In many respects it reminds me of Suanetia in the Caucasus, an analogy which also holds good with regard to the inhabitants, the Galchas of the Zarafshan being an older and purer race containing more Aryan blood than the Sarts of the plains.

Soon after Iskadar, where we spent a night, the valley begins to widen out considerably making room for a broader development of terraces. We have, henceforth, the canyon or notch of the second degree sunk into the conglomerate plain forming the floor of the great alpine valley. On a huge terrace several miles long lies the important village of Varziminar, which takes its name from a minar (minaret) built of mud-concrete in the shape of a tapering tower or obtuse cone. To get here from Panjikent had taken us two and a half days or twenty-two hours actual going. As the distance is about 54 miles by the road we had covered 2½ miles per hour on an average, while in the plains or on easy roads we were wont to reckon from 4 to 4½ miles per hour. The rest house at Varziminar belonging to a mingbashi was the finest we had yet seen in all our wanderings. It would seem as if the old Persian tradition had maintained itself in its purity here. The patterns and stalactites of the ceiling were exquisitely painted in green, gold, blue, red and white, while panels in relief stood out from the wall. Carved wooden shutters of beautiful design protected the three windows above which were filigree screens in stucco of lovely geometrical pattern. At both ends of the room and between the windows carved niches took the place of cupboards. The ceiling and frieze of the pillared entrance hall were decorated in the same manner.

Here we stayed for a day to rest the horses and also in order to make a selection from our outfit for a fortnight's excursion into the Fan valley. Moreover our native servants were clamouring for a tamasha after so many days of hard work. In its narrower sense "tamashà" means a feast, the solid and conventional basis whereof is a gorge of the fat-tailed sheep washed down with floods of green tea, while the beating of a tambourine coupled with melodious howls and the serving of rancid sweetmeats or other refinements are merely elaborations and variations of the fundamental theme. In its wider application the word signifies "fun," "something going on," "something exciting," as applied to a fallen horse, a railway accident, a fight, a law suit or a funeral; in fact any agitation or eruption on the placid surface of humanity or nature, giving pleasure to all concerned or, at least, to the onlookers. Thus tamasha is the great thing of Sart life as it is in the lives of all nations on the face of earth. Who is not fond of a tamasha? It means sensation in its widest sense as an interruption of monotony, it means a change from mechanical routine to conscious feeling; tamasha is any more or less pleasant awakening of our senses from the dull and daily round, a peal or even a discordance in the hum, anything new or old or odd, restless or bright or still, relatively to the qualities, movements and noises of ordinary life. Steady progress is a rhythm which, however slow, is the only means of rapid advance, but to the human machine tamasha is a necessary stimulant marking time, recalling us to an energetic resumption of the rhythmic beat at the same pitch, for otherwise it might imperceptibly become slower and slower. Wars are the tamashas of nations, whatever the beaten side may call them afterwards. Any intensification of vitality is a tamasha, which therefore, for healthy people, is always associated with movement, colour and noise. Those to whom absolute rest is a feast are overworked or overfeasted, and it speaks for the wisdom of the Oriental that the word tamasha is never applied to anything devoid of bustle or sound, while its highest realisation is combined with the smell of broiling grease and the digestion of mutton.

CHAPTER X

FROM VARZIMINAR TO THE ZARAFSHAN GLACIER

After our return from the Fan (Chap. XII) we left Varziminar on August 11, making straight for the Zarafshan Glacier 150 miles away. This we reached after seven days and a half, thus covering an average of twenty miles per day. Over the flatness of terraces and along the outrun of great slopes travelling was now easy, the only difficulties being those opposed by the fissures which tributary rivers had sunk into the vast blocks of Zarafshan conglomerate. Into and out of these we had to dive by means of corkscrew tracks scratched into the hard grit. Occasionally, when the water had wormed itself through hard rock, the cleft was sufficiently narrow to be spanned by a bridge and in this way we rode over the Chindon torrent 200 feet above its bed, our eyes vainly endeavouring to fathom the dark recesses of a gurgling depth (Fig. 89). Thus having to cross and recross the Zarafshan several times we were treated to samples of every conceivable variety of bridge (Fig. 81). Being innocent of railings and enjoying that springy elasticity which we sorely miss in the carts of the country, they afford good practice for beginners in the art of rope walking. Most people dismount on approaching these frail hyphens of the sundered road, but natives of more than ordinary fatalism or officials pervaded with proper pride remain in the saddle with unconcerned or haughty mien. We always got off and sent the horse on in front, for being weak in the faith and but poor travellers we were unable to look either unconcerned or dignified. One might as well ask me to show lofty indifference in a drawing-room with a cup of tea on my knees, and as to stately importance, why, a European's attempts are hopeless unless he be wearing evening dress, or uniform or, at least, a top-hat.

During our progress we had several cloudy days and an occasional sprinkle. The Russian forest guards told us that July was even a fairly rainy month. Thus now, in the middle of August, the driest time was only just beginning, for among the higher mountains the finest weather prevails during the latter half of August, September and October. The worst conditions we ever encountered were the three days of almost incessant downpour near and on the Zarafshan glacier. Moisture-bearing, westerly winds discharge their humid freight against the slopes of the Hissar and Zarafshan ranges. They cannot do so while blowing across the overheated plains, but on reaching the snowy mountains condensation sets in with the fall of temperature.

Above Varziminar (follow Map II) the valley narrows down to a gate only to open out again immediately, giving space to the wide shelf of Sasun. The picture (Fig. 77) taken just above the village of Sasun gives a good idea of the landscape. We are looking downstream, towards the gap, and the flat expanse of the terrace, nearly a mile in width, seems to merge into the slopes, for the river, on the left, remains invisible, being entombed in its trench. On the right a portion of the slate mountains, consisting of marl, stands out with lighter colour in the glare of the sun. Here a considerable landslide has occurred, resting its pale bulk on the plain behind the trees. As can be judged from the regular shape of the house and wall on the outskirts of the grove, there is sufficient clay or loess in this locality to admit of the same style of architecture as in the lowlands. But the majority of the Zarafshan kishlaks (i.e. villages) are built of pebbles. The tops of the garden walls are often crowned with a layer of rose thorns (Fig. 80), evidently serving the same defensive purpose as the broken bottles at home. That the soil is very thin in places is shown by the enormous number of stones gathered from the fields to make room for cereals. Sometimes, near a settlement, the road runs along the top of a broad jetty of cast-out rubble. Although the highest barley fields are left to a reasonable chance of rainfall, the more regular and systematic cultivation, especially that of vegetables, is dependent upon artificial irrigation. I wonder if it ever strikes the more philosophically inclined among the natives under what tantalizing conditions they live. They inhabit a desert[1] with vast quantities of water actually running through the middle of it; they see and hear the running well of life

[1] N.B. A desert remains a desert—in the climatic sense—however great the number of oases.

that is not for them. Indeed, the stony trough of the Zarafshan is liquid wealth collected and hoarded for the denizens of the plains, while the mountain people catch but a few trickles by difficult and devious means. As the tributaries have also burrowed deep down, joining the main stream far below the level of the arable surfaces their water has to be tapped high up the side-valleys. Where the long ditch comes to a cliff it is continued in a thick layer of grass-sods supported on props (Fig. 36). Evidently the water question spurs the natives to their best feats of engineering so that even tunnelling is resorted to. Such a pipe driven through several hundred yards of conglomerate is to be seen near Riomut.

Geology of the road can be studied at leisure, for the natives avoid constructive work, using as much as possible the formation of the ground. Bands of conglomerate, outcropping strata and nullahs are welcomed to connect the level stretches of the flood plain with those on the flat tops of terraces.

At Sasun we saw a mazar or holy grave combined with a small prayer-house. The offerings of passers-by were here disposed in a curious way. Hundreds of tiny pebbles, besides the usual coloured rags, horns and other trifles were laid on the cross-pieces of the ornamental wooden balustrade fencing off the portico. The village children at this place seemed to be very polite and not at all shy, greeting our appearance with loud salaams. We now begin to meet the first considerable brooks from the right or northern side whence we may infer a greater height of the Turkestan range and, perhaps, the presence of a few patches of snow. But the water is still tepid, having been drained from arid slopes facing the sun.

At Shavadki the snow peaks of the Zarafshan range begin to rise into the field of view and are henceforth often seen peeping out of side-valleys. On the right bank none are visible as yet, but streams are getting cooler and more plentiful. The landscape offers a continuous change from mountain steppe to little paradises. Just before Rars the path crosses a perfect desert of rocks, marl and loess where no tree or bush is visible. Between Rars and Pakhurd, as well as between Shavadki and Veshab the valley is uniformly bleak and monotonous, although full of interest to the physiographer. The character of our surroundings is here determined by huge scree slopes on the

88. Man and the Mountain; Dangari on the Zarafshan.

mountain flanks and by the canyon exposing the conglomerate which now consists of fragments of dark slate. Below the green and thickly tufted grove of Veshab the river, by strange contrast, has chosen to run through the blackest and most sinister looking of gorges. Between Sasun and Oburdon harvest is now going on and in the fields one sees neatly piled bundles of wheat and barley. Oxen are treading out the grain, the winnower flings his golden fountain to the wind and laden sleighs are marking their homeward tracks upon the dust. But the most glorious harvest is that of the apricots. Thousands of little yellow balls gleam like stars on the leafy dome of every tree. Women and children with frolicsome eagerness are reaping the sweet bounty and laying it in the sun to dry. Where screes are near, the fruit, like so many gaily coloured eggs, is put into nests among the burning stone, and when we first saw these vivid orange spots among brown slate at glaring midday, we thought of some sulphurous mineral, or colonies of rust-red lichens, or fungi, or the flowered cushions of a saxifrage. Enterprising boys had even climbed bold ribs of rock to store their mothers' property beyond the reach of two and four-legged animals. Being also less exposed to the filth of low ground I would prefer these cliff-dried goods to all others and gladly pay a better price for the "higher" quality. There is also a special way of preparing dried apricots by breaking the stone, extracting the inner kernel and sticking it into the fruit, thus producing a combination which reminds one of the familiar almonds and raisins. Most of these are exported to Ferghana across the passes, for the Zarafshan fruit is famous for its fragrant taste developed on an austere soil with mountain air and mountain sun. The trees occur in great numbers, often forming extensive groves. In size they correspond to the average of our apple trees, including specimens eight feet round the waist, but in shape and foliage they are much finer. The mulberry tree with its creviced bark and gnarled branches is equally characteristic and was also yielding an abundant harvest. Although to our taste the mulberry is of a somewhat sickly sweetness, it forms, in its dried state, an important winter store of the native population, for the sugar it contains makes it very nourishing. Remembering the old saying

> A woman, a dog, and a mulberry tree
> The more you beat them, the better they be,

I asked the people for their opinion on this controversial subject. They replied that they did beat their wives and threw stones after dogs, but that there was little satisfaction in thrashing a mulberry tree because it refused to yell. Above Veshab my favourite, the walnut tree, also came to the front, while lower down it had been rarer.

At Ustanak, where the valley is a little narrower again, the aspect of the landscape was slightly more alpine, showing how the shape of the ground influences the other features. The fields were inclined, barely irrigated, and a good deal of low scrub mixed with rank weeds occupied the untilled patches. From here we looked straight into the magnificent chasm of the Darkh river (Fig. 78). Oburdon again was like an oasis of the steppe, with flat, irrigated plots and no room left for useless vegetation (Fig. 79 represents this type). Just before reaching this place we had to cross a slope composed of large boulders. As horses cannot step over them, a deep and exceedingly narrow passage or defile of Cyclopean walls has been formed by lifting out the blocks down to the surface of cemented deposit.

Our quarters at Oburdon were more than ordinarily picturesque. The elders had installed us in a small courtyard situated on a terraced shelf and covered with lawn, and real, green grass, even if thin, is always a sight to be thankful for in Turkestan. Over a balustrade formed by the coping stones of the lower wall we looked down upon the road, while at the back another wall supported the upper houses of the village. Here, in a recess of the mountain slope stood a mazar built in the shape of a four-poster bed. Large blocks of glistening slate crowned with the horns of wild sheep lay at the head of the sleeping saint. In one corner an ancient mulberry tree cast a cool and penumbrous shade over this idyllic retreat, while a smaller one half standing, half lying, was supported by the little house over the grave. Near its roots bubbled a clear spring which, after flowing across the courtyard, lost itself under the slabs of the threshold. At both sides were verandahs with clean, hard floors, where one could lie down and let one's eye wander over the crooked lanes of the village, over groves, fields and mountains. A severe critic might have felt offended by an old calendar hung on the wall by some enterprising trader, but fortunately its brutal daubs were so faded as to blend with the quiet half tones of our surroundings. This holy shrine proves that even these rough peasants cannot be without artistic taste or appreciation, for even supposing that things were thus arranged without much forethought, they have, at least, avoided blunders capable of destroying the unsought effect. The intention of creating something picturesque may be rare, but the faculty of knowing what is pretty or beautiful seems to exist. The natives build and live under the immediate pressure of and in accordance with their natural environment which makes them less liable to err on the side of artificiality. This is, of course, speaking from our point of view

89. A Giddy Bridge.

and there remains the question if strange races have the same fundamental taste as ourselves.

On this point I should like to express a belief that nature forms the impartial basis of aesthetic valuation, if not between all mankind, at least between the cultured individuals of civilized races and all men. Thus we (i.e. the educated and thinking Europeans, Chinese, Japanese, etc.) form the subjective standard of beauty, because only we are able to abstract aesthetic impressions from utilitarian or moral considerations and because among ourselves only are such problems discussed to their uttermost limits. The objective standard then is nature and my axiom is this, that the nearer the work of man is to nature (the less differentiated) the lesser the risk of ugliness, the greater the chance of its being found beautiful or at least escaping adverse criticism. Towards nature we keep a distance, we never blame her aesthetically or morally. Nature may be charming, or desolate, or terrible, but nobody ever speaks of her as ugly, inartistic, in bad taste, or immoral, for although she can please us in varying degree she cannot offend us, leaving us merely indifferent when the worst comes to the worst. Man can effect pleasant changes on the face of earth, but always at the risk of creating ugliness. Something evolved from its surroundings in a simple and naïve manner is not likely to arouse a protest and keeps open for itself the lucky chance of turning out something supremely beautiful.

Has it ever struck the reader how unwilling we are to ridicule the unsophisticated attempts at ornamentation made by primitive tribes, while modern buildings or furniture bring the most scathing remarks to our lips? The more differentiated our work becomes the more it has to do with "art," the more difficult the production of something artistic as apart from the merely artificial. Hence our aesthetic sense is hardly ever incited to contradiction in wild or primitive countries, and we live under the happy illusion that the natives share our ideas of the beautiful, whereas in reality they have unconsciously clung to nature and the dictates of utility. Baubles from a sixpenny bazar excite their admiration and European trade can easily pervert their artistic innocence. An exception to my theory must be made in the case of man, his body and the decoration thereof, for man is too near, we cannot keep a distance. The objects of impartial judgment must be detached from the body and soul of man and what is equally detached for all mankind is nature in its original state.

Above Oburdon the river still flows in a canyon, but the valley as a whole is visibly wider. Screes are trumps and majestically drape the slopes with straightened folds from ridge to foot. Seen from the further bank their gradient appears enormously steep, and a quaint effect is achieved by narrow, white tracks drawn through the brown and seemingly vertical surfaces (Figs. 86, 87). They climb in direct line to the grass plots above and are used for dragging down the hay, which is left to dry low down among the screes. Its square stacks are visible from afar against the dark ground. At Riomut we found another forest guard with a wife and two children established in a vast apricot garden. They were just beginning to gather in their harvest, and some of it was already spread on the flat roof of the house. Two natives were employed in squeezing the stones from the fruit, but while one squirted them out very quickly and skilfully with his fingers, the other saw fit to apply the more scientific principle of suction. This undoubtedly added a nice taste to the otherwise tedious operation, and also gave an opportunity for shooting out a projectile on the principle of the air-gun. We particularly asked our host for apricots *with* stones, as we were rather fond ourselves of experiments in physics.

The canyon is beginning to lose its character of a valley within a valley, and the terraces are often dwindling into comparatively small shelves, leaving a wide flood plain between them where the river flows among numerous gravel islands. Only from time to time the bulky conglomerate dikes come together again like a vertical wall with a crack for the passage of the Zarafshan. Frequently the rubble slopes merge into the stunted terraces, smoothing them out into even gradients. The lateral rivers are still keeping their terminal chasms. We are entering the middle region of softer features found among all high mountain groups, that belt of gentler dips lying between the rugged shapes of ridge and summit weathering, and the zone of precipices and gorges due to river erosion. Whoever has crossed the Alps will remember how after the Via Malas, the Witches' Cauldrons and Devil's Bridges the landscape opens out again on the upper pastures near the pass. Here reign the flatter curves traceable to snow pressure and glacial erosion or accumulation and of which the Pamirs furnish an example on the largest scale. If we did not see the blackish water of the Zarafshan rushing towards us, and the snowpeaks which are now beginning to loom at the head of the valley, we could well imagine that the road was leading out into the plains. At Khairabat the illusion was all the stronger because nature had here seen fit to introduce an exceptional note by a large deposit of loess along the river which spreads in many shallow ramifications among quicksands and gravel beds grown with tamarisks. Here one seemed to be in between the fields of irrigated lowlands. On the path the horseshoes struck no stones but stirred up dust or splashed through water strayed from ariks, while at the sides the high

90. Fan Delta and Glacier of Sabak.

tufts of steppe grass raised their waving plumes. The houses are built of pure loess, which here however is grey instead of yellow.

At Madrushkat we found shelter in the mosque, a building erected with more than ordinary care and lately renovated (Fig. 83). The ceiling made of whitewashed timber was quite classical in its severe simplicity. The crops were less advanced than lower down, and at Tabushin we saw the last apricot trees. Thus, travelling upward we shall soon have passed through all the seasons, although the ascent is hardly noticed by the eye. The trees are gradually growing less and less until but a few hardy kinds remain, mostly poplars and willows. Amid the solemn and heroic landscape these solitary ancients made one think of pictures by Claude Lorrain. Plenty of fresh milk was obtainable, showing that the cattle cannot be far off, and that some of it has already been driven down from the summer pastures. Herds are seen on the slopes, and droves of sheep come down in tumbling throngs like rolling avalanches. Beyond the domestic animals, among which the impertinent house-fly has unfortunately to be reckoned, there are very few living creatures to be seen; some small birds, chiefly yellow buntings, grasshoppers, hardly any beetles, and very few butterflies. So far we have not seen any eagles, only an occasional white vulture. Rooks visit the cemeteries, smelling down into the open holes of old graves or sitting on the monuments, upright slabs without inscriptions.

The kishlaks diminish in number and the inhabitants are visibly poorer. Nearly all the peasants whom we meet stop still and salaam, placing one or both hands upon the stomach. Those on horseback dismount until we have passed. The women are less shy than in the plains, and we often get a glimpse of a face or even a full view, without haste. This is a result of economic conditions, for here the women have to toil in the fields, and practical work never leaves much room for conventional fuss. The children are sometimes strikingly European in their features, while the men are pleasant to look at, strong and sunburnt, clad in the dark homespuns of the country. These are the people called Galchas by many travellers and ethnologists. Undoubtedly they look very Aryan, more like Caucasian mountain tribes, such as the Suanetians, than like the other races of the Duab. That they are the remains of a former indogermanic population of Middle

91. Paldorak; Looking North-east.

Asia seems not unlikely. Like the Suanetians they have preserved memories of their old religion under the outer cloak of the later creed forced upon them. The sacredness of fire evidently points to Zoroastrian worship.

The Galcha element seems to increase as one goes higher, for the racial texture is apparently not uniform throughout the Zarafshan valley, nor are the character of the people, their hospitality and politeness fluctuating in a remarkable manner, though fortunately by easy stages which allow one to become accustomed to the change. At first I had a theory that the right bank was more genial than the left, or that the amount of our tips projected a mysterious influence in front, malignant or benign. But on the whole I should say that severer circumstances of life are responsible for a decline of outward suavity with the growing height above sea-level. The human voice also rises to the occasion, and the natives of the upper reaches of the Zarafshan have a conversational shout evidently pitted against the growls of the mountains and the rumble of the torrents. The villagers of Paldorak proved themselves hard and greedy bargainers, being also more independent of demeanour, not to say cheeky. But after all they are more honest than the discreet and polished Sart of the plains. As was afterwards plainly shown by our porters on the glacier, they are naughty children of nature and, like children, unable to disguise their motives or check their impulses. The women made but faint pretext at veiling themselves, but in vain did they thus exhort the devil of temptation. He refused to budge probably because he knew that the high-priced stock, more likely to further his aims, was kept under lock and key.

Owing to the scarcity of wood the dung of cattle is used for fuel. Its preparation, falling under the head of bakery, is given over to the kitchen department of the ladies. After mixing the raw material with chaff, they pat round disks thereof against the sunny side of the house. When dry these cakes fall off, being afterwards stacked on the roofs, where piles of various size mark the wealth or industry of their owners. As a source of heat these briquettes are by no means to be despised. The fumes and odours of the initial stage are no worse than those of damp wood. Once the fire is well started a pleasant, smokeless glow ensues, and in a proper stove dung-cakes would do as well as the peat bricks used in North Germany.

The houses of Paldorak are collected in clusters between erratic lanes and the only timber visible is that of the doors. Nor can the rafters supporting the flat roofs of beaten mud be very solid judging from the elastic swing as one walks across. The more ambitious homes boast of clay chimney-pots about a foot high. Here, more than anywhere, the roof serves as a platform for useful duties and the no less useful exercise of laziness. On it all housework is done from the winnowing of corn to the spanking of children; gossips congregate and from this point of vantage women, boys and girls gaze upon the sights of the world. A broad piece of greensward in the middle of the hamlet formed our camping ground and within the shortest possible time after our arrival the spectators had arranged themselves on the tops of their low stone hovels. They crowded and huddled together, intensely curious, a crew of brown humanity, the sombre tint of their rags relieved here and there by a gleam of colour. We constituted a free circus for the mob as well as for the more dignified citizens of Paldorak. Soon they come nearer grouping themselves on the ground outside our charmed circle of trunks and boxes which is to keep them at the distance of the longest flea-jump on record. Thus they sit expectant, young and old, often two deep, forming a crescent the concave centre of which is turned towards the door of our tent as being the focus of interest. Even late at night, looking up from our supper through the faint radiance of the lantern we espy the pale sheen of listening and peering faces, patiently waiting for new impressions, drinking in the glorious tamasha. Only of photography they seemed to be afraid, but one day I hit upon the brilliant idea of executing a war dance in front of the camera to attract the foreground necessary for enlivening the picture. The stratagem was crowned with success and while I made the exposure my wife kept up the spell by a continuation of the performance.

92. The Boys of Paldorak.

Camp life has by now found its groove of well-ordered routine.

On arrival I take off my nailed boots and heavy coat, wash the dust from my face and the greasiness of horse-bridle from my hands. While I bend over the diary Mac has the horses rubbed down, tethered and fed, then he starts bargaining for mutton, fowls, eggs and bread. Mac loves to hear his commanding voice resound through the village and all loungers who happen to be near are pressed into service, for the master mind must not be weighed down by such trifles as the lighting of the fire and the boiling of the kettle. He is very busy with his hands all the same, for the high-class cooking devolves upon him and his rice soup with chicken is a work of art. Unless overburdened with other cares he also attends to the laundry for he does not trust the native women on the score of cleanliness. Albert performs his duties of master of the tent and baggage in his quiet and unobtrusive manner. Ishankul the strongest and brightest of our grooms has taken Mac for his model, aping him whenever he gets an opportunity. Sometimes he is in charge of the photo-horse in front of the caravan and on arriving at a kishlak overawes the worthies with a grand show of authority.

At the western end of Paldorak is a graveyard combined with a clay-pit. Those occupants whose tenancy is deemed to have expired yield up their bones to a common pit while other parts of their former selves enter into the texture of the roofs which now shelter their descendants. My collecting instinct was roused by the sight of so many fine skulls, but owing to the nearness of houses I risked discovery possibly followed by a tremendous row.

An excursion to the small side valley debouching at Paldorak already revealed a glacier within easy distance and the panorama of ice-topped mountains gathered around gave significance to the fact. In this ravine we also found extensive spreads of red, or rather, violet snow. From a little above the village we obtained a fine view of the great basin (Fig. 91). This is the last widening of the Zarafshan valley which, after Yarom becomes a trough of more regular breadth (Fig. 93). It is also the last place where the terraces play a dominant role, their flat expanse almost filling the amphitheatre. Over the gently merging mountain flanks the fields and pastures continue without interruption to a great height. The cultivated surface of the enormous shelf is divided into numberless small terraces formed by the agricultural

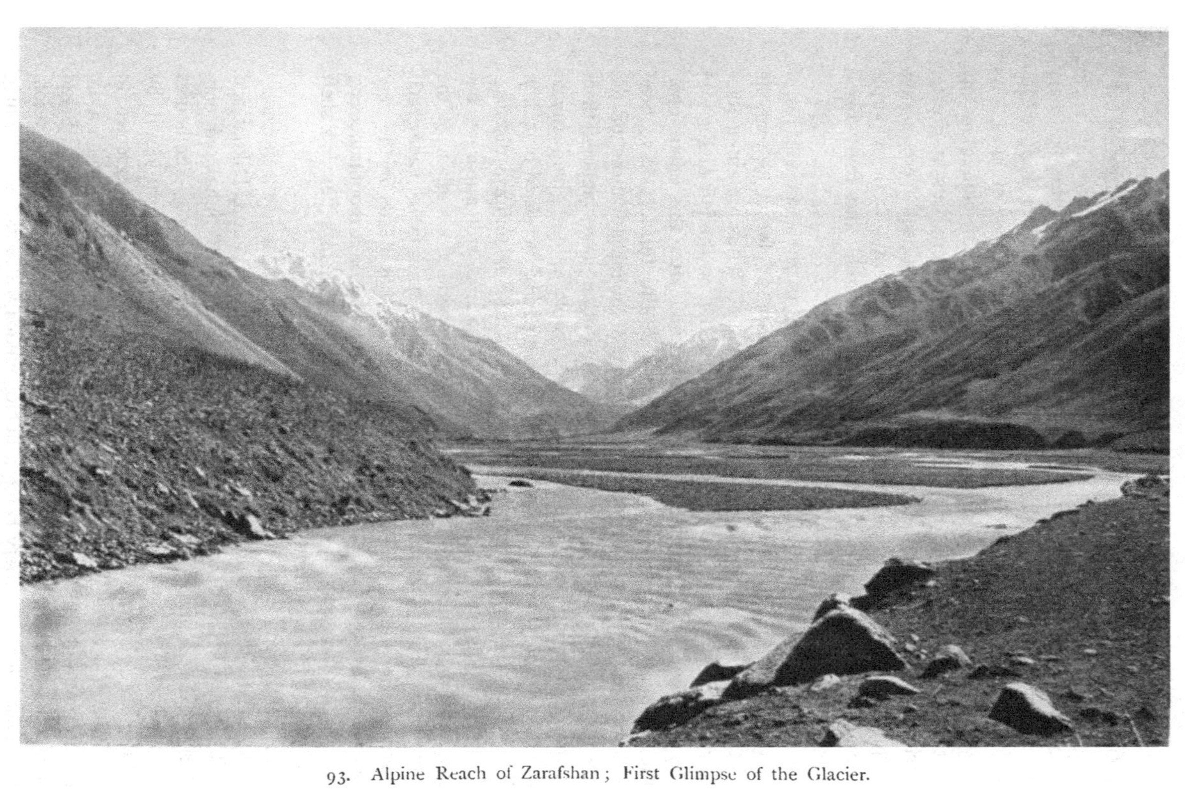

93. Alpine Reach of Zarafshan; First Glimpse of the Glacier.

activity of generations, by the collection of stones at the boundary of each plot and the levelling of its soil, by paths, irrigation canals and other work cumulating small effects in the course of ages. As may be seen by the more frequent occurrence of clouds in my pictures of this region the Zarafshan valley offered us a greater variety of atmospheric effects, and colours at sunset were always exquisite. The depths of the canyon were generally a deep purple while the western sun turned the heights to gold and flung black shadows over every glen.

Proceeding on our journey we now came to the twin villages of Yarom and Khudgif where a Russian sign-post informs us that the distance from Panjikent is 170 miles (280 km.). Thus remain another thirty to the glacier. At Yarom (in the shadow, on the left of Fig. 91) the great valley contracts for the last time and after this enters upon the alpine section of its career. Nearly all the terraces have been reduced to remnants owing to the more frequent advances of the glacier, and deposits due to stronger denudation have blended them with the slopes. Screes have to a considerable extent given way to slopes covered with vegetation taking advantage of increased humidity. The mountain sides accompany the river at almost uniform distance. Here and there the blunted edge of a terrace is still apparent, but generally the flanks descend evenly to the wide flood plain of the Zarafshan (Fig. 93). Snowy peaks overlook the main valley and many glaciers are seen to hang almost above our heads in the very numerous lateral ravines. Here one might carry on a mountaineering campaign on a sound mining principle using the main valley as a shaft and the glens as galleries.

After Khudgif, where we saw a picturesque mazar in the cemetery (Fig. 35), there are only four more villages, Lianglif, Vadif, Dikhab and Dikhesar. Above these are four lailaks or "alps" inhabited only during a few summer months when the cattle is grazing. The houses of Dikhab are built of angular slate fragments brought down by mudspates. From these stones of equal size have been selected to construct the walls. The layers are not flat but set on edge with a dip in alternating directions so as to form a herring-bone pattern. Bread is now hard to get, but milk is provided in all the states of aggregation, the most alluring variety being a homogeneous white salve. Running throughout this gamut of dairy produce was a harmonious admixture of hairs and dust.

An ever thickening carpet of plants began to drape the slopes, but always retaining the character of the steppe inasmuch as bare ground can be discovered in between the plants however close, whereas in our climate the interstices are filled up by smaller weeds, grass and moss. Sometimes we broke into a perfect oasis of wild-flowers; purple phlox, wild roses, yellow and white vetches, geranium

94. The Highest Fields on the Upper Zarafshan.

and other home favourites bordered the path in profusion, while from the mould of moister places sprang the leafy gardens of acanthus.

Opposite the lowest of the summer encampments we found the last bridge over the Zarafshan and also the last fields. These are of the unmistakable type which one may call the high or alpine field of the Duab (Fig. 94) and which occurs at heights between 6000 and 9000 feet. In this instance we find the highest barley at about 8500 feet above the sea, while the highest known has so far been observed at

Sarikol in the Pamirs. Loose stones cleared from the slope form islands or low walls among the irregular patches of the crop. On these crests and ridges, outlining them still more by their rank growth, thrive high, golden milfoil, willow-herb, spiraea, various umbelliferae, St John's wort, barberry, dwarf willow, and many kinds of the thorny tribe. On the patches between these stony hedgerows we find barley and lucerne, the former often mixed with oats to such an extent as to make us believe in an oat-field. But the natives say that they grow of themselves and they well may seeing that they are wild oats (*Avena fatua*) as Lipski affirms.

Wheat, which is perhaps the most important food-plant, also grows on the mountains but at lower altitudes. In connection with it Lipski mentions the curious fact that he often found the wheat mixed with rye and once even a field purely of rye. His theory is that the admixture is unintentional or, at least, not brought about methodically, for nowhere in the Duab is rye cultivated for its own sake (except by Russian settlers). Perhaps the natives of the mountain valleys where this mixed crop occurs are wont to regard it as normal, for the overbearing of rye in some of the fields proves, to my mind, that nothing is done *against* the impurity. Supposing the people really prefer the wheat—a problem still to be solved—this may be an instance where, in the struggle for life, a cereal takes advantage of the indifference of man. The highest barley does not ripen every year and much of it is eaten raw, after having been crushed between stones. We tasted this porridge and found it agreeable, especially when taken with cream.

As the Zarafshan valley runs due east to west the vegetation of the right bank was far in advance of that on the left, where the fields ended much lower down. The last poplars stood above Dikhesar, around the gay sepulchre of Hazrati Imam Musai Ash Ali the length of whose name alone is, indeed, deserving of much honour. From afar we saw a flagstaff covered with bunting and lines hung with coloured pennants, as if a yacht lay at anchor behind the trees. Our guide dismounted, we following, to tread the consecrated ground on foot.

Darkness was gathering before we reached the lailak on August 17. Grown, as it were, naturally out of the stony land these rude dwellings are nearly invisible from a distance. Lying close to the mountain the heaped up lairs, hugging each other for support, merge shape and colour into their surroundings. Only in front their square doors betray them, while from above the flat roofs might be mistaken for a group of mud pools run dry. We lay down in the open, too lazy to pitch our tent and too cautious to accept the hospitable offer of a rabbit warren with its charm invisible but strong that knocks a Feringi senseless at the door. Luckily it did not rain in spite of many threats. In the lower valley the setting sun pierced a veil of yellow and pink under the heavy canopy of clouds. Next morning the ladies had much difficulty in performing their toilet, as no retreat, however distant, seemed safe from the scores of ever watching eyes. The path now wound up the mountain side which was a perfect garden of meadow-sweet, flowering nettle, purple geranium and other blooms of many colours, amid which the dock with its dark reddish brown blossoms made a charming contrast. A few small juniper trees were dotted about the river plain. At 10 a.m. on August 18 we reached the glacier resting its sluggish bulk on the pebble flats. So here we were at last, 200 miles from Panjikent. On a high residual terrace of the left bank we found a shepherds' shanty for our men, and, turning the horses loose upon the knee-deep pasture, we here pitched camp for a longer stay. A few minutes further was the gate of ice whence issued the Zarafshan, a swift volume of slate-coloured water noisily greeting the light of day and carrying blocks of transparent ice.

Our next stage was the glacier itself, but we had to wait for our porters from the nearest village as well as for a gracious mood of the weather. When one awoke next morning one might have imagined oneself in the Highlands. There had been rain and strong gusts of wind during the night which sometimes threatened to blow our tent down. At 6 a.m. the scene was anything but reassuring. Grey clouds rolled down the Rama valley obliterating the distant peaks. Gleams of sunshine broke through those resting on the trough of the Zarafshan, and touching the turf turned it to emerald. A magnificent rainbow threw its arch for one moment over the valley and then we had such colours as one sees nowhere in such perfection as in the Scotch Highlands. Dear Old Scotland, how the heart sometimes hungers after a glimpse of your beauties, despite the fact that it rains with you three parts of the year and that sunshine is a thing almost known only by tradition.

Although our passage through the Zarafshan valley may have appeared uneventful it abounded in remarkable sights, and it may now be time to review the physical features of this section. I ask the reader to recall to his memory the three distinct portions of the Zarafshan's mountain track. From Yabon to Varziminar a rocky gorge in the shadow of overhanging heights; from Varziminar to Yarom the sides open out to a wide landscape of terraces and scree-slopes; finally there is the alpine valley plain, whence more or less grassy slopes rise in sweeping lines to the eternal snows. The middle course, being the most typical and important, we shall investigate more closely. Fig. 95 very clearly shows the arrangement of the terraces descending by steps to the present level of the river. On the left we see the edge of the highest and largest terrace which sometimes exposes its entire thickness in vertical precipices (Figs. 78, 79, 88, 90). This, the largest and most conspicuous of all the present shelves I shall henceforth call the Great Zarafshan Terrace[1]. There are the remains of older and higher ones above it, but its size and clearness make it a convenient starting point for the dating of events. The geological time responsible for its origin I propose to call the Zarafshan Period. Below the great terrace are the middle terrace and the low terrace. I leave out of count the insignificant minor strips below the last. What then is the genesis of these broad levels? Once upon a time there was a valley..., which means that we are beginning somewhere in the middle of things, for the existence of a valley implies that similar processes had been repeated many times before with endless complications almost unconceivable in their co-ordinate sequence. This valley was overwhelmed by a rapid inpouring of rubble from the upper reaches, side glens and slopes, filling it up to the level of the great terrace which thus became the floor of a great flood-plain. In other words, the river, or, rather, this long stretch of the river,—for the work of a stream is not the same throughout its entire length, excavation alternating with accumulation along different reaches,—received more than it could carry away, being kept busy in spreading out the shingle. It also contained more water and sent its floods rambling over its wide bed, evenly distributing their cargo of rock and sand. Then the supply of mountain waste grew less and the river

[1] No identification with the *Hochterrassenschotter* of the Alps is intended.

devoted its energy to taking away more than it received. Digging its way into the gravel, it made a canyon into the deposit, wherever it was held to one line, or swept out wide clearings where, by a frequent alteration of its course, it swung from side to side like the jet from a vibrating hose. A stream chiefly produces such changes of direction in flood-time, when old channels are stopped and new ones opened.

At first the quantity of water was still great and its force assisted by the steeper gradient of the gigantic gravel cone or fan which the mouth of the valley had shot out into the plains of Samarkand. Eating into the cake of deposit from outside, the Zarafshan, at the end of the accumulation of the Zarafshan period, endeavoured to clean out its old, rocky lair. But gradually its waters dwindled, the gradient had also grown less, and when a level below the surface of the middle terrace had been reached, the process of erosion was at its lowest ebb. Then things were reversed again by a fresh invasion of detritus, and in the same way as already described a new floor was formed exactly flush with the middle terrace which is the remaining fragment thereof. On the promontory of the lower terrace to the right of our photograph, one can still see the curved markings, untouched by man or beast, where meanders shaped their course, perhaps when Alexander came to Bactria, or long before. The reason why erosion must have progressed below the level of the next terrace before it was then raised to its definite height is that we cannot easily imagine the creation of so flat a surface by a washing-off. There must, at least, have been a stand-still, with a slight increase of imports over exports. How far down the subsequent inner valleys were cut below the great terrace one cannot say without painstaking and protracted investigation of the deposits. The relative size of the minor terraces is no safe guide to the comparative magnitude of the intermediate periods which followed upon the great Zarafshan deposit, there being other possible explanations than changes of proportionate importance in the upper reaches. One thing is certain, namely that whatever accumulations took place afterwards were inferior to the oldest of this series, i.e. the one with which we started. That the relative ages of the steps succeed from top to bottom is also shown by the state of their edges. Through longer weathering the drop from 1 to 2 has received a gentler gradient than from 2 to 3, while the sides are very fresh and steep in the entrenched meander of to-day (Fig. 95). Thus the surface of each landing denotes the end of a period of accumulation which, in the case of the lesser ones, may merely have been a short flagging of erosion of just sufficient duration for the even spread of a thin layer of gravel. I do not know for certain whether the middle and low terraces are *cut out* of the bulk of the great one or whether they were *filled into* voids from which the substance of the oldest terrace had been entirely removed, thereby boxing or telescoping each deposit into its predecessor. I incline towards the former view. Not having followed every level to the end I must warn the reader that the conditions seen in our leading picture may be only local, that is to say confined to a few miles of the river's length. The great terrace runs right through as a historical mark for the whole of the valley, but we dare not assume that the three regular steps of our example are an adequate expression of general causes during the post-Zarafshanic period.

Diluvial terraces are found all over the world, but each occurrence must be studied separately in order to trace the character and sequence of the stages. Although the general reason, namely alternation of transport to a site and from it, is plain enough, an enormous variety of combinations is possible, often making the problem one of the most difficult in physiography. Changes from accumulation to erosion and back are due to changes in the baselevel (base of erosion) or place to which the water descends and from which it eats upwards as long as the dip is steep enough. Every stream is intent upon making an even and gentle fall from source to mouth, by carving (degrading) where the land is higher, by filling up (aggrading) where it is deeper. For whole continents the base level is the sea; for mountains first the outer plain, then the sea; for every reach of a river first the nearest step below, then all subsequent stages down to the sea. Every increase in the difference of height between any two points along a river stimulates erosive energy, the extreme case being a vertical drop or waterfall. This can be accomplished in two ways, either by a lowering of the base such as a sinking of sea-level, a subsidence of the plains, the destruction of a barrier, etc., or by a heightening of the upper reach, such as a lifting of the mountains, an upthrow in the valley or an accumulation of rock waste. Increased transport is the paramount agent along this newly tilted stretch, erosion being most conspicuous by attack upon the

95. Terraces of the Zarafshan near Rars.

edge of the reach above (the ledge of the waterfall) and dumping into the reach below (beyond the foot of the fall). Conversely every approach between base and top leads to a decrease of erosion or even to accumulation. To bring this about the ocean may rise, the plain be elevated, a landslide obstruct a passage, a lower reach be filled with deposit, or, an upper reach may subside through a movement of the earth's crust. Having briefly reviewed these possibilities indicative of many nice puzzles and desperate complications, I turn to our narrower choice, being one general cause, namely the ice-age, and a local one, namely rocky steps. The great terrace must be attributed to an enormous increase of the glaciers during the Zarafshan period, while points of greater resistance are responsible for some of the smaller terraces forming independent sets confined to particular reaches and their little history. There may be minor levels running throughout and therefore symptomatic of slight glacial recrudescences, but we do not know yet[1].

During the great ice-age the Zarafshan glacier and its partners bore down a colossal amount of wreckage. After the climax the frozen cataracts receded, dropping or exposing the mountains of stone and mud which they carried on their backs or dragged along under their bellies. This drift was seized upon by the torrents gushing from the retreating snouts, washed together, packed and levelled out like the plain in front of the present glacier (Figs. 93, 100). This was the surface of the great terrace to be. The reason why we must assume glacial action lies in the fact that only a comparatively sudden filling up will meet the case, as a gradual deposit by water would have been balanced by a corresponding amount of removal further down. The denudation of slopes is insignificant in the Duab and almost equal to nothing on dry screes, while river transport, especially on the very gentle Zarafshan grade is small and, during the last glacial interval, cannot have been much more than it is to-day. All declivities are littered with the products of disintegration, with rocks and splinters blasted off the mountain by sun and frost. Then come snow-sheets and glaciers

[1] It is well to mention here that Machacek denies a fluvio-glacial origin to the terraces of the Western Tianshan which, more than likely, correspond to those of the Zarafshan. He believes them to be due to a lowering of baselevel by shrinkage of the old Aralo-caspian sea. But origin of substance and cause of gradation need not exclude each other.

pushing down this stagnating mass, cramming it all at once into the nearest hollows.

When the retreating ice had uncovered most of the length of the Zarafshan valley, when the tributary glaciers had withdrawn a little within their ravines, the filling-up process had reached its maximum, for the separating ice now withdrew its morainic freight from the central channel. The main river which till then had been overloaded, having done nothing but dump shingle over the fan it built from the gate into the plains, now regained destructive energy, for one cannot add and take away at the same time. It began from below with its last heapings, attacking the alluvial cone and flushing out the gorge above Yabon, in which work it was assisted by the Fan-darya. But this radical evacuation only succeeded as far as somewhere between Dashtibet and Urmitan, for, in the meantime the climate had still further improved, deglaciation progressed apace, and the stream, dwindling to a narrower thread, could no longer sweep out all former accumulations. Like a fine file it ground canyons into the gravel beds (later hardening into conglomerate) or, at best, cleared out a few basins in favourable places.

Now, as to some of the smaller terraces, we should keep in mind that the oldest and deepest valley is sunk into the hard substance of the mountains, into the bedrock. Its floor and sides are provided with irregularities such as projecting ribs, shelves, steps of waterfalls and great boulders, all of which were buried in the detritus of the great terrace. When the river while gnawing its trench into the conglomerate came upon an imbedded obstacle it did not always move aside, but being guided by a channel which it had already dug into the softer material at this spot, began to traverse the harder rock, just like a saw guided for the first cut by a finger held to its side or a knife cutting through almonds in a cake. Once the water file had gripped the bar it was permanently held fast at this point, whereupon followed two results. Firstly, at the moment of touching and until a gap of sufficient depth had been made, transport across the step was retarded and a short period of accumulation set in above, whereby a level floor was formed, let us say the surface of the middle terrace (Fig. 95). But concurrently with the deepening of the new gorge erosion increased again. Secondly, the river could no longer clear an opening or "outsweep" of the same size as before, having now less room for swinging from bank to bank.

The fixed points or thresholds in a valley divide it into reaches, and within each of these there is a maximum of gathering arc. If the length is enough, the mountain flanks are attacked, except near the entrance and outlet where the river cannot "get at every corner," because its angle of deflection is limited. When the distance between points gets shorter the stream cannot reach so far in its outsweep, cannot fetch so wide a circle. The nearer it approaches the bedrock the greater the number of possible thresholds cropping up from below, the greater the subdivision of reaches, the lower the terraces, the smaller and more numerous the outsweeps boxed within each other. For the same reasons there is clearly a lessening chance for follow-up terraces, traceable to glacial growth. Needless to say it can also happen that a fixed point is given up again. Permanent gates with permanent reaches between them are formed by the meeting of the mountains on either bank. The spacing effect of such a solid landmark can only be overridden by a complete choking of the whole valley to high level or by dislocation.

Less durable fixtures are those furnished by knobs and isolated projections buried in the middle of a great mass of sediment. Here shiftings frequently occur through floods, local accumulation, landslides, deflection of the river's angle of impact (when the obstruction was onesided), interference by an affluent, and various other causes. The Zarafshan valley is full of examples of such minor thresholds, most of which still exercise their functions as preservers of terraces, while some have been subjected to slight alterations and a few entirely superseded. The phenomenon can be studied at many of the side valleys turned into tributary reaches by an obstacle at the mouth. The chasm of the Darkh (Fig. 78) is closed by a barrier composed of slate (on the left of the picture) and conglomerate (where the slanting path goes up). Before the Zarafshan period, when the valley was empty, the stream passed freely to the right of the rocky spur projecting from a mountain ridge on the left of the picture (i.e. right bank of the Darkh). When, later on, the torrent carved its way through the fresh filling it happened to work further to the left, finally striking the knob. Instead of moving aside to the softer material as would seem "natural" to our mind[1], it stuck to the chosen line and sliced the slate bar which thus "inherited" its fissure from a superimposed canyon in the softer deposit. The guard of stone has chained the filament of water to this spot. By preservation of the great terrace the Darkh ravine, instead of opening like a funnel, is narrower at the end than in the middle. If any accumulation were to take place behind the barrier so as to raise the stream just above the top, we can well imagine how next time it might escape to the right, leaving high and dry the cleft witness of a former state. Such a thing has been done by the trunk river near Vardachit, only with this difference, that the new channel also lies through rock. On a flat terrace thirty feet above the present flood plain stands a short, waterworn canal of solid stone looking, for all the world, like the ruins of a disused lock. Close by is the new bed, already much deeper, where the compressed floods of the Zarafshan dash their uproarious turmoil below a bridge.

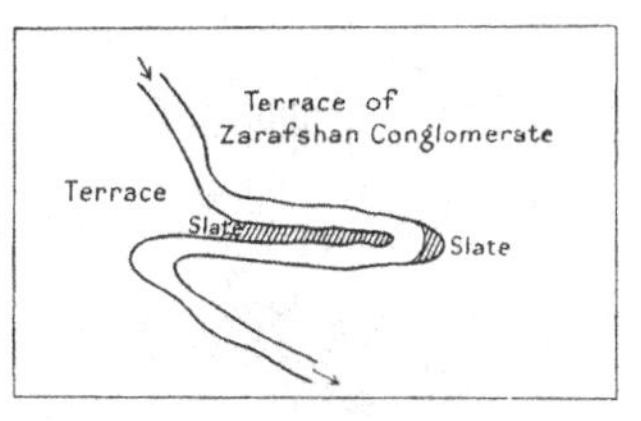

Fig. 96.

Near Oburdon the Zarafshan has touched a slate reef in its bedding. After dissection the result was an exceedingly sharp S-bend caused by deflection from the slate at the turn (Fig. 96). A similar thing happened at Dardar. It may here be mentioned that practically all the strata of slate or limestone thus exposed near the sole are in an upright position. Near Riomut another slate wall was more or less parallel to the general direction of the valley (Fig. 97). When flowing on the level of the higher terrace the river here made a loop indicated by the curve of the slope between the upper and lower tier. Keeping to its centrifugal swoop round the corner it ground a notch into the wall sidling down along the new edge. Then a change occurred somewhere (flood, accumulation, percolation, lessening of water and change of current, etc.), and the river, so to speak, slipped off its rocky perch, which now remains as a lengthwise screen. At Taumen and Yarom there are similar screens of tilted strata, serving at the same time as transversal barrier to an affluent which has severed it at right angles. These few examples

[1] N.B. Inertia is a universal phenomenon also familiar to the dynamics of our brain in the shape of determination, obstinacy and fixed idea.

show that terraces are excellent records especially in a dry and clear-featured valley like that of the Zarafshan. Each embankment tells a long or short story from the great terrace down to the inundation terraces (on the gravel in the foreground of Fig. 95), a foot or two in height, due to seasonal floods.

When the forces of nature have been at work on a regular and descending scale as on the Zarafshan during the postglacial (or post-Zarafshan) period the reading of the shelving book is comparatively simple. The smaller, inner terraces are the younger ones, and any local deviation, such as a re-covering attributable to accumulation above

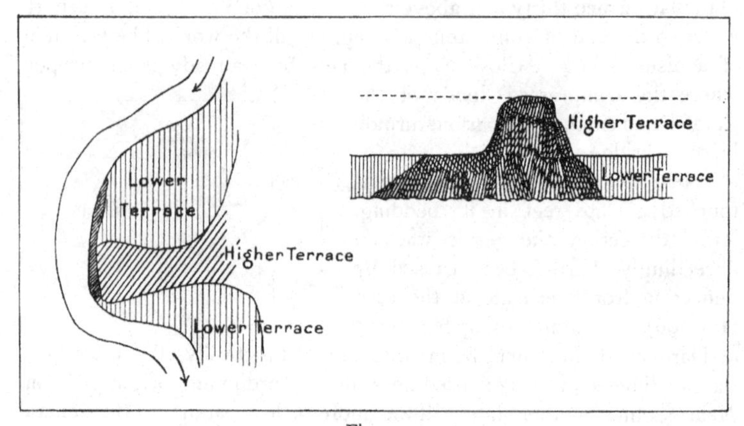

Fig. 97.

a landslide in a critical defile, can be corrected by the general rule drawn from many observations. But the question assumes another aspect, almost appalling in its intricacy, when we attempt going back beyond the last maximum of glacial and fluvial energy. It was preceded by just such a period of diminishing glaciation as the one we are living in to-day, and the returning surge must therefore have overwhelmed the delicate etchings and mouldings of a waning age. The glaciers of the Zarafshan period destroyed and the gravel of the great terrace buried the face of a valley which, in most respects, looked like the present one, but probably a little more evacuated of deposits. Blurred fragments of its terraces can be found on protected slopes above high level, while parts of its old floor are exposed beyond low level where the Zarafshan has touched and cut into bedrock. The piecing together of these remains is a task requiring months if not years of study, and I have limited myself to the nearest object.

In addition to the terraces we have to consider the testimony of moraines or first deposits left undisturbed by water action on the spots where the glaciers dumped them down. The most important of these

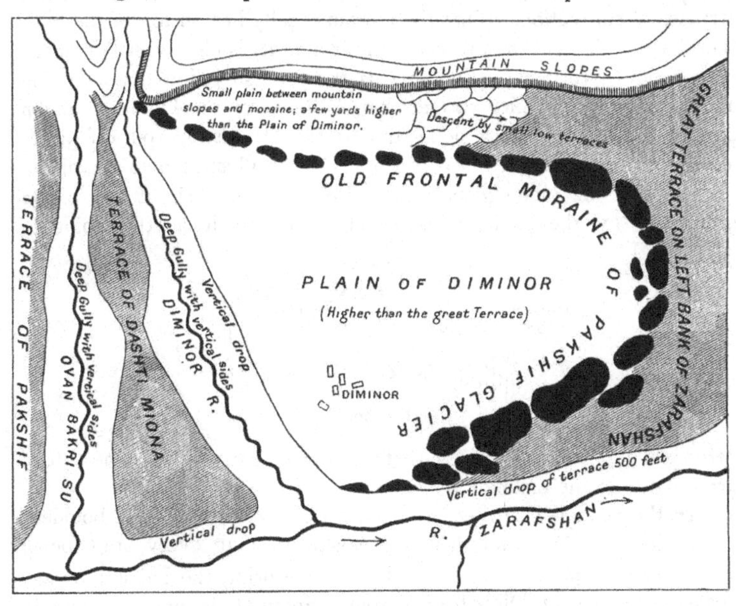

98. Old Frontal Moraine at Diminor.

is the amphitheatre of old terminal moraine at Diminor. Its shape and situation (Fig. 98) show the outline of an ice-tongue projected obliquely on to the great Zarafshan terrace from a side valley, namely that of the Ovan-bakri-su which comes from the Pakshif pass, entering the main valley at a fairly acute angle from the south-east. The glacier of the parallel Diminor river probably did not join the Pakshif glacier. The missing portions of the morainic crescent have been engulfed by the canyons of the Zarafshan and its two affluents, while a few faint traces of advanced outliers are still visible on the opposite or right cliff of the main valley. Among other things we notice how the humps get larger as their line curves round, showing that the snout has unloaded itself chiefly in a forward-outward direction. The plain of Diminor is somewhat higher than the rest of the great terrace owing to accumulation behind the wall, while the same difference in the protected nook under the mountain slopes has been shaded off by low terraces of cultivation. That the moraine really lies on top of the table (i.e. that the latter has not been built around it) is clearly visible at the edge where a vertical section reveals the reddish-yellow mass capping a conglomerate precipice 500 feet high. Towards the Diminor canyon part of the terrace has a slope, and on this are resting split-off pieces as well as an intact portion of the moraine as illustrated in the above vertical section. The glacier either found here a slight depression preformed by the Diminor river, or excavated it at the point of

Fig. 99.

strongest impact. During retreat the tail end of the moraine settled irregularly on the incline.

In the section of the valley from Yarom upwards erratic boulders, scourings, moraines and drift are conspicuous at every step, being chiefly the traces of the main glacier which, during the Pakshif period, may have ended at Dikhab or Yarom. Some of the present tributary glaciers afford interesting observations, notably the Sabak (Fig. 90), which demonstrates the importance of lateral contributions to the deposits of the trunk (as do also Figs. 78, 79). The enormous fan delta of the Sabak was level with the great terrace and is now sliced off in a vertical cliff descending to the Zarafshan which must be imagined as flowing several hundred feet below from right to left. The bulging snout of the glacier peeps from its snowy basin about 3000 feet (900 m.) above our standpoint (7800 ft.), while the peaks at the back are something like 15,000 feet (4500 m.) above the sea.

Without such indications it would be impossible to realise the vastness of the landscape, in which the village on the left, with its trees, becomes a mere smudge. To cross the flat alluvial steppe would take at least an hour of fast walking and another three are needed for climbing to the ice. A glacier with marked symptoms of retreat we saw in a side valley near Vadif. Its grey tongue, depressed in the middle like a mudspate or avalanche, stood out with wonderful plasticity from the white snow which surrounded it on all sides.

From what we have seen at Diminor there is no doubt that the Zarafshan period was followed by a lesser revival of glacial activity which we shall call the Pakshif period, because the Pakshif moraine lies on top of the great terrace.

Although the explanation of diluvial terraces and moraines and the fixing of their order or relative age is often more puzzling than my diagrammatic description seems to warrant, yet this is easy compared to any attempt at ascertaining their absolute age. To give an absolute date to glacial phenomena in different parts of the world means to give them an Alpine date, for the Alps are our standard of mountain time, just as the platinum metre walled up at Paris is the norm of all scientific length. So far the Alpine calendar speaks of four ice-periods or divisions of the quaternary epoch, the Günz, Mindel, Riss and Würm, the first and oldest being so hoary as to escape the legendary by a hair's breadth, at least for purposes of universal application. Suffice to say that the most ambitious traveller may pat himself on the back if he succeeds in proving parallelism with the Riss and Würm periods. I make no such claim, although I have certain suspicions. What I believe to have done is to have made a close limitation of our choice. Leaving to a final chapter a technical discussion of the snowline, I confine myself here to the peculiar traits of the Zarafshan valley itself. That the substance of the great terrace is the outcome of a glacial climax no one can doubt. The reasons in support of fluvio-glacial theories are especially strong in a dry country like the Duab, where only ice and snow can shove rubble down the arid slopes heretofore storing their waste for want of removal. In trying to assign an Alpine date, we had better begin at the end asking ourselves if the Pakshif period corresponds to the Würm or was merely a postglacial recrudescence. The former theory is supported by the data of Professor

Machacek who travelled through the Chatkal and Talaski-alatau in 1911 and who kindly wrote to me on the subject. He puts the snowline of the last main glaciation at a height of 9700 feet (2900 m.) and above. This would meet the case of the Pakshif glacier, while it would bring the Zarafshan ice-stream down to about Yarom or make it flow out into the basin of Paldorak.

If then the Pakshif and Würm periods are synchronic, the last main glaciation has made a poor show in Turkestan. On the other hand there is the Kemkutan moraine (cf. Chap. VII) which would indicate a depression of the snowline to below 6000 feet (1800 m.). According to the premisses furnished by the views of authorities on ice-lore (Penck, Hess, etc.) the lowest snowline in the Alps belongs to the last of a series of glaciations. With such a climatic descent of the permanent snowcap the vast accumulations of the Zarafshan valley could easily have been brought about during the Würm period. But as Kemkutan, although supported by the Kughi-surkh moraines (6000 ft., cf. Chap. VIII), is an isolated witness against the twelve valleys visited by such an authority as Machacek, I shall treat it with caution. Yet it will clamour for a hearing. If the moraine is not one, I shall blush; if it were supposed to have remained from a time when the granitic bastions of Kemkutan were 4000 feet higher, that would be absurd; if the present hypothesis of the relative values of glacial periods require revision, we may yet learn much from the diluvial history of the Duab.

As to the great terrace, we must, until further elucidation, make a difference between its *substance* and its *surface*. The substance may be older than the surface, may possibly be the same as that of some fragmentary shelves *above* it, so that the great terrace may be the floor of a notch cut into a larger one. The name of the Zarafshan period I apply to the substance of the great terrace and any higher ones which may, later on, prove themselves as being of the same origin. Its surface, of course, need not (unless it *is* the uppermost storey of the same glacial set) be as old as its substance. That the surface cannot belong to the Riss period is shown by the insignificant amount of detritus which denudation has thrown upon it. Even where slope wasting is so small as here it must have attained a higher value than that visible on the shelves of the Darkh gorge (Fig. 78) and elsewhere. The Darkh, by the way, offers a direct, graphic demonstration of the difference between river action and dry weathering. Both started with the same new floor, the one to excavate, the other to heap up, but while the torrent has emptied out a hundred million cubic feet of conglomerate, the slopes have barely strewn the top with a hundred thousand cubic feet of scree, or a thousand times less in bulk. But in spite of this resistance of the slopes we may well doubt if they could have kept the terrace so clean and flat in various places through one interglacial, one glacial and the postglacial time. The surface of the great terrace is older than the Pakshif moraine, but it need not be much older, in fact it may have been levelled by the Zarafshan just before the Pakshif glacier descended upon it.

We have therefore the following choice. Either the Zarafshan period (represented by the substance of the great terrace and higher tiers possibly forming part of it) is to be reckoned parallel to the Riss, the Pakshif corresponding to the Würm; or the Zarafshan period (if the surface of the great terrace is also the upper level of its substance) coincides with the Würm, and the Pakshif was a postglacial revival. I confess to a predilection for the second alternative, but owing to doubts which need revision, we shall only speak of the Zarafshan and Pakshif periods as of the two last glaciations in the Duab, leaving open the question of Alpine identification.

The greatest Zarafshan glacier—leaving aside all question of period—probably reached as far as Samarkand. Chapan-ata (the hill immediately behind the sepulchre of Fig. 57) looks as if it might be a moraine, but when the idea occurred to me I had no opportunity of examining its internal structure. I only remember that the railway runs along a rocky exposure, but that may just be the buttress preserving from destruction part of a huge morainic crescent, if there ever was one. The conglomerates found at Guzar may be remains of a cone delta projected into the basin of Samarkand.

Shape and position of the Zarafshan valley fitted it eminently for the generation of a mighty glacier propelled far out into the plain. With its opening towards the humid west, and straight as an arrow, it runs like a gutter between two roofs, tapping majestic ranges, the higher one of which faces north. Moreover at Varziminar debouches the Fan-darya with its widespread ramifications, especially the Yagnob, built on the same lengthwise plan under the northern declivity

16—2

of the main ridge, so that a basin measuring 4000 square miles (10,500 sq. km.) is drained by one outlet. The basin of the Rhone covers about 2700 square miles (7000 sq. km.) on the horizontal projection. Thanks to the perfect herring-bone arrangement the trunk glacier of the Zarafshan received its tributaries by the shortest and steepest line, thus ensuring coalescence of the ice-streams even if the snowline was barely low enough. Akin to a long probe it is thrust straight into the heart of the mountains, so that by this shortest route a minimum of melting was assured for the journey from the summits to the plain, while snowy chains uniformly maintaining their great height promised additions up to the last. The altitude of the valley floor above sea-level was another favourable condition.

A few rough figures of comparison with the valley of the Rhone which sent its glacier down to Lyons, may prove interesting. The valleys are compared between the glacier and the lake of Geneva and Dashtikazi respectively, their length being measured in a straight line, without every bend of the river.

Length	Rhone	90 miles (145 km.);	Zarafshan	150 miles (240 km.).
Drop	„	4600 feet (1400 m.);	„	5500 feet (1700 m.).
Drop per mile . .	„	50 „ (15 m.);	„	37 „ (11 m.).
Mean height of floor above sea-level .	„	3500 „ (1100 m.);	„	6000 „ (1800 m.).

Along the middle stretch of the valley—Brig to the Lake and Pakshif to Dashtikazi—the distance of the culminating ridges on each side from the river is 14 miles or 22·5 km. (Rhone) and 7½ miles or 12 km. (Zarafshan), while the height of the crests (say 11,500 ft., 3500 m.) is about 8000 feet or 2400 m. (Rhone) and 5500 feet or 2800 m. (Zarafshan) above the valley floor. These figures show how the Zarafshan cuts into the mountain flanks much nearer to the summits, its bottom and sides thereby losing melting surface, but gaining glaciation surface. Hence by its topographical configuration a Zarafshan glacier can make up for many climatic drawbacks.

The Zarafshan is a beautiful example of a comparatively mature, well-graded river, for even the side streams have, without exception, worked down to accordant grade. All join at low level, no waterfalls tumble out of hanging valleys. For a long time undisturbed by upheavals the Zarafshan has cut or filled the irregularities of its bed and all tributaries have been able to keep up with it. A long, gentle gradient rises from Karakul, passes through gorges, over flood plains and under the glacier, emerging on the divide of the Macha Pass. Above, sometimes slightly dipping below it, are the lines, more or less broken, of all the valleys which have gone before. As long as no important convulsions happen, as long as a valley remains practically identical, all these past and future slopes converge in an ideal point somewhere above the present watershed, where once the highest summit rose still higher into the heavens. Here meet the continuations of all channels, terraces and glaciers which ever formed the longitudinal profile of the current. The great terrace becomes lower and lower, losing itself in the alluvial gravel of the alpine reach, while the bedrock underneath rises to the surface from under the neve of the Macha Pass. Downhill it is the same with regard to base-level, be it the ocean, Amu-darya or the landlocked delta of Karakul. On the way between these limits of opposite convergence there are often points of intersection where the travel-paths of shifting matter, of water, ice, rubble and silt cross each other. In the plains the sediments of three glacial ages cover each other, the oldest being at the bottom, while their corresponding terraces or morainic smears along the slopes of the upper regions are ranged with the oldest on top. Somewhere in the middle the greatest complications occur, due to continuous dismantling and re-stacking on the way, to local interferences or to smaller climatic changes incapable of leaving a great impress throughout. Here the extremes meet in confusion on a shifting battle ground of the mountain wishing to come to rest and the plain aspiring to a height.

Of other things to be seen in the Zarafshan valley museum of physiography a few are worth mentioning. The discordant superposition of younger formations upon the tilted bedrock so frequently exposed at the mouths of lateral ravines, is particularly well shown near Guzaribad. The upright layers have been smoothly abraded by the old river and are now overlaid, at right angles, by a bank of soft conglomerate and several feet of loess. Looking downstream from this place there is also observable a fine sweep of mountain folds with a radius of about 1500 feet.

Side valleys of very regular formation can often be surveyed from top to bottom, their steep but even gradient sliding in ever broadening snaky curves from a jagged crest to a fan dump at the end. The talus cones of Fig. 87 have been worked out of the side of an old terrace by dry disintegration combined with the severing influence of rain gutters. The share taken by a dry atmosphere and glaring light in bringing out the sculptured features of the land is strikingly illustrated by the welted furrows and "lions' paws" of Fig. 85. A regular supply of small mudspates is furnished every season by the two gullies, building out cones of mud breccia, easily distinguishable from the river pebbles. Its exposed ledges of a yellow and reddish colour form the front part of the claws. The two steps correspond to terraces, being notches cut into the mud cones and partly overrun by gravel which can be clearly seen as a layer of different texture on the fissured lower bluff. Lions' paws are a recurring trait of talus breccias and solidified mud shoots undercut by streams. They are also evident at the foot of the great twin cone of Fig. 86 overlying an old terrace. Besides two typical mudspate sluices this same cone reveals a surface of interlaced welted furrows akin to a design of tangled worms. Discharges which missed or overflowed the main gutters have traced their paths in various directions, often interfering with each other. It is they who have really left a deposit and are thus responsible for the growth of the cone. The frequency of steep dejections at the mouths of ravines is among the characteristics of dry mountains where disintegrated slopes are flushed by intermittent gushes of water dropping their cargo when the force is spent.

Below Dikhab we found the very flat spread of a recent mudspate covering nearly a quarter of a square mile. After having been belched forth from a lateral ravine the rocky mass had fingered out into many distributaries, some of which, after a run of 900 yards, dropped over a terraced edge into the Zarafshan, while others ended in depressed avalanche snouts on the gently sloping ground. The branches looked exactly like broad, badly paved roads between dams from two to six feet high, thus representing welted furrows on a large scale. While the stones of the roadway were plastered with grey mud, the boulders of the flanking walls were clean and black, having been washed by the frontal wave of water riding on the surface. Older tracks betrayed such outbreaks as endemic in this place.

The action of wind on rocks can be studied at certain corners of the valley where eddies are likely to play and where polished patches of slate glisten in the sun as if they had been freshly tarred.

Fig. 88 gives a good example of the headward growth of erosive channels. Separated by a perfectly horizontal line we have the remains of an old terrace overhung by a slope of whitish grit. As can be seen towards the left, no furrows generate on the upper slope as long as it enjoys a support or a flat outrun, where rain can lose itself. This also was the case in the middle as long as the least little shelf remained of the old terrace. Finally however this was worn away and as soon as the gullies that were biting its edge had touched the grit they began to "draw it in," sucking, as it were, threads of gathered surface water from the upper incline, now robbed of its dividing ledge. Hitherto independent orographic individuals became locked together into one. Gradually these grooves will eat their way still further upwards. The contrast between the wider shapes of the conglomerate and the sharp, close-set flutings above is explained by the finer and more homogeneous material of the grit. One can carve cement more delicately than concrete.

We are never left in doubt that the geological face of the Duab revels in design and colour. At one place on the Zarafshan there is a big slope composed of innumerable thin, horizontal streaks of almost every hue of the rainbow, mainly green, red, brown, yellow and grey, all as straight and parallel as if drawn with a ruler (probably sandstones, clays, schists, marls and chalk). At another time we watch alternate layers of grass green and red amid screes of a rich brown. Whatever the features of the land lack in the tints of vegetation they retrieve by the brilliant painting on the soil itself. Again the reason is not far to seek. Most of the brightly dyed materials of outer earth are friable rocks and soils which in an alpine or English climate are either washed out or covered with plant life, taking advantage of their softness.

On all sides there is apparent the rule of a typical climate over scenery and dynamics. There is little precipitation, but rain sculpture is more conspicuous than with us. There are dry ravines, yet like the wadis of the desert, a menace of destruction by floods. In arid regions the "turning on" of a watercourse by sudden rain assumes a catastrophal aspect for what was dusty and quiet before becomes a raging cataract. There are gullies where it means almost certain death to remain in them even a few minutes after the outbreak of a thunderstorm, so quickly the waters run together on the bare, upper slopes and so terrific is their onslaught down the narrow chute.

Yet contrast of extremes is a preserver of those features which would suffer by transition. Perishable banks stand close to the path of destruction and oases thrive in the midst of the desert; the language of remains is as eloquent as the graven line of the destroyer; they do not mingle but reveal each other. In muggy climes the tearful heavens make the outlines merge and cloak the tumbled surface with a living green; their new creations are ever half in ruins and what they preserve they hide.

Between the mountain gate and the glacier the valley of the Zarafshan is a grand and simple theme without the shaggy fur of the woods, the tangled folds of copses or the mural carpet of sloping greensward. It is a landscape of classical asperity and sculptured nakedness where the village oases are like gay frescoes amid the plastic massiveness of terraces and canyons.

For the purposes of description the world may be imagined as composed of types, concepts, ideas, things in themselves, influences, definitions, abstractions, or whatever one likes to call them. No concrete thing is ever a pure type, the definition is never wholly realised. This can best be seen when an individual object is itself raised to the dignity of a type, for no other object will then be exactly equal to the original. None are quite the same as this one, all contain an admixture of other types. The types of knowledge, particular or general, elementary or compound, are constructions to which the phenomena of nature are approximated in order to name or explain them in terms of type. The more certain types predominate, the fewer their number and the less they interfere, the more we call a thing "typical." Thus a landscape is composed of geology (i.e. the objects and scope covered by a definition of that science), geomorphology, weather, vegetation, animal life, human activity, etc.

In the genetic aspect of the Zarafshan valley two types of influences rule supreme, tertiary geology and quaternary physiography. Those of organic life are practically negligible, while "climate" owing to its

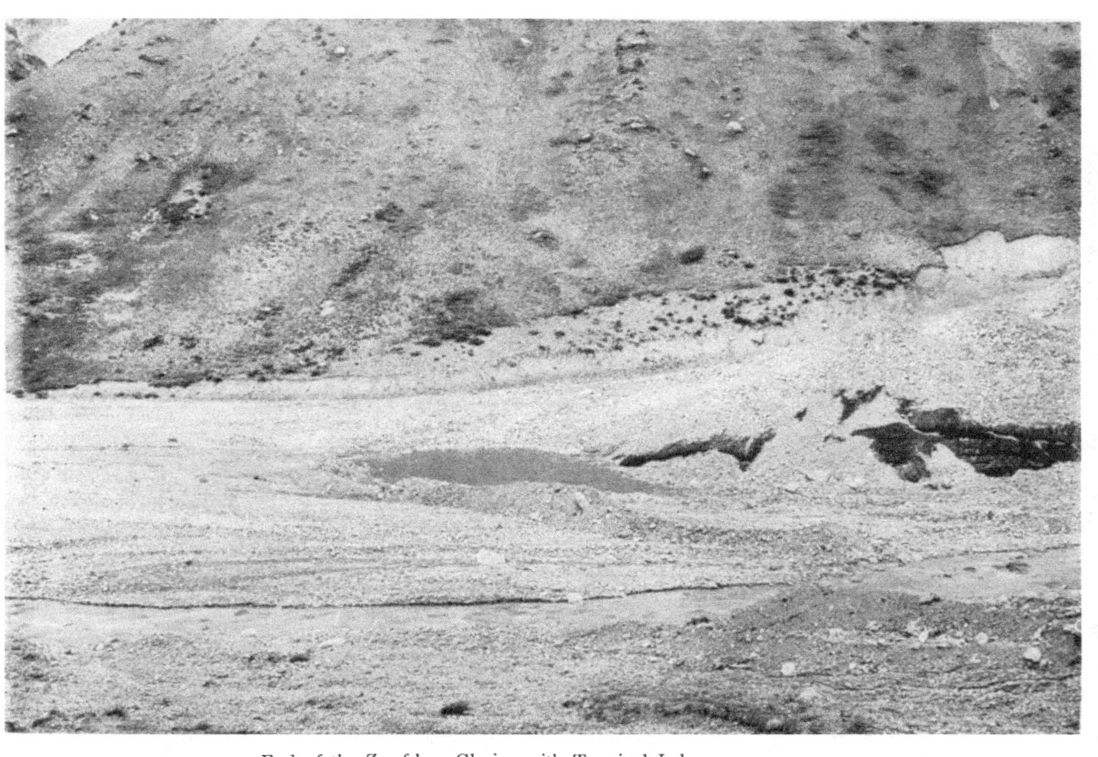

100. End of the Zarafshan Glacier, with Terminal Lake.

dry and non-generative qualities (chemically and biologically rather inactive) merely figures in its capacity of a constant dynamic factor or medium of inorganic physiography. Hence the text-book simplicity of this valley, the characteristic purity of its features composed from the typical traits of a dominant order of phenomena. To this must be added its great length and straightness providing an extended scale on which differences of degree are more accurately registered. In my opinion there is no object more worthy of minute exploration, none more eminently fitted for serving as a standard of comparison for the physical history of the Duab.

CHAPTER XI

THE ZARAFSHAN GLACIER

THE nice, homely Highland weather lasted for several days which we devoted to a thorough examination of our surroundings. Almost without transition the great glacier rests its foremost point upon the alluvial plain (Figs. 100, 101). A fringe of low terminal moraine bears witness to the latest and most rapid stage of recession, which was the work of a season, for, as the natives say, there was ice a year ago where now the water is. After showering a hem of rubbish from its slippery sides the extreme tip melted back very quickly, leaving a round lake in the place it formerly occupied. This pond is fed by a fairly strong source from the glacier, the overflow escaping through a gap in front to the wash plain where it ultimately joins the Zarafshan. In the morning the terminal lake is almost dry, presenting a glistening surface of black, sandy mud, only covered with a foot or two of water during the heat of the day. Traced on the level plain this circular dam which measures about 120 feet across, is also conspicuous for its reddish colour against the grey of fluvial shingle. That the disorderly heaps of stones in front of us still form part of a glacier can only be seen from the shiny, black facets of dirty ice appearing here and there. An irregular wing of the frontal wall is sent out towards the left[1] where it is breached by the main river. Its continuations flock together with other lines or mingle with the morainic deposits of the Yarkhich fan (Fig. 101). Higher up the shrinking of a lateral bulge of the snout has led to the formation of the mud lake, so called because it is entirely silted up. It receives a filament of water through an inlet from the Zarafshan, losing it again by percolation, there being no

[1] Unless otherwise stated 'right' and 'left' are always used in the orographical sense, meaning right or left as one looks downstream and downhill.

opening in the lower rim. As can well be seen from the diagram the ice-born Zarafshan once hugged the edge of the glacier, whereby one curve of the great S-bend was forced to encroach upon the morainic cone delta of the Yarkhich valley.

The principal glacier gate lies well back on the extreme left, a

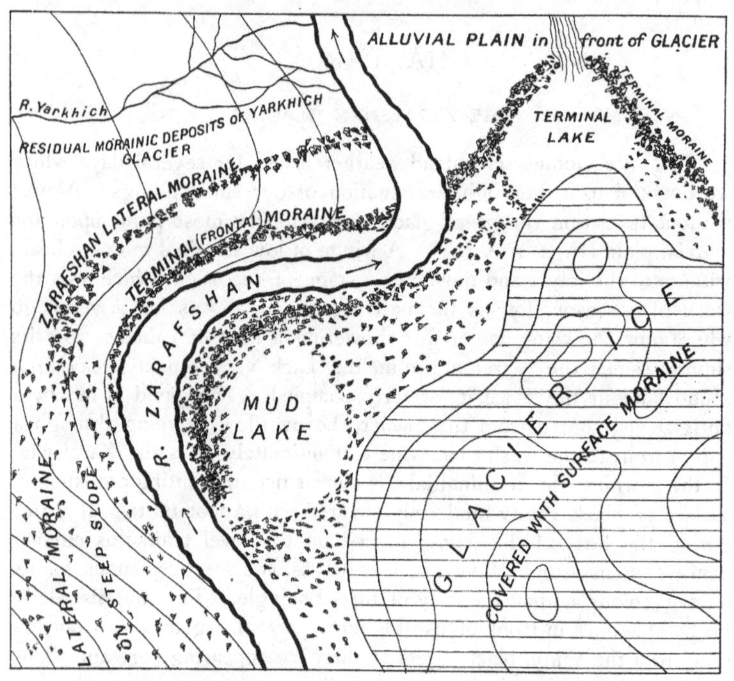

101. Snout of the Zarafshan Glacier, August 1906, showing left half of Glacier with extreme tip and stream.

third of a mile from the tip of the tongue and almost in touch with the mountain slope. This is but one of the many symptoms of the lopsidedness of the Zarafshan glacier, which must be ascribed to the greater influx of rock waste from the right. A glance at the map (Map II) will tell us that the tributary glaciers from the northern range are longer and have a wider catchment area. The mouth of the tunnel (Fig. 102) is 40 feet broad and 20 feet high, while the shape of the ice wall suggests that until recently it was a pocket-lake, now broken out. From this vault issues the Zarafshan, a spout of turbid broth, a solution of pulverised mountain. Clambering to the top we find ourselves on the surface of the glacier, if a scene of utter collapse can be called a surface, and cold storage of shivers and smithereens a glacier. One has certainly to recast ones Alpine and Caucasian memories, learning, among other things, of how much variety a defini-

102. Glacier Gate of the Zarafshan.

tion is capable. What from afar seemed a moderate roughness of the sluggish worm resolves itself into the gigantic and glorified reproduction of a desirable building site on lease for 99 years, with the brick heaps of former houses and water-logged pits. We are swallowed up by the heights and depths of a landscape, almost grotesque, contained within the embracing vastness of the mountains. Balancing over loosely piled stones that never settle, we scramble over ramparts of confused blocks, climb to the summit of pointed cones rising fifty or a hundred feet above us, and then dive down again into sunken valleys, with nothing around us but blasted granite and mangled slate making it difficult to remember that there is ice underfoot. Sometimes we nearly stumble over the sharp and sudden brink of a pocket or sink-hole (Fig. 103). One of these, by no means the largest, was a crater 300 feet in diameter, and at one point its steep, greasy incline measured 150 feet from the apex of a sliced cone to the lake at the bottom. In parts the ice was yellow and red, in others of a slaty grey. In the red ice gleamed a blue grotto larger than the gate of the Zarafshan and with a sheet of frozen water in the chill of its further shade. All around from fissures and arches water drips incessantly, adding its weary music to the dull thuds of single stones or rattling cascades of whole batteries churning the pool with joyful splashes. The number of these funnels and vats is very great indeed and their size and shape as varied as the colour of the water which most of them contain and which often looks like a very unappetising peasoup. Occasionally one even meets with extensive defects in a longitudinal direction.

If the underside of the ice—in this lower portion of the glacier—rests upon a floor somewhat lower than the wash plain in front, then its thickness cannot be much more than 300 feet at the highest or about 150 feet on an average. This condition prevails for many miles up the glacier and makes me think of a lump of wet sugar falling to pieces. All these many pocket-lakes would hardly hold their water so well unless close to the ground moraine.

Seen from a height the back of the squashed and warty reptile shows irregular stripes (Fig. 104) of different colour. The light streaks are granite and gabbro, mostly from the Turkestan range, the darker ones slate from the left, while intermediate tints represent a mixture of the two. This is all that remains of distinct ridges of medial moraine, so characteristic of energetic glaciers. Here they have been broken up into cones and partly vanished into the gaping traps of kettle-holes. These are also evidences of a very slow movement, for otherwise the melting of their sides could not keep pace with the forward pressure of the ice. They are mostly old crevasses opened out by an excess of melting over advance, and many of them are in connection with sub-glacial water-courses. Hence they offer an analogy to the dolines or funnel-shaped sink-holes of the limestone regions of the Karst. An enormously increased wasting surface is thus produced and atmospheric disintegration of the ice is accelerated by its dirty and therefore heat-absorbing condition. One cannot expect this honeycombed, not to say rotten, mass to be endowed with any sort of impetus. With a glacier deep-reaching dissection must be as symptomatic of reduced flow (not only increased melting), as the separation into detached pools is with a river. All the same there is a possibility of imagining a continuance of the flow in spite of the cellular structure. One can assume the

103. Pocket Lake on the Zarafshan Glacier.

surfaces of each pocket to be a kind of glacier termination where the new ice is melted at the same rate at which it draws in, threatening to close the hollow. In that case every funnel wanders down the glacier. Still, the fact remains, that movement must be very slight. Obviously the section of the glacier which has numerous and large cavities is "ending all over" by intussuscepted snouts, as one might call them. This cankered and corroded state obtains to about midway between the Farakhnau and Tolstov glaciers, though, of course, a very gradual transition takes place. It is this lower section which I have in mind while discussing the special features of the Zarafshan glacier, for higher up, as one ascends, it changes into an alpine glacier of the ordinary type. This lower portion can be looked upon as practically stationary, although there must be a little flow enabling it to extend beyond the junction of the last lateral glacier. Nor can the longitudinal sections behave in the same way, owing to lateral influences, and I have already mentioned the lopsidedness of the glacier, which is further accentuated by a very deep, valley-like depression on the left, over the subglacial current of the Zarafshan.

The Gargantuan load of detritus under which the glacier groans heightens the moribund impression it leaves on the spectator. At first one may feel inclined to accept this gigantic burden as evidence of brisk activity, but in reality it is only a show of hard work and prosperity. A miser is not necessarily productive, and the Zarafshan is hoarding instead of going in for a large turnover.

Every glacier is a living individual. The upper snowfields are the seat of vigorous life gradually relaxing as the end is reached where the dead matter is cast off, just as organisms drop their nails or branches. Like a bulb it may lie dormant, shoot out afresh or be annihilated altogether. In most glaciers with which we are familiar the last stages of dissolution are marked by a gradual tapering and flattening of the lower course to a long, pointed tongue which, although deeply fissured, retains the appearance of a consolidated mass. Movement is retarded, but kept up steadily to the very end. The Zarafshan glacier however remains uniformly broad, and increased melting surface is attained by deep-seated defects in the mass itself. Hence its aspect of senile decay, especially in the portion below Farakhnau junction, which looks more like a deposit or abandoned position left over to the solvents of sun and air. The two Nazaryailak glaciers are very flat at their mouths, the Porak barely touches the trunk, and the Farakhnau does not look much healthier than its recipient, so that one gets an idea as if this lower section was nothing but a welding together of the fag ends of the main glacier and its last affluents.

Undoubtedly the lower inflows give off more water than ice, which means that they add destruction instead of enlargement. The inert and congested mass is simply roofing a system of rivers, doing much of the work and transport which the glacier can no longer accomplish. The frequent occurrence of round pebbles among the surface detritus is but further proof of the very considerable co-operation of water. Owing to the many holes and the slippery gradients debouching into them, most of the surface litter is turned into ground moraine by falling to the bottom. Here it is seized upon by the subglacial streams changing much of it to water-worn pebbles before the end is reached. Moreover the ice is so full of fragments and rockflour that we may speak of an uninterrupted transition from surface moraine, to interior moraine and ground moraine. As long as the Zarafshan glacier is thus stationary its cargo is trimmed into the floor, so that nothing is left over for the formation of frontal dams, except during rapid retreat. The real moraine of blocks and angular material lies *under* the alluvial plain in front of the glacier, having been closely packed by water and covered over with a layer of smaller fragments converted into smooth pebbles by the subglacial brooks. If a glacier retreats so slowly that the water can keep pace, a flat shingle bed will be left in front of the receding snout. But it will be otherwise when melting proceeds on a grander scale, when the mighty snail quickly withdraws its back from under its load, dropping the morainic landscape as a whole. Thus we can understand how a certain type of old moraine is formed, such as seen at Tupchek (Figs. 139, 145, 151). To all intents and purposes this kind of glacier is a conglomerate of rocks kneaded into a matrix of ice and mud. Let the glacial cement be turned to water and the bulk will settle into a wavy plateau with rounded hummocks which once were ice-cones. If one measures all dimensions and volumes of a lower section of the Zarafshan glacier and finds out the maximum rate of melting at which no terminal moraine is found; if furthermore one measures an old moraine like that of Tupchek and estimates the quantity of ice which must have corresponded to it, one can, finally, calculate to within a decade the minimum rate of recession necessary for the formation of such an old moraine.

Judging from photographs this type of smothered and lacerated glacier-tongue (i.e. "dry" glacier, below the névé line) is also found in the Himalayas. A still further reduction of ice would turn it into an Alaskan rock glacier, while a more miry variety was introduced to me by the "dirt" glacier in the Sagunaki mountains.

Apart from the signs of lesser fluctuations already mentioned and centred in the crescent of terminal moraine, the Zarafshan glacier also offers evidences of a long and sustained recession during the latter half of the past century and up to the present time (1906). This is the shrinkage belt or waste-mark clearly shown on several photographs (Figs. 100, 105, 106). To understand its origin we must remember what goes on at the sides of a moving glacier. Between the sole of the ice and the valley floor a mass of boulder clay, the ground moraine, is pushed along. As the plastic belly of the glacier fills its bed and is moulded to the banks, the ground moraine is also brought in touch with the mountain slopes on either side. At the same time the rim of the glacier's upper surface sheds rubble upon the slopes, so that here the two kinds of moraine are brought into contact. When the ice draws away, latero-terminal accumulations are left behind consisting of a mixture of ground and surface moraine, and greatly varying in shape according to the nature of the ground. In the case of the Zarafshan where the mountain flanks are not too steep and fairly soft owing to a mantle of older deposit, the ground moraine is pressed into the sides. This adhesive strip is crowned with a greater or lesser fringe of lateral dejections from the surface. The whole of this one may call a hanging moraine because it clings precariously to a declivity. When the glacier dries up it exposes to daylight the hanging moraine which, being composed of freshly ground rock and devoid of vegetation, generally shows up as a pale border. Where the hanging moraine cannot remain in position, because the slope is too yielding, or its base is undermined, or too heavy a crest of surface moraine weighs upon it, there it slips off. This most easily happens near the end where the conditions just mentioned are most likely to occur. On the right bank, opposite the terminal lake (Fig. 100), the ground has been thoroughly flayed. This waste-mark begins a hundred yards from the junction of the Rama valley and, passing through various forms, is continued some distance above the Farakhnau junction. It is broadest just below the Farakhnau, thinning out each way towards a recent limit of the glacier and the present limit of desiccation. Some of the older parts of the waste-mark are beginning to show a light down of vegetation. In Fig. 100 three stages are observable near the middle of the picture. Above is the large slope covered with the darkest, best established vegetation and strewn with stranded boulders. Below comes an intermediate strip of boulder clay preserved in the recess of an older gash in the mountain side, made by a previous short advance of the glacier and left filled in after retreat. On this the black bushes are in evidence, while more exacting herbs have formed a thinner coat, much lighter in colour, owing to the shorter disintegration and therefore lesser fertility of the grit. Finally there is the newest flaw boasting, here and there, of tiny specks where the hardy weed of the mountain steppe has gained its first footing. Further to the right the intermediate deposit is overridden, the rent cuts into the slope with a raw surface and clean edge ripped from the turf.

Higher up the glacier comes a more irregular portion of the wastemark, because here the hanging moraine is half clinging, half slipping. This is due to the greater mass of moraine and more complicated conditions: chiefly to the great volume and weight of superincumbent lateral surface moraine, partly to the broken nature and fitful shrinking of the glacier in this part, least of all to the formation of the slopes (Fig. 106). Over a stretch of the left shore, for about a quarter of a mile upwards from the water gate, the process of shrinkage assumes a catastrophal aspect (Fig. 105). Here is the long, caved-in hollow over the subglacial tunnel of the Zarafshan, here all is disorder and destruction. Robbed of its base by a too sudden lowering of the ice-level and overweighted by a ridge of outer moraine the whole mass of drift has given way, dragging down everything in its wake, raking the hillside, tearing off sods now vegetating as forlorn islands in the midst of chaos. Bits of masked ice can be traced proving that the frozen substance of the glacier is still reaching some distance up the sides. But there also appear many black patches of slaty bedrock, uneven knobs and lumps which cannot have been subjected to much abrasion. Only lower down, near the Yarkhich junction, we found a hanging moraine still glued to the flanks. Its top was formed by the remaining half of the dam of lateral surface moraine running along the slope like a ledge with a small valley on the mountain side. This formed a convenient alley which we often used in our excursions. The outer part of this ridge, after losing the support of the glacier, had collapsed and in union with the uncovered ground moraine formed a precipitous boulder gradient down to the river 200 feet below.

But the most interesting portion of hanging moraine is preserved further up, especially on the right bank (Figs. 104, 106). Ground rubble and waste-mark are one, forming together an inlaid moraine, a most beautiful piece of glacial marquetry. From a distance this white band is set off in a perfectly straight line from the darker pasture land. A closer view reveals the fact that this dado follows every contour of the slope. Practically flush with the ground it curls round

104. Zarafshan Glacier with Macha Pass in the Distance; View of Right Bank and Farakhnau Junction.

every rib, in and out of every gully, so that all the bulging or grooved features of the slope above find their natural continuations obeyed by the waste-mark below. A mosaic of concrete has been firmly pressed into the flutings of the mountain by the plastic ice. Avalanches may have contributed to a smoothing out of the gullies, for nearly everywhere we found vegetable remains, the flotsam and jetsam of winter, precipitated out of cones of melting snow.

The minor period of glacial growth anterior to the present shrinkage cannot have lasted long, for otherwise the folds of the mountain side would have been abraded.

Wherever the mountain has been violently scored or the covering been stripped away, exposing a waste-mark, an edge or bench results, because the new ground has a steeper gradient than the older slope. The broken configuration of the lower slopes (Figs. 93 right, 106 left and middle) is due to the great number of small benches formed in bygone times by minor fluctuations of the glacier. There also exist great benches, namely those notched into the rump of the mountains and ascribed to overdeepening by modern glaciology. The ridges coming down into a wide valley have a profile of angles (Fig. 109) which can generally be found to correspond. Going from ridge to ridge and from bank to bank it is often possible to connect corresponding angles by a gently graded line probably representing the floor of a former valley when the mountains were not yet so deeply eroded and everything flowed at a higher level. Thus the valleys of three or four succeeding ice-ages are set within each other, their formation being analogous to that of river terraces. Roughly each angle marks the bottom of an older valley and the rim of the younger, "overdeepened" trough below.

The Russian geologist Mushketov reports that in 1880 the Yarkhich glacier joined the main trunk. Since then it has lost 600 feet in vertical height and a third of a mile in length. This is all the more credible as the ravine is very narrow. Moreover small glaciers are apt to be jerky and capricious, their movements being liable to fits and starts. High up one sees the Yarkhich nosing about in its lair, and over the forehead of its black face one espies the white séracs of an icefall. The waste-mark is very small, while the old moraines have been partly worked into a cone delta. Owing to the softness of the rock, glacial scorings are not much in evidence, the walls having been bruised rather than polished. Altogether the smoothing influence of ice is not so clearly demonstrated as in the Alps, the rock having been either splintered by weathering or covered with detritus. In the Duab I have not yet met with extensive fields of roches moutonnees or with huge granite bosses such as those of the Grimsel.

We also paid a visit to the Rama valley, using the Zarafshan glacier as a bridge, as I did not wish to risk our horses in fording the tumultuous river. Descending to the alluvial plain we found it well spread, none of its asperities being more than two feet in height. Willow bushes came up quite close to the morainic ring, often dipping their pink roots into the icy current issuing from the breach of the terminal lake. In drier spots tamarisks were thriving. The Rama torrent, which adds quite one-third to the volume of the Zarafshan, flows through a bed somewhat different in character to that of the main valley. There are low, dam-like streaks of cobbles as well as sharp stones. Between

105. Slipping Moraine on Left Bank of Zarafshan Glacier.

them they often comprise dry channels, on the sands of which we followed the tracks of shepherds and their flocks. The whole place was like a stony fallow overgrown with a profusion of blue sage spreading its blaze of colour through a straggling grove of juniper, birch and willow, and exhaling the aromatic perfume of its blossoms.

The glacier has withdrawn about a thousand feet from its last position, but loss of thickness is far more apparent than that of length. The blanched waste-mark, indicating a recent loss of 300 million cubic feet at the lower end, follows a straight line along the slope 200 feet above the valley floor, ultimately joining the frontal moraine by a steep curve. Hence the glacier must have had a bulging and precipitous head instead of the present, dirt-laden humps. On the right bank the pale zone of shrinkage presents a noteworthy feature. Here descends a face of rocky slabs, bare and ice-worn as far as the waste-mark of to-day. But below this line their gradient is continued by a slope of drift, of which enough has remained to hide a slight lateral overdeepening, a mere scratch, by a flowing outrun of the upper incline.

Mushketov does not give specific data enabling us to measure the reduction of the Zarafshan glacier since the time of his visit. He states emphatically, however, that the Yarkhich joined the trunk and that there were undoubted signs of retreat. This and a few other clues derived from his narrative may possibly point to the fact that the major portion of the recent recession has been effected since then, that is to say within the last thirty years. An intelligent native told me: "I am 35 years old and I have seen the Andarak (a small glacier similar to the Yarkhich, below the Farakhnau) joining the big glacier, but my father told me that he saw the big glacier as far back as the Farakhnau." In the Alps there was a period of distinct retreat between 1855 and 1880; in 1900 retreat was still observable; in 1902 the glaciers of the Bernese Oberland were fairly stationary.

The Zarafshan witnesses eloquently to the different character of ice and water action. The broad back of the glacier smothered under broken mountains fills the whole valley, pressing against the sides and crunching over the floor. Underneath issues the slim ribbon of the stream tumbling in haste and saturated with the finest dust. The glacier is a beast of burden, carrying heavy loads over short distances; compared to it the river is a scientific machinery rapidly conveying suspended and dissolved matter over great distances and in enormous quantities. All things being equal the energy of a river and the corresponding glacier will be about the same. Also both are modellers and transporters rather than destroyers. The greatest havoc to the skin of earth is done by atmospheric water and temperature. Destruction decreases as the channels of collected water grow larger and larger, while finally the sea is practically water which has done its task as a surface destroyer. The first action of water on touching earth is destruction, then transport is added, then accumulation, until at last the building up of the bottom of the ocean becomes the ultimate action of water before leaving earth again.

In view of our expedition to the Macha Pass we had at once begun to draft porters from the summer villages, it being impossible to obtain them within a day, still less at a few hours' notice. So we kept them on board wages and installed them near our camp where they built themselves a shelter of stones and branches. This part of the programme seemed to please them very much. Of their ability to do the work expected of them—apart from the question of good will—I had not the slightest doubt, for all the mountain peasants of the Alai-Pamirs are accustomed to the crossing of long and dreary and sometimes even dangerous glacier passes. Also many of them are redoubtable hunters who smite the wary ibex with a matchlock rifle. Their field of operations is however fenced in by hidden dangers, crevasses, steep snow slopes and difficult rocks, to defeat which they lack the necessary outfit and systematic training. That the lower glacier does not frighten them was proved one day by a man whom we saw descending the moraines with a horse, ass and wife, all packed with heavy baggage, save himself. Often we watched fires gleaming through the night far up the valley where shepherds seek pasturage close to the snowline.

On the 19th of August the weather was quite alpine, wet and gusty, reminding one of a bad day at a club-hut. Over the glacier a yellow Egyptian darkness and an icy cold, spitting rain kept us indoors, for we had become as sensitive as shorn sheep. Next day there was a promise of betterment, and we started on the 21st. Fifteen Galcha porters answered to the roll-call and, all told, our party numbered twenty-two, for besides Albert and the indispensable Mac we had also taken two of our Samarkandis, Ishankul and Palevan. The porters formed a motley crowd ranging from a man well stricken in years to a boy of fourteen who carried their supply of spare dressing-gowns made of brown homespun and apparently very warm. Though they looked extremely ragged, their clothes seemed sufficient considering that they were a hardy lot. Most of them had nothing but pants and a long coat (khalat, chapan), the nether folds of which they tucked in at the waist, thus girding their loins for the march. My chief anxiety was

106. On the Zarafshan Glacier; looking down.

regarding their boots, and I saw to it that only such men were taken as were provided with sound footgear in the shape of the rough leather stockings of the country, perhaps best described as primitive Wellingtons with soft soles. Their Aryan, sun-bronzed faces were mostly pleasant to look at, betraying a character against which one could not bear a grudge for long, in spite of childish behaviour. Before starting I admonished them again that they must feed themselves for their daily wage, but this merely as a saving clause in case we should run short of provisions, for I knew perfectly well that in practice it was a farce, just as with many Swiss guides. On some of them I found loaves of pease-pudding, seemingly the only travel-store which the poor wretches could muster, while a few carried linen bags with sour milk in a semi-solid state. It can be diluted with water. Half the sack of bread which I had sent up from Paldorak they devoured during the first day, for here obtains the sound principle of the camel of the desert: eat while you may, to meet the unforeseen with all bunkers full. While tramping over the first two miles of morainic convulsion I felt somewhat uneasy about their comparative silence until I saw that they were munching all the time, only emitting fragmentary talk between the last swallow and the next bite. Their manner of walking was against all rules of climbing. Notwithstanding their forty pound loads they went very quickly, but took frequent rests, that being to them preferable to a steady march with halts at long intervals. Among them was a certain Karaul-begi whom the gang had named as their foreman, probably because he was titled and a rich man, as wealth is reckoned near the snowline. We did not like him at all, for he was an impudent braggart and later proved himself at the bottom of much intrigue. So we called him Pig-bashi, teaching the other men to address him by that honourable Feringi title, laying stress on the pig. This gave us innocent pleasure, while flattering him greatly.

The Yarkhich torrent bore down enough water to make a crossing from stone to stone fairly difficult. Not wishing to wet our stockings we had ourselves carried across by a stalwart native, for these men are accustomed to wading through turbulent waters, and unbooting is a simple matter with them. For a mile or so we kept to the larboard lurch of the glacier where a spread of frittered slate afforded soft going. Not far from the bank we found an enormous sink-hole, 500 feet in diameter. Its bottom was a gravelly plain over which meandered a small brook, coming out of, and disappearing into, the ice. Whether more ice was underneath (it cannot have been much) or whether the ground moraine was thus laid bare, we had no means of ascertaining. Near the Porak valley the surface was again deeply sunk in, probably the first stage in the formation of a new funnel by erosion from the subglacial Porak torrent.

Presently we wheeled to our left in order to cross over to the Farakhnau side, and now lost ourselves amid the mighty dunes of a rough-hewn desert. We took nearly an hour to traverse this landscape of petrified energy, with its large, rolling hills sometimes rising two hundred feet above hollow vales in which the rest of the world was blotted from our horizon. Here and there a conical peak turned a sheer declivity to a gaping and glaucous crater. Upon this scene of tortured topography is showered a deluge of spiky rocks, a jumble of splinters and cubes. All this is full of snares and pitfalls, for the melting ice allows no settlement, and all disorder is disordered worse. Our feet are in constant danger of getting pinched between blades of sonorous granite, and whole embankments, delicately poised, are waiting to come down at a touch on the trigger-stone. The jumping stride of the practised mountaineer is oft of no avail and must be changed for a cautious blundering through this most diabolical place for locomotion. It was, indeed, a sermon in stone, but punctuated with the oaths of man. None of us can resist the allurements of jutting blocks on the edge of a great vat with its circus of toboggan slides and watery ring. But what we do in passing, with a dab of the ice-axe, Ishankul cultivates as a sport, sparing no exertion. Once he succeeded in tipping a three-ton boulder over the ice-shoot with that satisfactory result of swish, thunder and splash so thoroughly enjoyed by every healthy human being, young or old, savage or civilised.

Life still finds a footing in the wind-blown sands of Karakum and is not extinct on the ice-drifted desert of the heights. A modest little plant opens its tiny, red blossoms and on the ice huggling masses of the "glacier-flea" (*Desoria glacialis*) may be discovered. On some stones I found the houses of insect larvae. They resembled small sea-acorns, were built of sand and still contained the cocoons of former occupiers now on the wing.

After having passed the Farakhnau we took to the mountain slope. Between the stony waste of the glacier and the grazing ground lined with earthy sheep-tracks, our delighted vision fell upon beds of beautifully tinted sorrel shaded from red to pinky green, its blossoming spears planted in thick bunches of sappy leaves. In between the St John's wort raised its golden heads, and islands of yellow cress throve in the moisture of dying avalanches. We could now increase our pace, but distances were more than doubled by innumerable bights into and out of an endless succession of gullies. Remembering to be thankful for small mercies we were quite glad not to have to do with Caucasian pasture slopes where a thick, knee-deep and juicy vegetation worries the foot of the traversing climber with a relentless inward bend.

The waste-mark is now lowering and the inlaid moraine is beginning to grow a small crest, which gets bigger and bigger until a ridge of lateral moraine is detached from the slope. Still higher we shall find that it is continued on the ice, that it was a medial moraine, now stranded on the bank. Half-way between the Farakhnau and Tolstov the glacier's visible spoils assume their grandest proportions. Apart from a general increase of quantity, there is a remarkable augmentation of large boulders. None of outstanding size are to be seen on the lower part of the glacier, while here giants of 5000 and 3000 cubic feet are surrounded by a bevy of satellites ranging from five to fifty tons. These boulders and the major portion of rubble in bulk have vanished down the hatches into the hold, long before the end is reached. The number of pockets and placid pools is still very great. One of them was 200 feet deep from the highest part of the rim and its sides were pierced by a gothic arch of transparent blue. Elsewhere a pinnacle of ice attracted our attention and was found to tower 300 feet above the bed of a small burn. Falling stones unfortunately warned us not to examine the floor of this pit, although I was eager to know if there was any ice beneath.

By pointed groans and distressful breathing the porters began to hint that they were working very hard, that the hand of the Tura lay heavily upon them in the shape of sundry packages, and that a good, long rest was a question of life or death. But the Tura was ever a fiend and turned a deaf ear, especially as his glance caught one of those fixed points where immemorial tradition condemns the passer-by to sit, stand, rest, rope, breakfast or spend the night, regardless of his wish. In this case it was the snug lair of shepherds, with a litter of hay and a battery of cooking pots. But half an hour later rain began to fall and we sought shelter under two enormous blocks meeting like the roof of a house. The fine drizzle changed to sloppy snow; we sucked the lemon-drop of consolation and smoked the cigarette of despair, and finally had to stop for the night. There was plenty of wood about, for the porters brought dead juniper trunks which exposure had turned to a silvery grey. One of them was nearly two feet thick, another was thinner but fifteen feet long. As there are no live trees of any size about here, these dead ones may date from a time when the glacier was much shorter, all the more so as juniper wood is very hard and durable. Huntington (*Pulse of Asia*, p. 53) mentions dead trees on Kashmir glaciers. Drinking water had to be fetched from a lugubrious tarn at the bottom of a crater which might have served as a picture of the entrance to Dante's inferno. Between patches of black ice looming through the gathering mist a staircase of toppling boulders led down to what, in the creeping twilight, seemed like abysmal depths. Very cosy was our cave with the fire playing on its walls and the crackling of the wood mingling with the patter of the rain outside. Later on it was discovered that our abode was not as rainproof as we had supposed. Ominous streaks were showing on the ceiling, where fat drops sucked themselves along with flat feet, then to let themselves drop where least wanted. Their steady drip continued throughout the night and in the morning the mackintosh sheet with which we had covered ourselves showed deep pools with several pints of water.

On August 22 a dull morning lifted its leaden eyelids. A sea of cloud wadding lay in the Zarafshan valley and a grey heaviness overhung us above. Mountain peaks loomed spectrally out of the vapour. Their summits and sides covered with fresh snow betrayed what had been going on through the night. We had, however, been kept very warm by the glowing embers of juniper encased in stones built over them, and acting like a stove. Later some rays of hope were shot from blue openings in the sky, and three of us started in advance. After two hours we reached the Tolstov glacier where the grass slopes come to an end. We still heard marmots whistling (11,000 feet) and a small duck rose from a pool. The moraines of the main glacier are beginning to form parallel ridges out of chaos. Here we waited for the others, before crossing the labyrinth of the Tolstov junction. When Mac came up he told us that there had been an attempt at mutiny among the porters who had thrown down their loads and even threatened our men with stones. Mac then fired off two shots which only perforated the atmosphere, but made a strong impression upon the minds of the Galchas, thus closing the somewhat theatrical scene.

On the side of a shallow crevasse I observed a rare and curious phenomenon. A glacial brook had managed to undercut the ice in such a way as to form a spiral loop in the manner of a mountain railway crossing its own track at a lower level. At half-past two it began to snow and we were forced to camp on the left bank of the Tolstov glacier. On the way we had met Kirghiz shepherds and bought two sheep from them, one of which was killed at once with a wonderfully soothing effect upon the feelings of the porters. Towards evening the spongy heavens parted wide and in the cold, clear sunset we had a magnificent view of Mount Akhun (Figs. 18, 109), one of the finest ice-peaks I have ever seen. Its snow-white crest is streaked with the sharply chiselled grooves and spines made by descending avalanches. Looking through a telescope one can estimate that these fine carvings are often from twenty to fifty feet deep.

Next day a glorious morning broke and we left early, reaching the Skachkov glacier in two hours. We walked over the middle of the glacier which now presented an ordinary alpine appearance such as that of the Gorner glacier in the neighbourhood of the Riffelhorn. The grey and gritty surface of the ice is loosely strewn with fragments, here and there gathered into lines of medial moraine. Where the stones are spread singly they have melted holes into the ice thus producing a honeycombed condition called, by Böhm, a "sieve-moraine." A few glacier-tables are seen and water is now running on the top of the compact ice into which it has cut the channels for numerous rills and streams. Still further up the medial moraines lose their ridge-like aspect, turning into flat strips, until we come to a place where they are just beginning to melt out of the glacier which is, all the time, getting whiter and whiter. We have reached the névé line, where the supply of fresh snow is in excess of waste by melting. The Skachkov glacier contributes more detritus than the main trunk, and altogether the right branches supply four-fifths of the whole of the Zarafshan moraine.

The porters murmured and mumbled again, but a brisk pace on the easy floor of the glacier does more than words or threats in keeping them up to the mark. Without Mac's help we should never have got these people to the pass, for he had lured them onward bit by bit. Last night they stoutly refused, but by an Oriental system of warning, intimidation, flattery and persuasion he induced nine of them to follow us, the others staying behind, with Ishankul as guardian of the camp. Mac is the only Eastern dragoman who understands the aspirations and needs of mountaineers and it is now seventeen years since I first introduced him to the type of mad European who plods up tiring slopes and dangerous cliffs in search of neither women nor gold. Like all men Mac has his faults, but there are only six honest interpreters in the Near East and he is one of them.

This was one of the most delightful glacier strolls of my life, with an easy road underneath and a cool breeze to temper the heat of the sun. When we got to the snow and the first crevasses I also made the porters rope together, which in addition to the goggles I had distributed seemed to please them very much. At two o'clock we stood on the Macha pass the low saddle which one sees at the end of the straight perspective of the Zarafshan valley. Looking west the eye was held between an avenue of surpassing grandeur, where soaring peaks lined an endless vista with their broken ridges and frozen straits. The warden of the pass is Mount Igol (The Needles, Figs. 107, 108); then comes Bielaya (The White One, Fig. 109), then Akhun, named after one of Mushketov's porters. Opposite Bielaya, Mount Obriv (The Precipice) forms a corner round which the upper continuation of the Zarafshan glacier turns northwards. The climber who here refreshes his memory with Alpine and Caucasian panoramas is struck by the absence of those rounded ice domes which, like bulging octopuses, uplift a heavy mantle of névé, trailing their fringe of tentacles through the gaps of the lower rock. This type of mountain is very rare in the Duab, even where a formation of the ground and sufficient height above the snowline should make it possible. On the whole one cannot help feeling that, other things being equal, the ice coat is thinner than in the Alps. Hence the frequent occurrence of ice-falls and broken masses of névé on slopes or ridges. The eastern view from the Macha pass is hemmed in by a near circle of great mountains from 16,000 to 18,000 feet high. As they show us their south-western flanks we only see a few small glaciers, while the peaks are merely sprinkled with snow, reminding one of the dry sides of the Pennine Alps.

The names just mentioned were given to the peaks by Mushketov who reached the Macha pass a quarter of a century before us. Although that was a hundred years after the ascent of Montblanc, yet the Russian traveller and his companions were still in the age of Saussure, as far as mountaineering goes. The most difficult pinnacles of Switzerland had already been scaled, the Alpine Club was a hoary institution and the science of climbing laid down in hundreds of printed volumes. But here were the Russians who had never seen a nailed boot, never heard of ice-axes, and did not even know how to walk on a moraine or among crevasses. On the other hand they had equipped themselves with sleighs and dogs as for a Polar expedition, while for moral support they relied upon the proverbial pluck of the Cossacks whose names are immortalised by the Tolstov and Skachkov glaciers. No doubt they did show themselves worthy of their fame, but these horsemen of the plains of the Don are as much at home on a glacier as Albert Lorenz would be on a yacht. Considering that any good walker can cross the Macha pass with a third-rate Chamonix guide, Mushketov's tragic description smacks of the incongruous. This however does not detract from the merits of the undertaking in itself and the honour due to the courage and perseverance of the travellers. They approached the unknown mountain world with grave misgivings and without that experience which would have enabled them to laugh to scorn the bloodcurdling tales of their porters who gave exaggerated

107. Approaching the Macha Pass; Mount Igol on the Right.

108. Mount Igol with the Macha Pass just below it.

descriptions of the dangers lying in wait for them. How they fared with their native crew I can vividly imagine after the comparatively harmless samples which had fallen to our share. We only feared inconvenience and loss of time, for desertion could not terrify us, as we would have felt as safe on the Zarafshan glacier as on the Theodul. But the Russians went in constant fear of losing their way or perishing in the event of bad weather, thus proving excellent objects for the levying of blackmail. Of this windfall Mushketov's porters availed themselves thoroughly, and on the third night even left the camp, although after a while they came back to negotiate. The Russian party traversed the pass entirely, descending on the other side into the Zardalia valley, where a somewhat steepish snow-slope made their hair stand on end. Here also night surprised them on the ice, and in order to keep themselves warm they burnt all the wood they could lay hands on; from the sledges to the handles of their instruments.

The Macha pass is not the true head of the Zarafshan valley, but a lateral gap, at least hydrographically, for the main glacier curves round Obriv towards the N. and N.W. On the other side of the saddle we found an artificial platform which seems to show that native travellers sometimes spend the night here. Climbing some way up the granitic ruins to the north of the pass, we made the porters build good wind-screens for themselves and for us. Then I scrambled to a point of vantage on the gigantic piles of sharp-angled granite (Fig. 108, left corner) and from there took a series of photographs (Figs. 108, 109), Mac in the meantime boiling tea. The starry night was wonderful. Before us the black, unearthly wall of Obriv stood like a mighty gnomon past which the moon and stars tracked their silent course.

At five o'clock on August 24 we greeted the sun lifting his dazzling orb over Alai's jagged crests. On our blankets lay the hoarfrost of frozen breath. Sending the others straight home, the two ladies, Albert and I set out towards the upper glacier, taking the Pig-bashi as porter for my camera, and also to keep him away from the others. We trudged about three miles in a north-westerly direction over undulating névé nearly free from visible crevasses. At the end of the first mile the Pig-bashi collapsed or pretended to collapse. Dividing his load between us we left him there with strict orders not to budge from the

109. Mounts Bielaya and Akhun; looking down the Zarafshan Glacier from the Macha Pass.

spot, until we came back to take him on the rope again. When we did return two hours later, he had vanished, but somewhat lower down we caught up with him, forlorn and helpless, in a maze of fissures. We enjoyed his crestfallen appearance while he was rather glad to have some of our refreshments.

Skirting the flanks of Obriv whose avalanches have imparted a heaving swell to the glacier, we finally came close to the foot of the last wall. At this point the glacier has accomplished a half circle from N.E. to W.S.W. round the gable of Obriv which has a N.W.—S.E. position. Here, in an amphitheatre of granite pillars, we saw a very difficult ice-couloir leading up to a sharp notch near a gothic roof with three small spires. This then, is the still unclimbed Zarafshan pass cleaving the divide at a height of about 14,400 feet (4400 m.). The total length of the glacier to the bottom of this ice-gully may be reckoned as 14 miles (23 km.), or two and a half miles less than the Great Aletsch. As our present knowledge stands it is the longest glacier of the Duab.

Thus we have stood at the beginning and at the end of the Zarafshan. It is the middle line of the Duab and may serve as a symbol for the whole. On its banks are greater cities than on the Amu or the Sir. The Zarafshan is history; it runs through the very heart of the country, it is in the midst of its events. Its waters cleave and fissure the rock and grind the mountains to dust; they blast the boulders from the spine of the land and carry them down on a back of ice; they cut the valleys and chisel the cliff, build terraces with villages and gardens; they level the plain, making it green with trees and golden with corn and busy with the throng of men; they have been the life-blood of an empire; for them the glorious temples, for their sake the clang of many battles. A new empire now watches the waters of the Zarafshan giving their last drop to humanity, before dying in view of the western horizon; they are in the full throb of nations, they have seen the glory of Samarkand and the greatness of Tamerlane. The Zarafshan is history, history of landscape, history of man. Will you hear the softly falling snowflake of conception, the majesty of ice-gestation, the thunder of a river birth, the youthful rush, the manly flow and the last dying murmur, then go and listen to the waters of the Zarafshan.

CHAPTER XII

TO THE MOUNTAINS OF THE FAN

WISHING to draw a picture of the Zarafshan valley in its unbroken length, I have kept until now our fortnight's excursion to the Fan district or Hazrat-sultan, as this part of the Hissar range is sometimes called (see Maps 110 and 115). We left Varziminar on the 26th of July, and soon found ourselves on the dangerous cornice-paths of the Fan-darya. The Russian soldier, fond of pet names even for pet aversions, has given the name of "balkonchiki" to these giddy ledges. All bridges and paths are emergency structures, for our friend the native only builds emergency roads and emergency houses, his whole life being evidently one long emergency in view of a more permanent state in paradise. The "balconettes" are characteristic of most Asiatic mountain roads, and reflect great credit upon the improvising skill of the inhabitants, having been made without the use of a single ounce of powder for blasting or a single inch of rope for tying. We have more eye for their defects, but explorers and Cossacks have horses, whereas the humble pedestrian praises Allah for having caused such a comfortable passage to be made by man. Steep rock walls rise from the foaming torrent and along their face runs the narrow shelf stuck together out of crooked sticks and rubble. Twisted trees and branches are jammed into clefts or supported by friction on pads of brushwood; little walls are raised on tiny ledges, or alternate layers of blocks and fagots formed into a coping. On this projecting scaffold is spread a mixture of stones and bits of wood, the surface of the overhanging road. Where smoothness of the rock defies building up from below, long timber is used to reach the nearest hold, with here and there a crazy prop to bear the weight of tipping slabs placed on the rafters to offer foothold to beasts of burden (Figs. 111, 127). How these jutting eaves cling to the precipice is always something of a mystery, but they do, by a cunning employ of the shape and balance of each stone or bough. Winding, climbing, descending, the cornice-path seeks the best means of support, along the towering cliffs which close in the

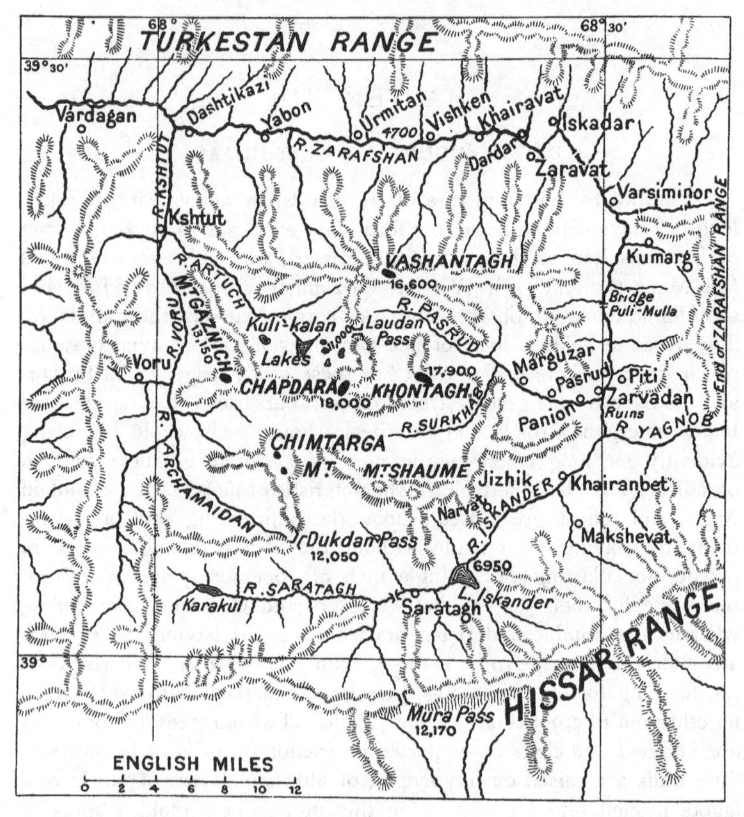

110. The Fan District.

sombre gorge. One must admire the ingenuity capable of such makeshifts, while regretting the want of public-spirited effort to create a more lasting piece of work. For generations untold these contorted balconies have enabled the villagers to reach the plains, and all through these generations they have been nothing but an unbroken series of daily repairs. Every horse will do some damage at one point or another along the line, and his master is obliged to replace branches or stones dislodged by those who went before. Thus it is mostly the travellers themselves who close the gaps and pitfalls on the way, unless they find themselves confronted by a blank wall where a long portion of the gallery has slid into nothingness. Then gangs of men are brought from the nearest hamlet, where the caravan patiently waits for days until the damage is made good. Pegs are again stuck into crevices, cobbles banked up, bundles of juniper or poplar branches pressed down and the whole finished off with stones as flat as can be found near at hand. Hence these structures resemble the famous beer-mug which had been in a German family for six hundred years. First the metal handle was broken and renewed, then the earthenware jug suffered the same fate, and thus, by a continuation of the process, the same vessel outlived many centuries. On these paths there is only one place at every step where the hoof may rest more or less safely on the wobbly boulder or between two holes; there is no margin for saving time or worry. Such a cornice represents the exact ratio between a natural difficulty and the human will opposed to it in this particular case. Nor does the native grumble because his road is made up of two halves: the work done by others, and his own ability of using it without danger to himself. For may not the traveller of to-day be one of the road-makers of to-morrow, pressed into service by the powers that be, whether beg, mingbashi or aksakal?

Only he who cultivates suicide as a sport will ride over the worst sections of the airy stage, for besides stumbling into a void one can also be knocked out of the saddle by an unexpected bulge of the mountain overhead, wherefore a dreamy state of mind and the composition of poetry are strictly at owner's risk. All the same it must be admitted that weird shudders are best felt from on high when gazing over the brink of perdition with eye of man and groping for safety with foot of horse.

In spite of precautions we lost one of our best horses. We had just left the saddle in order to walk, when the grey mare got her foot caught and fell across the shelf, only three feet wide, with her hind legs dangling in space. Now, the horse is an intelligent animal, but with an intellect sadly marred by nerves. We always had a rope ready for emergencies, and this we managed to throw around her neck and thereby to help her up. But instead of keeping quiet to await further developments, she struggled, fell again, and went right over the edge, almost pulling down to death Ishankul who had stepped into the coils of the slack. Down she went: a light grey spot, a yellow saddle-felt in the slaty, gurgling waters of the Fan-darya, six hundred feet below, and that was all. She died with the honours of the mountaineer, with thirty yards of Alpine Club rope that could not save her, though Albert held on till his hands were raw. Perhaps, some day, in the desert of the plain I shall pass, unknown to me, a stately camel the burden on whose back is tightly gripped by twisted strands with a red vein running through their heart, the mark of noble origin.

The cornices of the Fan defile are the most horrible I ever trod. Fortunately, the Mingbashi of Varziminar gave me ten men to help us over the worst places, and sent ahead an even larger number to hide the most glaring defects. This in order that the great Tura may report favourably at Samarkand, and, better still, shed the golden rays of his countenance upon his humble servant the Mingbashi. Of course, the Tura's countenance is in his pocket, and silver is also taken. I never disappoint those dignified worthies, for a stately head keeper of the West (who expects paper) easily carries five mingbashis in the hollow of his palm. But in spite of the solemn assurance that the moon will reflect upon the meanest creatures the glory of the sun, I advise everyone to shed direct a furtive ray, however small, upon the half-naked wretches toiling at the road, so as to make sure of the reflection which gratitude may kindle in their eyes.

The deep ravine of the Fan, which keeps us imprisoned between lofty precipices up to the mouth of the Yagnob, is but one of the many instances so often met with in these mountains, where a river has to cut through intervening ridges before running into the main valley or the plain. Accordingly, many valleys are quite inaccessible by what should be the chief entrance, being reached by passes over a lateral watershed. Thus all traffic between the province of Samarkand and the villages of the Yagnob, Pasrud and Iskander rivers goes by way of Panjikent, Kshtut, and the Laudan pass in order to avoid the execrable roads of the lower Zarafshan and Fan. Communication through the Fan gorge is, however, important as a very short cut from north to south, and that is the reason why a way was forced through it at all. Here goes the direct line from Hissar and Karatagh in the Khanate (Bokhara) to Uratiube in Ferghana, by way of the Mura pass, Iskander-kul, Varziminar, and over small passes in the Turkestan range. A hardy traveller can cover the distance in five or six days, whereas a fortnight is the estimate of a journey by Denau, Derbent, the Iron Gate, and Kitab (Shakhrisiabs). Of course, the big merchant who can calculate to a nicety how much he saves in wear and tear of man, beast, or material by following the line of least resistance in commercial topography, will always send his goods by the caravan route. If the consignments of his Russian products are very large, he can send them from the railway station of Charjui up the river to Termez, and thence by camel into the eastern provinces of Bokhara, in which manner he avoids the customs duties of two or three local frontiers, for each beg levies a tax on merchandise and cattle passing through his domain. That is, among others, a reason why English goods via Afghanistan until lately had such a good chance in East Bokhara where Russian wares could not penetrate without having to pass many of those barriers, more formidable than wall of rock or water-ditch. Moreover smuggling across the Oxus formed a valuable asset, since then much depreciated by Russian guard posts. Were a Russian to take his transports himself he could pass unmolested by native excise, but native solidarity should prove more than a match for him. In Baljuan or Kalaikhumb he could not possibly rival the patience and capital of time which the local trader so richly possesses. A native wholesale purchaser, on the other hand, is a doubly welcome prey to the beg, for he cannot keep his stores a secret in the bazar. There are, however, people, it seems, who have found the middle road, and to this I attribute the appearance, during late years, of several Caucasian Mountain Jews (calling themselves Ossetes) and Armenians in a small way of business. These have enough cheek to assert their supposed privileges as subjects of the Tsar (such is the power of European prestige), while yet having the qualities of the native.

For the reasons set forth above, all humbler business prefers the high mountain road, and during the summer months flows over it in a constant stream, paying only one tax on entering the begship of Hissar. There we meet the small trader with one or two pack-horses, the man with a donkey, or the pedestrian in search of work. To all these a shilling is a real entity and not a mere cipher of wholesale arithmetic, and to save it is well worth a quick, if exhausting, struggle with the dangers of the heights. Eagles and vultures are kind-hearted creatures who only eat the dead, but the rapacious officials of the Amir skin you alive. Moreover, on the longer track the money spent at hostelries must run up to a pretty sum in the end, for one cannot endure for two weeks what one can easily risk in a short sprint of six days,—pace and privation. Thus we meet them; the man with the girded loins striding forward with an energy which strikes us as unusual in his race; and the horse arching his back under a heavy load. They hurry on all day, only snatching forty winks under some eaves of rock or a noble karagach tree during the heat of noon. At night the wanderers sit shivering around a small fire, eating a crust of dry bread; the animal gets a few handfuls of barley brought from home, and a bunch of grass or hay which the master has collected during the ascent, then it stumbles about near the encampment to supplement this scanty diet with what it can find between the stones. In this wise they push, and pant, and struggle, and starve, until reaching the other side, where they rest and feed up again. The reckoning is all right, and generally works out well, at least for the man, for who sums up the horse's profit in the long run? Frequently the result is disappointing, because the preparations were inadequate. With true Oriental improvidence, some owners feed their horses insufficiently before the start, or overload them, so that many collapse near the top of the pass.

Among the sensational items on the Fan track there is also a narrow passage leading up between two rock walls, like a steep corridor. Here my chestnut slipped, rolled over and nearly squeezed me flat against the wall. Those coming behind were involved, and for a few seconds the sounding alley was a turmoil of kicking horses and shouting men. But all went well and nobody was hurt. I am particular about mentioning these events, because all our little adventures happened in the course of ten days, while the rest of the three months' journey was entirely bereft of those narrow escapes forming the credit balance of live explorers against dead ones. I might have spread these incidents over the book more evenly to satisfy the dictates of harmony, but I have so often failed as a liar that I am forced, much against my will, to sacrifice artistic effect.

Bridges are the weakest as well as the most important spots of the mountain road, for they connect the easiest slopes. If the path comes to an impossible precipice, it must look for the other side, and I have counted five bridges within two miles on the Karatagh-darya. Such as are above the level of the spring floods enjoy a great reputation, being anchors of faith on a pilgrimage of uncertainty. One of this kind is the Puli-mulla, which lies halfway to Piti, and marks the end of the worst stage of the road from Varziminar. Whoever has learnt of its existence from the map is at first puzzled, until, having shaded his eyes many a time, he at last discovers a slender streak lying across the sunlit canyon. I have seen many giddy bridges in the East, but this one is the giddiest of all. Innocent of railings it lies gingerly over a chasm at least thirty feet wide. The usual cantilever of overlapping sticks and stones (similar to Fig. 89) cautiously balances on this side, reaching out for a hold in mid-air. The gap thus shortened is spanned by three poplars overlaid with rough boards, fastened by means of bast and withes. This diving-board sways and trembles 200 feet above the thundering turmoil of the whirlpools, between limestone walls into which water has wormed its queer concavities. Such a rickety piece of furniture is a fitting termination to the cranky wickerwork cornices of the Fan. Uttering a prayer of thanks, we take a well-earned rest on the other side, where saucer-shaped and rounded rocks have been hollowed out by the eddies of diluvial times. They offer convenient seats glistening with a high polish imparted by countless travellers among whom there may even have been the great Timur, to add his mite of shine.

The road now becomes very easy. Owing to the steepness of the mountains and the subsequent want of spaces for irrigation, Piti, where we spent the night, is the only considerable village of the lower Fan. It hugs a forbidding cliff serving as background to the sloping garden of apricot trees and the terraced roofs, bearing such close resemblance to the pueblos of Mexican Indians. Rising tier upon tier the boulder-walls and platforms culminate in the mosque with its pillared portico. As the approach is over the graveyard, the horseman, whose gaze is upon the golden apricots in the foliage above, may find himself upset, if his stallion's foot intrudes upon the inmates who enjoy but scant privacy under yielding vaults.

On the left bank of the river, opposite Piti, are outcrops of slate with a strong admixture of coal. The same coal, in moderate seams, occurs near Kshtut, where a Russian firm is trying exploitation on a large scale.

111. Cornice Path and Bridge.

Also on the left bank, below Piti, there is a smooth rock covered with rude drawings of the ibex (the ahu of the natives) reminding one of the mural paintings of the Bushmen of South Africa. Like the rag-trees, to be mentioned in due course, they are wayside offerings made by the passer-by. In the Duab every remarkable spot on the road is made an occasion for such offerings: the beginning or end of a dangerous passage, the top of a pass, a parting of the ways, the grave of a saint, a freak of nature, waterworn rocks, curious shapes, a cave or other place attracting the eye, impressing the mind, or hallowed by tradition. Or it may be that someone began something at a certain place, that someone did tie a rag to a conspicuous tree, outlined an ibex on the rock, deposited a couple of horns, or buried a relative in holy orders. Hence by force of imitation the first offering may become the cause of repeated offerings in the shape of rags, drawings, and the deposition of anything curious, such as oddly contorted stones, very round pebbles, fossils, or that most typical of all sacrifices, the frontal ornament of horned animals. It is in the nature of the ibex pictures that they only appear in the mountains, the home of the animal, and on smooth slabs offering a writing surface of sufficient size. As is shown by the caves of bushmen and carved bones left by prehistoric man, the human being has, since time immemorial, had the instinct of representing things. In its lower and most unconscious stage this desire of painting something is called into action by any surfaces which invite scribbling, such as snow, sand, rock slabs, and paper. Any large smoothness seems to be endowed with irresistible temptation to cover it with drawings or writings, from the crude and illiterate to the highly artistic. The seashore, sandstone, dried mud, the first snow of winter, and the author's blotting pad covered with mysterious hieroglyphics straight from the hoary soul's unfathomable depth are proofs thereof. Likewise soft or plastic materials invite sculpture and modelling. The uneducated of civilised races betray this invincible desire by covering the walls of public places, often unmentionable, with texts or symbols of ever-recurring type, the writer's own name being, of course, the one that most readily suggests itself, for it must be remembered that the act is spontaneous, not preceded by meditation. Hence habitual things, such as the proper name, hackneyed sentences, doggerel reiterated to nausea, mannikins and childish or indecent sketches are constantly used. The same with these ibex slabs. Originally there may be an ancient cult connected with the practice, but now the thing has become a routine of the traveller, who simply imitates his predecessors, just as we do in the West when we must eat whitebait at Greenwich or write a picture postcard from the top of the Eiffel Tower. That explains why the diagrams of ibex are more or less alike in one place, but often quite divergent in different localities. Near Piti the figures had this shape and were drawn with bits of coloured sandstone. At Tabushin, in the Zarafshan valley, they were dotted into the rock by making chips with a hard stone. Thus the wayfarer simply follows the type and method of work set by his predecessors who again have mechanically copied the symbol started by the first man. The impetus may have been given centuries ago, in which case one will find innumerable faded traces; or the series

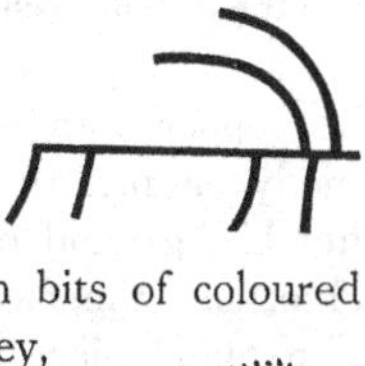

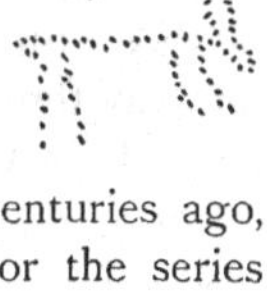

may have started quite recently, for new mazars, fixed places, and painted walls may spring up anywhere at any time.

In full view of Piti there is a very fine talus clearly showing the formation of slope breccia, its lower end having been undercut by the river thus exposing the layers underneath. The slope is the result of dry weathering in spite of which its interior looks as if caked with yellow mud, while the surface consists of clean, sharp and arid screes in the spreading of which water took next to no share. In this country one must ask oneself if a surface of angular stones is not just as capable as grass of catching loess dust floating in the air. Rain washes it down into the interstices below, while disintegration of the rocks undertakes a renewal of the rubble. Undoubtedly the stones themselves provide a good deal of dust, but I dare say that the scree slopes collect a fair amount from outside.

Above Piti the Fan, for a short distance, passes through a fine and regular canyon of solid rock, where the river pursues a gentle flow. To the perpendicular wall on one side there still remains attached a narrow strip of old river terrace, supported at its base by a broader talus of its fragments. Undoubtedly it corresponds to the great Zarafshan terrace. The perspective of this gorge also shows a fine profile of three successive overdeepenings, in the clearest manner possible. This profile is very steep, the benches or steps being very narrow in comparison to the sides of each successive trough[1].

About an hour from Piti the river opens out a little, and riding becomes still easier. Standing near the ruins of Zarvadan, perched on a bold promontory of limestone, we survey the basin in which the upper Fan (Iskander-darya), Yagnob and Pasrud unite. It is chiefly a picture of streaks, of the dipping strata which dominate the background of the scenery. It is the desert still, and as if to advertise the fact by some striking sign, nature has here laid out a genuine piece of sandy waste, a few acres of pure, dry, drifting sand swept together by the eddies of the valley winds.

We now take the Pasrud for our guide, and on the left bank pick our way through a maze of huge blocks of a ruddy, brecciose concrete fallen from the cliffs above, where one observes alternating layers and

A photograph of this view is to be found in the *Geographical Journal*, November, 1907, p. 497.

transitions from pure sandstone to rough fragments embedded in red cement. Presently the smile of a more gracious mood flits over the face of the earth. On the broad floor of the valley are villages among green trees, pastures and yellow fields. A pleasant hour of rest is spent in luscious grass under the shade of the weeping willow by the stream. This is more like some Swiss prospect. The swift current dashing against the round-backed boulders lifts a cool spray, and when we raise the eye it scans the long ridges of high mountains, here and there dotted with snow, or penetrates into the depths of some wild corrie, where the light begins to fade from the tiers and screens of purple crags festooned with icicles.

After passing Marguzar, the highest village, the desert claims us again for a while, because white limestone screes, reflecting the glare of the sun, descend to the river's edge. But plant life is bravely disputing every inch of ground, as shown by the bushy fringe which partly hides the stream. Close to the water is the thirsty willow, holding and hemming the rim with a web of roots, while the more ascetic barberry, hawthorn and similar shrubs line up behind as an advance guard to stem the flood of stones. Some juniper scrub also begins to appear on the slopes.

Then suddenly two buttresses come close together, and we see a mountain gate, the foreboding of a change, of something new, for the passage is not long, and beyond we perceive the massive build of a great peak, Khontagh. On approaching the neck we find it choked with a dam of huge blocks mixed with smaller detritus, the ground moraine of an old glacier. Under one of the largest rocks the water has dug a subway, thus forming the natural bridge of which previous writers have spoken. After clambering over the obstacle, we land upon the level where the upper Pasrud falls into the Surkhab. The latter is much stronger, though the valley of the Pasrud, being wider, longer, and more important in every respect (save that of water supply), claims the honour of the principal name. Pausing to think, we realise once more how tradition scoffs at nomenclature based solely on hydro-arithmetics. An impression of stateliness, a more engaging aspect, or greater length, the presence of villages or pastures, or the direction of the road to the pass, are more convincing to the native than a cubing of the water power. Quite well can we imagine how the first shepherds and hunters were attracted by the inviting openness of the Pasrud (named by them after the long familiar stream at the village, for all naming proceeds from below) before they ventured into the darker recesses where the Surkhab glacier lurks in a lair of blackened slabs.

We have now reached, at about 7000 feet, the region of high grazing land laid out on the round-back ridges and spreading slopes which form the pedestal of towering pinnacles or ice-clad pyramids. The view is alpine, but at this time of the year we see greyish-yellow

112. Bent Strata on Khontagh.

tints of dying grass, and hard soil is discovered between the tufts of grass or the single stalks of coarser weeds. Then we realize that the air is dry, that drought has not stopped short on its ascent from the steppes. But if we were to come in spring, we should find a beautiful emerald lawn thriving under many showers and the water from the melting snow which soaks the ground. And the conclusion is that this region owes the preservation of an alpine aspect (in summer and autumn) to its cool height, enabling the truly alpine conditions of spring to preserve their main features, to save their face, so to speak.

By the time the winter snow hides this face again it has become quite shrivelled, but, thanks to the proximity of the glaciers, it is able to keep up the outward appearances of freshness, while the caked and crusty stretches of the lower Fan cannot hold the soft caress of any rains. We are here on the line where the mountain air begins to gain on the desert air, just as the gardens of the lowlands mark the ground where mountain water proclaims a victory over drought. For the desert is hot air against which the mountains wage war with ice, cold air, and water.

The place on which we stand, and where the upper Pasrud flows, is what the Tyrolese peasant calls a Boden, an Alpine floor, meaning a flat (often alluvial) plain in a high valley. A peaty sod hides most of the underlying gravel, and the brook meanders peacefully in a setting of greensward. Grassy undulations rise towards the sides, studded here and there with dark-green junipers, solitary or in clusters. The birch is also seen, with silver bark and feathery crown. Cattle, mainly fat-tailed sheep, browse the pastures, under the watchful eye of uncouth shepherds.

We find that agriculture has not stopped short, for we see fields. Yelling women, seeking shelter from the male gaze, and squealing children, frightened in their rustic games by our approach, proclaim the neighbourhood of a lailak, or summer resort, corresponding to the alp of Switzerland, the yaila of Pontic Transcaucasia, and the ailak of the Hindukush. The dwellings are tumbledown hovels of stone or rag-clad wickerwork, which is to represent an apology for yurts or kibitkas. All the old men, the women, and the children are sent up here, to live on milk and to prepare the winter store of cheese, while all strong hands remain in the villages to reap the crops. Later they come up for the higher fields which sometimes are found at an altitude of 8000 feet. Once started under the auspices of a mild spring, the barley has a fair chance of ripening.

Just off the north-eastern face of Khontagh we climb an old moraine, dividing the valley floor into an upper and lower level. To the traveller who has topped the block strewn elevation, Khontagh here bids majestic welcome. Beyond the door of a side valley it shows itself in the manner of a tremendous red block. Only the very head bears an ice crown, while the unassailable front, though breasting the north, is barely specked with snow. The opening which frames it has a bold, earthy cliff of boulder clay on one side and a softly puckered, tree-studded pasture hill on the other. When we thus saw the view great clouds were caressing the sky, causing it to shed a gentle light and to blend its eager glare into a deep perspective.

For a long way the Pasrud now meanders peacefully. Its banks

113. A Dead Juniper Tree.

are skilfully rounded off by a natural process, the river having washed out gravel from under the grass, causing the unbroken thickness of the turf to hang into the water with a graceful bend. Presently we branch off from the Pasrud for a while, and climb the Laudan valley by a good path leading through fine groves of ancient juniper, gnarled, twisted and weather-beaten (juniper in Figs. 113, 114, 116, 118, etc.). Two kinds of this conifer are chiefly met with, the more tree-like *Juniperus excelsa*, and *Juniperus pseudosabina* which is smaller and often crouching. As the Latin name shows, science classes it as a juniper, but the general appearance and the shape of its leaves are those of the thuya (*arbor vitae*) and such it is often popularly called by the Russians. Its native name is "archa." Cypress, thuya and juniper are united in one botanical suborder, nearly all the members of which one readily associates with Mediterranean countries having a poor soil or a dry climate. The numerous species of these three families range across the whole of Eurasia from Gibraltar to Port Arthur, and most readers know the juniper whose berries are used for flavouring gin. Our juniper is generally a small bush, but in certain localities will grow into a good-sized tree. In the Lüneburger Heide, that primaeval heath of North Germany, one can see specimens which, in size, shape, and colour remind one of the cypresses of Italy.

Emerging upon pleasant downs, where hidden glens still contain sweet herbs for the horses, we make our camp at a height of 9000 feet, this being more or less the upper limit of the juniper tree, the good friend who supplies us with fuel. Grass ascends to about 12,000 feet, and in valleys of the Pamir type even higher. A stiff breeze, blowing from the head of the valley, teases the flapping canvas of the tent; the noise of camp life is answered by the shrill whistle of the marmot as it dives into its burrow. These little red-coats are very plentiful, and so is the eagle which spies them from above.

Next day we are impatient to reach the top of the pass, in order to enjoy what we have been longing for: a mountain view in bracing air after those tiring marches through the ovens of sun-baked stone. At first the rise is gentle along the middle of the trough. The sun is up and the traffic is astir, coming and going,—horsemen proud of purse and bronzed wayfarers with heavy packs. Stout staves in hand, they have risen from behind their wind shields, low walls blackened with the smoke of many fires. Bethinking ourselves of the fleshpots, we await the progress of long streamers of dust, the pennants with which a cattle-dealer heralds his approach. We pay him his presumptive price and pick a strapping yearling from the herd which is to be sold at Samarkand. Yesterday we saw this flock sleeping in the sun at noon, closely huddled together and motionless, like a monster bale of wool. To-day they have begun to move at early morn ; three hundred well-nourished sheep, black, brown, yellow and white, fine, strong animals, with curved, Assyrian noses, and with the short, curly fleece that seems to speak of wild forefathers, as does the wiry leg. But domestic is the blubbery bump of fat wobbling on their buttocks as they walk. In front marches a small guard of clever goats, long-haired and dignified, as if aware of their responsibility as leaders. Behind are black calves, slow and clumsy. Then the drovers, urging forward with stick and stone, and blowing through their teeth a sound like that of the spitting cat ; on their heels the faithful dogs, slinking drowsily. And last of all, arrayed in blue khalat and white chalma, the owner, enthroned upon a pair of carpet saddle bags that almost hide the shaggy pony.

The conditions and rights of pasturage are somewhat complicated in these mountains, for not only are there the animals from the near villages, but also large herds sent up from the plains, in spite of the mountain peasants whom one would imagine to be the *beati possidentes.* Curiously enough, as Lipski remarks, those living farthest out in the plains go up highest into the mountains. But what is more remarkable still is the fact that those who have come so far have such an undisputed claim to the grass that the mountain villages dare not take advantage of the pasturage in their immediate neighbourhood. It would be worth while following up the history of this interesting variety of alpiculture.

We push on. The pasture-mat, now almost brittle straw, is notched into long strips by the feet of grazing animals. This makes the slopes look as if scored with an infinite number of irregular contour lines. But the thin, silvery fibres of the watercourses are bordered with a fringe of lawn, sprinkled by the splashing stream, and cropped close by every beast. The gardener of a tennis ground would turn green with envy.

Soon we meet the first patches of snow, remains of avalanches covered with a film of slimy earth. And here the juniper begins to leave us, only sending up those crooked dwarfs which can squat under weighty snow, or crouch in hollows from an icy blast. Some drought-hard plants now wear their nuptial garb ; the tough broom, the tangled clusters of yellow roses, and the royal thistle in its gorgeous ruffle of glossy grey leaves drawn out into long spines. Near the water nasturtium and a tiny dandelion are in flower. Presently our attention is claimed by the steepness of the path, which climbs a long and dreary incline by endless zigzags, and we restrain our curiosity until the top of the Laudan pass is reached (about 11,000 feet). Then we lift our eyes to look around, and our silence and our talk speak alike of wonder and awe, and of the eagerness of inquiry. First comes the overpowering nearness of those gigantic piles Khontagh and Chapdara. Khontagh is a colossal pyramid, with flanks carved out to make tiers of vertical precipices surrounding the corries at their base. Where have I seen such walls before? Ushba, the Terrible, in the fastnesses of the Caucasus, comes back to memory. On its western side there is a sheer drop of 3000 feet. The shape of Khontagh is not so bold, but its granite quarry shows an unbroken face of well nigh double that height. Did I say granite? The square cut and the reddish colour of the blocks lead us at first to think that the two mountains which we have compared are built of the same stone. But later we learn that the material of the great Pasrud peaks is limestone covered with a thin scale of shining rust.

In Chapdara, drawn out into a long frontage of tremendous cliffs, we easily recognise the mountain visible from Samarkand and Panjikent. It cannot boast of isolated prominence, like Khontagh, but impresses by the heavy broadness of its northern bluff, now deep in the shadows of the declining sun. Its upper shelves are loaded with ice, sending down avalanches thousands of feet into a transversal valley-ditch running along the base. Immediately below us, so as to appear quite flat from this height, is the morainic floor of Kulikulan, specked with a multitude of black tree-dots among which the tiny lakes flash merrily like so many pieces of broken glass. To the north is Vashantagh, a long embankment of slanting masonry, sending round to our stand-point the curve of its main spur. Further west are dry ridges hurrying away to the wide expanse of Turkestan. Into the eastern distance retires a row of black towers riding, as it were, on an ice-clad crest which recedes towards the headwaters of the Yagnob. Down there, in the far haze, a fearful monster lurks, a crocodile pointing straight up to heaven its gigantic, tapering snout, with jaws slightly asunder.

Studying the Russian 10-verst map (1 : 420,000), we find a good record of everything that has some practical interest: cultivated valleys, villages and hamlets, roads, bridges, and passes. It is, in fact, a splendid and valuable piece of work, covering, as it does, thousands of square miles of rough country. A very large staff of topographical officers have been busy for many years, with the result that we have before us a complete survey of the main features between Fort Pamir and Bokhara city. The realm of ice and snow they have filled in with a few located summits surrounded by caterpillars of uncertain contortions, and dotted lines. That is not a fault, for such detail had to stand back in view of more pressing needs, and has been left to the mountaineer. Even time permitting, the region of glaciers and forbidding rocks would have received but scant attention owing to the want of experienced climbers. Deep down in the heart of every true Russian there is a curious, unconscious hatred of the mountain, which is not surprising when we consider the fact that one can walk through his country for 1700 miles without finding a place to fall down from. It would be nonsense to call it want of courage or laziness; it is a question of training and surroundings. The officers of the Tsar have ridden and walked and mapped in storm, rain and desert sun, and they will certainly begin to survey the glaciers as soon as the still young Russian Alpine Club has carried its mission into the army. Moreover Russian officials, nearly always accompanied by a mounted escort, have more or less suffered from the *idée fixe* of the Cossack. Where the Cossack fails no one can penetrate. Unwilling to separate from his animal the doughty horseman, almost incapable on foot, soon comes to a standstill among the mountains. Wherever he says that progress is impossible one believes him. Now the Cossack is a splendid fellow, willing and plucky, but he hates walking, and off his saddle he is only half a Centaur. The courage is still in him, but not the power or experience demanded by all undertakings among the high regions. My own map does not claim to be a marvel of accuracy; it only gives the most important of the higher ridges, and its chief interest lies in the high valley of the Pasrud sources, never before visited by Europeans.

As to names, I have adhered to those of the Russian sheet. It mentions a big mountain called Chimtarga to the south of Chapdara, but owing to loss of time I have not been able to make sure of its existence, whereabouts and size. The name of Chapdara is therefore given to the peak whose position coincides with that of the ordnance, and Khontagh is the great mass which seems to have escaped the notice of the Russians. It is always better to keep to the names on the first good and comprehensive maps (government surveys), unless they be flagrantly unreasonable or misspelt. One must not listen too much to the inhabitants of different valleys, for in each the same mountain has another name, generally that of the pastures below it. As the names wander upstream, and as many valleys culminate in one elevation or one group of peaks, all the names arrive together on the top, where they begin fighting, so to speak, as soon as the topographer heaves in sight. Shepherds, hunters and guides only learn from the tourist the habit of choosing one and the same appellation for a mountain, no matter from which side it is seen. Topographic nomenclature is therefore more important to us, the tourists, travellers, writers, and users of written records.

To the native his haunts are surface places, not points. Only mathematical surveyors, European science and exactitude, strive after dots pricked into the landscape, while the records of sport also require fixed points, determining arrival at a goal, apart from the general Western craving for finality. Thus the appearance of the surveyor makes things "come to a point." Our sheets of white paper with their limited space are probably at the bottom of all this desire for confinement. For everything beyond his immediate neighbourhood the native has only vague names accompanied by a large sweep of the hand. Hence many rivers have several names along their course, as instanced by the Voru (Map 110), which is only a short torrent and yet can muster three names: Kshtut river, Voru, and Archa-maidan. The greater example, to whit the Kizilsu-Surkhab-Vaksh trinity (of Karategin, etc.), would probably have dozens of names, if its banks were not inhabited by peoples of higher culture, traders, government officials and others making it important in commerce and politics. Lipski mentions three names for the Laudan pass, which is also called Fon (i.e. Fan) and Pasrud; while Krafft knew it as Lailak, and Kossiakov as Rasi-rabat. Many villagers know only one pass and one river, *the* pass and *the* river. Lipski throws an extremely interesting sidelight on these cases. According to him the Little

Russians are in the habit of saying, e.g., "Our Psiol is bigger than yours," thus using their specific name instead of the generic "river," just as if a Cockney were to say to a Parisian, "Our Thames is bigger than yours." He rightly remarks that the Londons and Yorks of America might almost be looked upon as generic equivalents for "town." Certainly a point worth remembering in the psychology of peasant geography.

On a map, which is seen from above, from one direction only, a point is an absolute entity, while the horizontal etymological relativity of a point can only be thrashed out in a book. Perhaps some day the author of a map for tourists will print on the margin the most important facts in the history of the map, such as previous maps, etymology, doubtful points, changed names, references to books, etc. The literature of civilised nations is so complicated that traditional names are havens of rest from which to survey the ocean of undefined spaces; they are the fixed points enabling us to trace general outlines. A striking case is that of the Central Caucasus where Dykhtau has been changed into Koshtantau, and Koshtantau into Dykhtau, so that I can only remember which is which when I have books at hand and can spend at least half an hour over the puzzle. This was done with the estimable but ludicrously belated ambition to satisfy the science of folklore. But a mere statement in a book, of native usage, would have sufficed, whereas it was quite out of date to change ideas, not at all erroneous topographically—for a name becomes a map symbol—which had obtained the sanction of a famous history of climbing. This change affects a generation of mountaineers, and the thousands of readers who identify the two peaks with the tragedy of Donkin, and the conquest of Mummery. Therefore, firstly, let us find out the surest names from the beginning, or give a number; and, secondly, let not the newcomer be too eager to "correct" well-established landmarks of recognised pioneers. If the pioneers know their duty to what is beautiful in geography, they will avoid the bad taste of enlarging the dreary topographic edition of *Who's Who*, or remind themselves of that Gargantuan *testimonium paupertatis*, a list of American towns. Even in uninhabited regions the language of the nearest tribes might be used to coin short and pithy fancy names which do not clash with local colour.

With our view from the Laudan pass we may combine a short sketch of the geology of the district. I quote from the report of my dear friend Albrecht von Krafft, whose death is still lamented by all who knew him. He passed through here on a rapid journey from Karatagh to Kshtut, and was only able to gather some of the more prominent facts. To his vivid description of the stately summits I owed my resolve to visit the Fan district. He says, " The northern slope of the Hazrat-sultan alps is far richer in sedimentary formations than the southern side. Near Saratagh begin the red and grey dolomite, the black limestone and calcareous limestone, with a thickness of over 3000 feet. These masses of dolomite and lime extend to the northern foot of the chain, and all the great peaks are formed out of them. Between the stocks of dolomite, the substratum, composed of phyllites, is occasionally laid bare. Dislocations traverse the region, so that the lower horizon of the limestones is found at different heights. Marguzar is in the midst of a dolomitic landscape, whereas the Laudan pass cuts into phyllites."

Next day, on July 29, we made an attempt on Vashantagh which offers a broad flank to the south where lay our camp at the mouth of a short ravine, opening out above. The appearance of Vashantagh from this side is that of an Alpine limestone peak with a gentle slope composed of smooth slabs, long screes, and a noduled network of snow patches. A couple of hundred feet below the summit we were just tackling a chimney, when I felt a smart tap on my head, caused by a small stone dislodged by the rope. Being occupied at the time, I paid no heed, but a few moments later I wanted to rub the sore place and shifted my cap. The result was a perfect waterfall of blood, about half a pint of which had collected under the cap. This was rather alarming, and I shouted to those above to pull me up quickly. The order was obeyed with laudable promptness, but on the way the open artery at my temple spurted merrily, and I was wondering how long it could go on at that rate. Luckily the task of stopping the flow proved easy to Albert's first-aid experience. A small stone was pressed against the artery with a bandage and I was able to walk down. The shock had however sufficiently unnerved us to give up our quest, especially as there were a few difficulties in front. Several days later we also tried to scale Khontagh, but returned at the foot of a long and smooth chimney leading up from the south-western ridge to the summit slope. I was still somewhat weak from loss of blood, to which must be added a wholesome fear of falling stones, a risk eminently peculiar to untrodden mountains. The ascent of steep rocks is dangerous enough in the Alps and in distant lands should be avoided as much as possible.

Near Laudan camp we had our only real adventure. It was one afternoon when I had finished my work and changed for a lighter jacket the coat with six heavy and bulging pockets. After tea I took the photographic camera and, accompanied by my wife, climbed a thousand feet of grass slopes, in order to watch the glaciers crawling in their cold ravines and the sun blushing on the heights. Turning round a corner we saw Khontagh, the tower of blood. The tent had vanished from our view, a deep silence was about, disturbed by neither man nor beast.

I set up the apparatus, and for some minutes the world was lost to me. Later only, I seemed to remember as if distant and muffled sounds had struck my subconscious ear while my head was under the black cloth. Then I packed up, and as we were retracing our steps, voices began to attract our attention. I gave a cursory glance in their direction, for what else could those be, but inquisitive shepherds. We move on. Suddenly a raucous yell roots me to the spot; unwilling, a thought begins to shape itself: danger!—Danger?—Nonsense!—Who in these mountains would ever dare raise his hand against the mighty Uruss and his vengeance, the sotnia of Cossacks with death and rape and plunder in their wake?—The idea is absurd. Yet, all doubt is dispelled by the hoarse shouts of two ruffians bounding up the slope almost on all fours, with three savage dogs at their heels.

Thus troats the stag on the autumnal moor, and howls the jackal of his hunger in the night. These are not curious peasants, but gentlemen on business. I thought to myself that these fellows, if not careful, would conduct further inquiries from a graveyard, for had I not my revolver? My hand dived into my pocket and a sunny smile began to sketch itself upon my features. My hand went down, down, down, and my smile froze before it was quite born, for the pistol had been left in the other coat. My first word was short, yet wrapt within its pithy substance were all my feelings of anger and consternation.

Short as it was, it gave me time to pause and think, and thus it may have saved our lives. In order to sound the intentions of our interlocutors I raised my hand and lifted a stone. Immediately they swung their long and heavy staves and made ready for action, also with stones. Knowing that since David's time shepherds are good shots, I opened negociations, for against two men with clubs and three

114. Awaiting Judgment.

dogs I felt hopelessly outnumbered. At this moment I was seized with poignant regret for having once tossed into the waste-paper basket the alluring letter from a professor of jiu-jitsu.

I became very polite and diplomatic, for diplomacy is when you daren't strike. Swallowing my rage and humiliation, I called a hearty greeting, something in the how-do-you-do-, glad-to-meet-you-style.

They swung their sticks, aiming at my still bandaged head, and growled, "What are you doing here? give us the woman and your things." "Of course, my dears," I answered back, "I shall be perfectly delighted; but patience." I spoke as many words as I knew of their language and made engaging signs, all to gain time. With almost superhuman effort I kept cool, collecting my innermost thoughts, while outwardly I made merry with these blackguards. I made them sit down, gave them cigarettes, teaching them how to smoke them, emptied the contents of my pockets, unpacked the camera, and did many other things. All the while my thoughts wandered towards the ridge, sixty paces further distant, whence it would be possible to see the tent. In order to reach this goal of hope, I got up, took them by the arm and under various pretexts lured them onward for twenty steps or so. Then they became impatient, shouting "The woman, the woman," and threatened to crack my skull. "Do you not see the glow of our fires, down there, on this side? come let us go."—I saw, and said it was very nice, but my thoughts were in the other direction, where my camp stood, a thousand feet below.

After a polite interval I got up again, saying that we would now follow them, but by a roundabout way over the ridge, "for I cannot walk here, it is too steep, I am afraid." As long as I live I shall not forget those few paces, separating us from the view into my camp. A false word, a false movement,—let us not picture the consequences, for I was not alone.

At last we did reach the corner of the slope, and furtively I gazed into the valley. But here the patience of the brutes seemed at an end. Again the one raised his stout pastoral staff and, shouting "Khatun, Khatun," prepared to annihilate the Gordian knot of adverse energies centered in my brain. But then the spell was no more. I had made a gesture of defence, he had hesitated a second, and in that second he had lost. In that very moment his accomplice uttered a cry of warning; both turned and fled.

Looking down the slope I saw Albert coming up with his swinging stride. He had been strolling about in search of marmots, had by chance espied us on the ridge in doubtful company, had whirled his rifle and burst into a yodel that rent the air from Vashantagh to Chapdara.

The *deus ex machina*, the great barber of the close shave, is a favourite figure of the novelist, but that fraction of a second by which we were saved was about the keenest edge he can ever have put on. What if Albert had not obeyed the ruling passion of the Tyrolese, or if I had not set my mind upon that view round the corner?

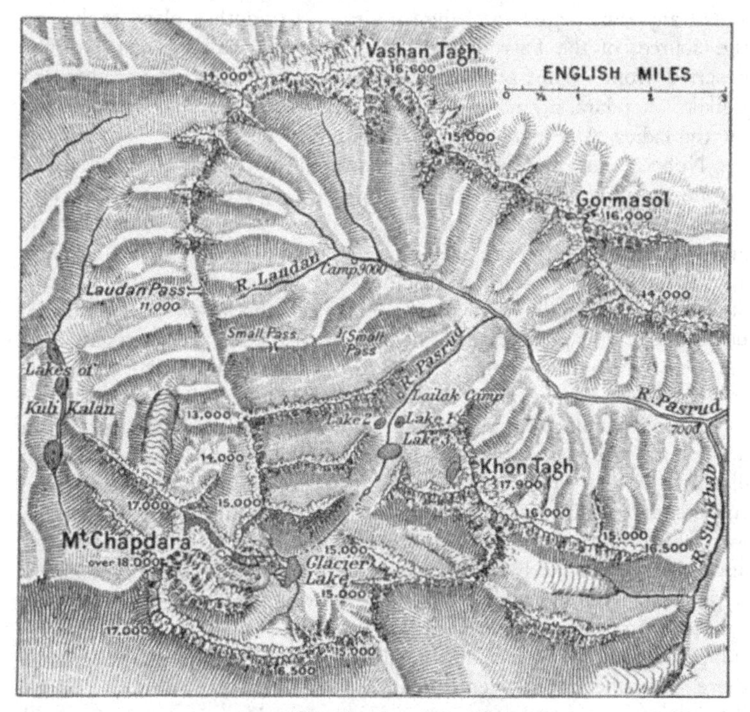

115. The Chapdara Mountains.

I was with difficulty prevented from snatching the rifle and letting fly after those rascals. But Albert said, he had only one cartridge as his intention had been to shoot only one marmot. But after I had given sufficient vent to my suppressed wrath I said, "To-morrow." He is the stronger man indeed, who, though beaten, can say "To-morrow."

At night, seeking the sleep of the just on my couch, I recalled the books of my youth, and the Red Indian, and his stake of torture. Sweet dreams of vengeance were balm upon my burning soul.

In the morning twilight I saw two figures against the skyline far above us. They vanished when our armed force began to move. I had anyhow wanted to change my camp to another place to-day, to the sources of the Pasrud in a wild and lonely valley which we had descried from a point of vantage. So I sent the caravan ahead to the Lailak Chapdara, myself starting on a punitive expedition, accompanied by the ladies, Albert, Mac, and Ishankul.

Nobody was, of course, to be found on the upper pastures, but, lying in the morning sun, I saw the woolly bales of a drowsy flock of fifty sheep. My soul began to smile. We drove down the valuable herd, and when we got into the deep gorge two scolding men appeared high above us, following at a safe distance. It was no longer the troating of the stag resounding through the forest, nor the jackal's hungry howl, but rather the yapping of dogs behind a fence. My soul was gay.

Our new camp we found at the summer huts called Lailak Chapdara on the upper Pasrud, about 8000 feet above sea-level. I sent a letter to the lower villages, and next day a Yuzbashi carried into our green and sunlit grove the gloss of his multi-coloured silk and the snowy dignity of his turban. Men were at once dispatched to catch the culprits, and in the meantime our friend began to eat mutton, for mutton was cheap to-day. One way of cooking these doubly delicious saddles and chops particularly struck my fancy. A roaring fire was made in a pit until stones thrown into it were glowing hot. Pieces of mutton wrapped in juniper twigs were placed into the hole, covered with more juniper, more hot stones, and finally a layer of earth. Six hours later there issued from this oven legs and chunks of tender mutton, in which the juiciness of Southdown commingled with the aroma of juniper and the sweetness of the spoils of war.

One man was caught, the one who had threatened me most. He still had his murderous staff which now stands in a corner of my study. The Yuzbashi bade him welcome with a heavy riding whip, which he drew across his face. He was then bound to a tree, in front of my tent, to let the vertical rays of the midday sun fall on his shaven head. Allah is great! He lets the sun shine on the just and on the unjust,

116. Moorland Scenery at Lailak Chapdara.

but the just may sit in the shade and praise the mercy of Allah. Our friend meditated deeply on the motives and necessities of human actions; abortive attempts and things left undone passed through his mind. The logical conclusion of his musings made him collapse more and more. It may also have been the sun and a certain void in his stomach, enhanced by the rich flavour that rose from the mutton-pit near by. I feel sure that next time he will strike the blow at once, and I shall always carry my pistol. The world is richer for two determined men.

The gloaming sank into the vale. Khontagh was red; like a gory stone of sacrifice, grand and terrible, it stood against the sunset's ruddy glow. Behind the tent of the Yuzbashi I hear a swishing sound, well known to our youth, which we hear, but cannot see, but which experience connects with the seat of pain. And I say to myself, "I hear what I cannot see, but the seat of pain is revealed, yet it is not with me; fulfilment is without me. Barakallah!"

A long wail now followed. But not the troating of the stag, nor the howl of the stinking jackal. My soul is filled with peace. Omin, habullah!

Lailak Chapdara is one of the rare sanctuaries which the water nymphs have been able to hide away in the barren stone pit of the gods of the Duab. The turfy glade, swampy in parts, is strewn with lichen-covered blocks, between which flow shallow brooks of clear, cold water amid banks and islands of beautiful flowers. Confined to a small space, and so prettily laid out, this charming landscape is just like a garden. The many little burns with their winding courses, tiny waterfalls and miniature lakes, with the tufts of long grasses and groups of leafy plants, all seem to be arranged with loving skill among the romantic boulders which build up the perspective of the general scheme, hiding new surprises round every corner (Fig. 116). Dark groves of juniper surround the clearing on the level, the last and lowest step of this morainic scenery, which rises towards the background in a series of steps, forming a graded transition to the rocky wastes beyond. Such was our moorland glen, alike unto the ice-born dales and fells of Caledonia. Why does nature wear this kindlier face? Because here the ground holds water, is soaked with water, not merely traversed by a ditch. Old moraines have been left across the valley in waves and

117. Khontagh and the First Lake.

heaps; great falls of rock from the precipices have added their share. In the bowls and pans thus formed the glacier water has collected, held back by the grit and clay in the moraines, or by the silt which it had itself carried down and deposited in the clefts of the lower dams of loose boulders.

The process works from bottom to top, thus explaining why we find the more finished landscape on the lowest step. The site of Lailak Chapdara must have been a shallow lake, gradually filling up, to be changed into a bog, and then luxuriant moor. It represents the last filtering of the glacier stream. But hardly has it left this place of purification, of peace, and beauty, than it hurries on to join the common rush, and, like a reckless youth, throws itself into the troubled career of the Surkhab which is full of mud.

Let us find the sources of the Pasrud. We go through a thick forest of juniper, birch and honeysuckle-tree, intertwined with clematis. Some of the sombre junipers are hung with festoons of lovely roses. Small buttercups blossom on the humid soil, while drier slopes produce *Ephedra equisetina* and *Artemisia*. After a climb of about 300 feet we discover the first lake (No. 1, Figs. 115, 117), a shallow, mossy tarn lying in deep shade below the sheer wall of Khontagh, garishly illuminated by the sinking sun. It receives and discharges visible streams. A hundred feet above it we meet three strong springs running down to supply the two brooks of the Lailak. Another 300 feet or so, and we stand on the grand piles of tumbled blocks forming the lower shore of the largest lake (No. 3, Figs. 115, 118). Overhung by high mountain walls and embedded in the cataract of titanic boulders it may well recall the sterner and wilder ones of the Scotch lochs or the highest Alpine lakes. When the light of midday brightens the gloom, we may revel in shades of blue and green, from the rich sapphire of the depths to the gay emerald of algae in the shallows near the bank.

To make our report quite accurate, let us also mention lake No. 2, which lies in a clean, stony bed held between the sharp-angled screes of a recent landslide and a hump of moraine. It has no visible outlet, but receives the left and weaker branch of an open escape from the great lake. The right prong of this forked stream only sees daylight for 100 feet, and then sinks away Attempting thus to trace the surface streams, we have obtained an insight into the complicated

118. Third Lake of the Pasrud Valley.

hide-and-seek of the water. Some of it leaves the big lake ostentatiously, diving away suddenly, or in the wider surface of a pool; some of it thrusts itself forward from underground.

Valleywards the large lake is fenced off by a ridge of rocks rising to 60 feet above its level. Towards the interior is a moraine dike 200 feet high and spotted with stunted juniper. Tracing an inlet at the eastern corner, we proceed on our search. Like a noble queen, the torchweed raises its golden sceptre above a host of tenderer plants already gone to rest, for we are now leaving behind the oozing sponge, that honeycomb of moisture between the third lake and Lailak camp. Behind the dike just mentioned the wanderer will march across a sort of irregular swelling, half a mile wide and long, which is also the width of the valley. This gentle rise is covered with numerous mounds or round heaps of astounding regularity. Small trees still flourish, and in a hollow near the screes on the right side is a colony of aspen. Close by shines a mirror, to which the stream, still above earth, will guide us. It gleams and glitters with the many hues and the brilliancy of a faultless opal, or like a bowl lined with mother-of-pearl and filled with pure water. This is not merely a poetic simile, but an attempt at very accurate description.

Onward now to reach the higher and blacker dumps of the retired ice. Passing over a shoot-cone from Khontagh, we imagine ourselves walking on billiard balls, so rounded are the edges of the stones, and that only from dry grinding. The stream, which is getting thinner and thinner from constant losses in the rubble, here issues from a spring amidst a litter of sharp and striated limestones of many tints. Then a low rampart, and then a very high one, dark, forbidding, crowned with a giant boulder that hails us from afar. Between these two morainic waves is a lake, a basin of thick, black ink in a hollow of ash-grey shingles. It is fed by a spurt issuing from the coarse rubbish heap in front. Here sprout a few long grasses, geranium, ephedra, crippled juniper, fading wolfsbane, and the stalks of rhubarb with their red, green-edged seed vessels.

Hence we do not see water for a long time, but have to climb patiently for many hours over what may be described as a monstrous outpouring of blocks disgorged into the narrow trench between the mountains, a veritable pandemonium of broken stone. To our left we obtain a glimpse of the south-eastern aspect of Khontagh, an interminable flight of moraines and screes without a single speck of snow. Nothing but loose rubble, on which next day we spent six hours of steady plodding. In a moister climate, even under the same latitude, there would surely be névé.

At last we reach our goal at a height of 12,000 feet. We stand on comparatively level ground in the focus of a magnificent panorama of

119. The Pasrud Glaciers with the Glacier Lake.

glaciated peaks. Before us a grey lake. Its opposite shore is a gravel delta covered with a network of countless channels. On our side is a miniature Norway, formed by the labyrinthine rocks of a terminal moraine that fall away into fjords of deep water. Fig. 119 shows the receding background of the amphitheatre. Its sides become steeper as they close in towards the entrance in which we stand. On the right the snout of the principal glacier is just visible, while still nearer to us, beyond the right rim of the picture rises one of the pillars flanking the valley gate. A small crescent of new frontal moraine below the snout of the inclined glacier speaks of recent retreat. Half of the lake is already filled in by the delta, a small example of the "sandr," as one calls the wash plains in front of Icelandic glaciers. It will be interesting to watch the rate at which this lake is being filled in. The fragments composing the fan are still very angular compared to the rounded pebbles cast out by the subglacial Zarafshan. Here the ice is short and steep affording little scope to water action under its body. While there are quite a number of large blocks near the end of the glacier, only two or three can be seen in the middle of the wash deposit, showing that heavier fragments are quickly worked under the surface.

From the gullies of Chapdara descend the twin glaciers, the sources of the Pasrud. Their waters unite in the glacier lake, secrete themselves in the block caves of its lower coast, and there begin their fanciful journey to the world below. The chaos of stone, of mountains firm and mountains shattered, is but thinly hung with that mantle of snow and lacework of ice which so royally clothes the shoulders of Swiss and Caucasian monarchs. Seeing what I see in this place, I am not so sure that a covering of snow has not a preservative effect, as contrasted with the weathering of the naked ribs of the Duab. The question of glacial erosion is, however, not one which can be answered in a general way. All depends upon the formation of the ground determining the shape of the deposit of solid precipitation, and the lines leading to points of concentrated dynamic agency. Very roughly one might say that névé preserves, glaciers mould or scoop, water destroys.

But these are second thoughts, the outcome of reflection which compares, not of immediate vision. This means that the effect of climate on the scenery begins to pale before a wider influence. Vegetation we do not expect in any case; the most familiar factor of comparison is ruled out, and that makes the lofty mountains of the world akin. Two things only are seen—rock and snow. Therefore the landscape is alpine; a cloudless day in a high limestone range.

We know that the great snow domes as seen from the valley present the same appearance at all seasons, and if we let our thoughts soar higher still, to regions where no human foot has gone, or higher yet, where even mountains but aspire, there may be said to be a realm of air beyond all climates and all seasons, like the bottom of the sea.

120. Boulderclay of the Iskander Valley.

There the shifting scenes of desert and garden, of the battle of the stones and of the wind-borne crystal, fade and merge into the cold of the outer sphere.

We left Lailak Chapdara on the 4th of August. It had been raining during the night and fitful drizzles accompanied us to the meeting of the rivers where the crystal clearness of the Pasrud was soiled by the loamy swell of the Surkhab. Humanity ever feels recalled to deeper meanings at cross-roads and partings of the ways. Accordingly this fork boasted of a rag-tree in the shape of a fine juniper hung with thousands of shreds and tatters of all colours and of all degrees of cleanliness. Whereas the ibex diagrams only occur in the mountains, the rag-tree is found all over the Duab at spots that mark the progress of the journey by sundry associations, such as I have already mentioned: holiness, danger, relief, wonder, etc. A tree, more particularly, apart from being the nearest convenient peg, may itself be the object of veneration, as a giver of shade in a glaring desert, or having been planted by some departed saint. Whoever passes by, rips a piece from his shirt, or from the lining of his robes, and ties it to a branch. With most people it is a matter of habit, but many a pilgrim, trader, or childless woman utters a silent wish. The ragged poor are the most pious observers, for I have noticed that officials in new and untorn clothes do not add their mite to the ancient sacrifice. Schwarz thinks that it may be an Indo-germanic custom, drawing attention to the holy trees of the ancient Germans and to our decorated maypoles or Christmas trees.

We camped on the banks of the Surkhab, finding its water so full of muddy sediment as to be hardly drinkable, but none other was to be had. Rain was the chief cause, for next day the water was merely grey with glacier-milled rock scales. On the right or south bank of the river there is a big limestone mountain discharging a scree cone of most amazing dimensions. It is like a mountain of its own, like a volcano of regular shape. Measuring a mile and a half round the base its even surface runs together in one tapering swoop to the mouth of the feeding gully, 2000 feet above the bottom of the valley. Looking at it from a higher position, somewhere upstream, one appreciates even more the imposing proportions of this shoot-cone. Scree slopes are often a striking element of landscape, but this one dominates it. Maintaining a uniform angle of 35 degrees, the straight line of its profile cuts through the view from the high mountain tops at the back down to the depths of the foreground. The horizon of the sea or desert is also straight, but such bee-lines as this one, rising up towards the sky, are very rare in nature.

This mighty talus affords a vivid picture of mechanical disintegration so prevalent in continental climates. In the Duab, the region of the most potent surface destruction must be in the middle belt of the mountains where vegetation gives scant protection and steep gradients allow transport by gravity alone. Above the snowline the difference, as against the Alps, cannot be great, for while Alpine peaks have larger shields of protective névé, their rocks also receive more moisture, freezing in the fissures. The lowest belt of the Alps enjoys the benefits of vegetation which holds together the products of weathering by forming a living or self-renewing coat against the impact of the elements. In the Duab, on the other hand, it is the want of sufficient transport which causes accumulation of débris on the lowest slopes. Forms of physical extremes are best watched in extreme climates. A pronounced condition of cold and wet oceanic weather reigns in Spitsbergen, where low temperatures, liquid moisture, exposure and poorness of plant-life are combined in an efficient manner. Here the phenomena of denudation assume a high intensity. It is cold, yet warm enough to expose large slopes, in summer, to running water during the day and to blasting frost at night.

In search of the glacier we wandered past graceful birches overhanging the torrent and through clumps of juniper, many of them dead or weather-beaten, exposing the corkscrew fibres of their wood. Macerated bark, like old ship's hawsers, frayed and blanched, clung to the petrified muscle of their trunks. The brook of the main glacier proved to be clean. My wife and Albert, whom I sent into a side valley, reported that the dirty water came from a small glacier, whence it issued quite red, turning yellow later on the way. Returning to camp we found a visitor in the person of the Russian chief of the district, the Pristav of Panjikent. He rules over 45,000 people, and, what impressed us still more, over thousands of snowy mountains. Being on a tour of inspection he had heard of our whereabouts and was now awaiting us, surrounded by his staff of native officials, various volostnois, mingbashis, yuzbashis, onbashis, and aksakals. Most prominent among these was a portly, moon-faced Yassaul-begi, the Sart police-master of Panjikent. It was a very agreeable surprise, for the Pristav proved a boon companion as we sat talking round a blazing fire till late at night. Grigoriev took great interest in the welfare of his villagers and from his baggage he produced native-grown potatoes, the first fruit of his agricultural propaganda.

The Pristav also persuaded us to accompany him to Lake Iskander. On the way we inspected a native alum factory, where the mineral is treated in primitive retort chambers undistinguishable from the ordinary hut. Near the junction of the Fan I marvelled again at the rippled inset in the mountains of a piece of sandy desert. This sand, coming from the boulder clay and sandstones of the neighbourhood, is proof of selection by wind. Of all wind-borne matter it has remained nearest to its place of origin, being unable to rise out of the cauldron, as floating dust can do. A boragineous weed shows special preference for this locality. The old castle of Zarvadan reminds one of European or Caucasian mountain strongholds, surroundings and building material having caused a deviation from the usual type of Muhammadan architecture. These ruins, quite imposing in a country so devoid of rocky citadels, tower on a promontory 300 feet high and overhanging the Fan, so that a pebble drops into the river straight through the air. On the breast of this cape the swirling pools of older Fans have left their polished pot-hole marks. Even the sloping side, turned towards the sandy triangle, is so steep that the structure seems to hold on by friction alone. The walls are built of irregular blocks of the same red sandstone which composes the background, so that at shadowless noon it becomes almost invisible from a longer distance. Little cement shows between the joints, and the unstable impression is thereby heightened. The top layers often consist of crowns and patches of baked mud squares, evidently later attempts at repair. Its brick-red hue gives a somewhat sensational effect to this tumble-down and romantic citadel of mediaeval knighthood.

Between Khairanbet, where we rested under a noble karagach (Fig. 71), and the lake the valley of the Iskander-darya is like a quarry of pastel chalks, yellow and brown, with many dabs of red and green. The Iskander glacier has simply choked the whole place with ground moraine now rising 2000 feet or more above the stream, especially

121. Fluted Boulderclay.

on the left side. Rain has sculptured the gritty substance with a close and delicate network of innumerable grooves and flutings, but genuine earth-pillars do not occur, probably owing to an unfavourable arrangement of the boulders in the matrix (Figs. 120, 121). Here and there planes of inclined bedrock have been laid open. From above, the carved moraine overhangs them in the same way as séracs balance over the "hot plate" of an icefall; at the lower rim some moulded strips remain, like plaster ornaments glued to a stone slab. In places one notices a pure, white, limestone ground moraine sharply separated from a superimposed layer of darker material, probably a mixture with surface moraine or slope detritus from the sides. Everywhere among the Alai-Pamirs we come upon such huge quantities of till, constituting important elements of landscape, owing, partly, to their preservation, partly to the absence of grass or forest. In a humid climate the soft deposits get soaked, settling into gentler gradients, whereas here they break off steeply, revealing their geological nature by vertical exposures.

Looking at these drift banks one cannot help noticing their likeness to loess in general appearance, especially as regards colour, the yellow of the loess being, so to speak, the composite tint of all the rocks of the land. Boulder clay is just the thing that should most easily produce loess when the finer particles are carried away by wind, leaving behind the blocks, coarse fragments and sand. Undoubtedly such accumulations of ground moraine strike one as the last or nearest birthplace of the yellow earth. Although to-day the "aerial rock," where freshly fallen, is largely self-shifted and mixed with dust from all processes of denudation, we can well imagine how, at the end of a glacial surge, the enormous masses of drift were able to supply the largest amount of ready-made material.

Along the hillside and on the river we saw poplar, birch, willow, tamarisk, roses, caper shrubs, barberry, bladder-senna, meadow-sweet, sea-buckthorn and a bush of the sloe tribe. *Hippophaë* (buckthorn) with its rows of orange coloured berries is also a familiar sight on dry grit slopes of the Alps.

The last gorge, where the river descends in a waterfall discovered by Lipski, is circumvented by steep zigzags, made uncomfortable by stones detached under the feet of two and four-legged animals in front. Then one skirts the outlet, a regular canal of even width, already at

122. Lake Iskander.

the level of the lake. Iskander-kul lies in a cauldron of barren mountain flanks. Along the shore stand a few juniper trees of great size, their gaunt and tangled boughs covered with spare, moss-like leafage stood out against the water like the silhouettes of Japanese fancy. The delta of the inlet at the western end, however, proved an idyllic grove of poplars, willows and buckthorn, with meadows of various grasses, eyebright, orchids, sedge, horsetail, cinquefoil, clover, geranium, lousewort. Here gleamed a gorgeous tent on a plot of greensward, at the foot of a picturesque rockery of boulders, overgrown with lichens. From under this bubbled a powerful spring of limpid water. Behind us was a screen of trees, leaving an open prospect upon the lake (Fig. 122). After a long ride it was refreshing to see a camp pitched there and to find the Pristav's retainers already prepared for our reception. There was no waiting until Mac had unpacked our provision cases. Cakes of dried cream, milk, bread and tea were served the moment we had dismounted. Later, as darkness came, a fire was lighted. The customary sheep having bled for our wants, supper, consisting of excellent bouillon with potatoes, boiled mutton, and pillau was served; after partaking of which we retired to rest under the stars.

Next day, having escaped the eyes of curious Sarts, always on the look-out for enlarging their knowledge of the habits and customs of Europeans, we found a secluded spot by the lake, overgrown with tamarisk, where we had a delightful dip. The water was delicious and not in the least cold. The sky was flecked with soft, fleecy clouds, the lake an exquisite shade of eau-de-nil, the silver grey of willow and eleagnus contrasted with the darker green of the poplar, while the sun brought out in all their fulness the different tints of purple, red, green and grey of the mountains. Over all brooded a stillness broken only by the occasional neighing of our horses in their gambols with one another. I here report the experiences of my wife connected with this morning bath. "I believe I got through the operation of tubbing unseen, but when halfway through my dressing I discovered that a regular army of women from the neighbouring summer camp had assembled behind the trees and that Cenci (similarly engaged a little farther on) and I were the objects of a most intense scrutiny. Going up to them afterwards, we were immediately surrounded by a shouting

123. Looking towards Iskander-kul from the Saratagh Pass; Ephedra Steppe and Meanders.

and gesticulating crowd. Thinking it might please them to see the contents of my rucksack, I began showing them a few articles. On producing my needle-book, it was torn out of my hands, rifled of its contents and finally retained by one member of the party the rest of whom would probably lead her a life in consequence."

An old shoreline cut into the rocks runs round Iskander-kul at a height of about 200 feet above its present level. It is best seen on the northern side, but a careful study of Fig. 123 will reveal parts of it on the deeply ravined rocks in the upper portion of the picture, and at about the same height as the low saddle behind the water in the background. The saddle or dam just mentioned is, of course, the old moraine which ponded back the river from Saratagh, maintaining the lake at the level of its lowest outlet (a little below the portion of the dike visible in our photo) for a considerable time, so as to allow the accumulated water to make an impression upon the surrounding mountains. Then came a comparatively rapid fall of the liquid store. Perhaps an earthquake made a crack offering a convenient purchase to the stream for eroding its outlet more quickly, while to-day it has found a layer of lateral bedrock arresting it at this stage, for the time being. The various causes underlying the formation and changes of lake levels or shorelines in mountainous districts are fundamentally the same as those applying to river terraces. All the high lakes of the Duab are morainic or glacial. The lakes on the southern side of the Alps, the fjord lakes of Norway and Scotland, on the western coasts of North America and Patagonia are dammed up by moraines. But what we miss in the Duab as well as in the Caucasus are those great sheets of water for which Switzerland is so famous. Speaking in a general way this phenomenon is due to the fact that the old Duab glaciers did not reach sufficiently far out into the plains.

On August 7 we walked up to the Saratagh pass. Among the boulders near our camp an animal akin to a vole disported itself. Its fur made it almost invisible, being shaded to the surroundings of greenish yellow with reddish tints thrown in. Two marmots also greeted us, standing upright at their door like a couple of penguins. Beyond the influence of the lake delta we entered a new speciality of the steppe, never described before, as far as my knowledge goes. It is an ephedra-steppe entirely composed of *Ephedra procera* and well shown as a large,

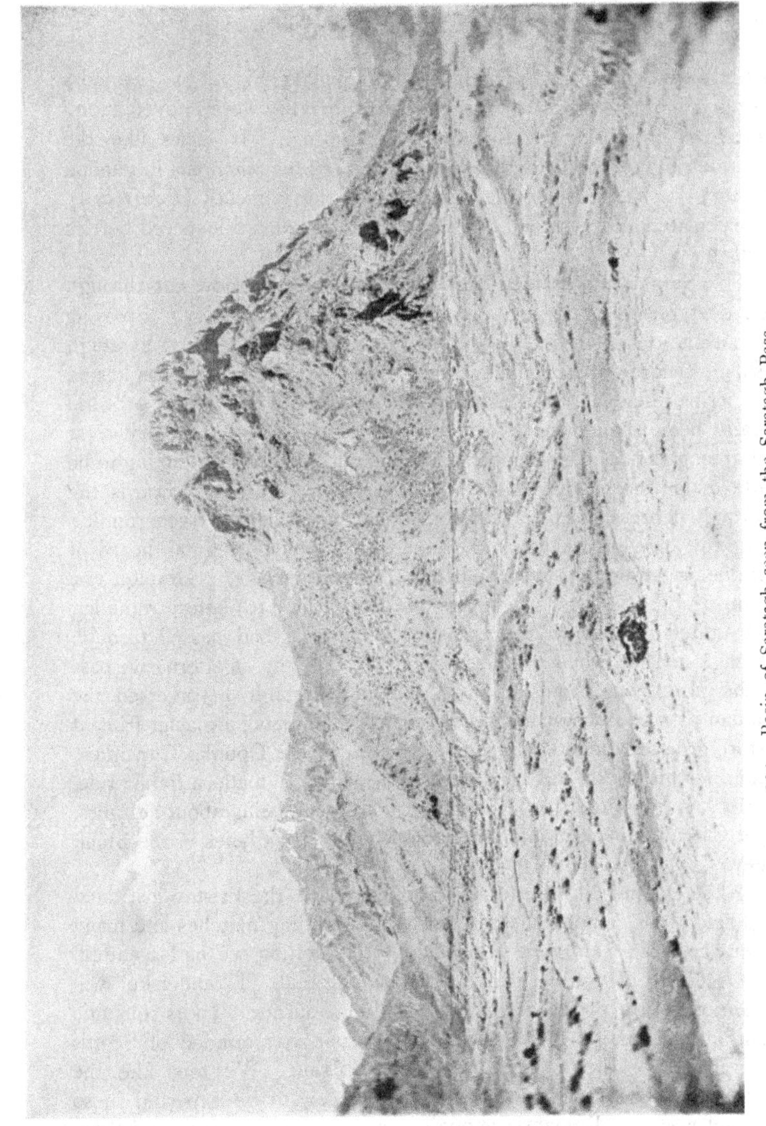

124. Basin of Saratagh seen from the Saratagh Pass.

spotted surface, in our view from the Saratagh pass (Fig. 123). Ephedra is a plant closely related to the yew and not very far removed (comparatively) from the cryptogamous Equisetinae. It looks like the prostrate pine of the Alps (*Latschenkiefer*, *Pinus montana*) in general bearing; its twigs are similar to those of the horsetail (*Equisetum*), i.e. articulated and sheathed like the tail of a lizard, while its red berries recall the yew.

The Saratagh pass is a dam of frontal moraine now cut through by a deep gorge on the left or northern side of the valley. From its summit one obtains a fine view of both the Iskander-kul and Saratagh basins. Before reaching lake Iskander the Saratagh-darya has leisure to describe a few languid meanders (Fig. 123), and one of the arcs has already been abandoned, clearly marked by the crescent of dry river bed among the ephedra-steppe. Another cut-off is just beginning to be made under the shadowed rocks to the right. Turning towards the Saratagh depression (Fig. 124) we notice an old lake-terrace running along the base of the gaunt limestone peak, while irregular heaps of moraine are straggling over the floor of the former lake. Saratagh was the meeting-place of many glaciers (cf. Map, Fig. 110) before ramming their united bulk down the Iskander trough. I had passed through Saratagh in 1896 when crossing the Hissar range by an alternative pass to the Mura, *via* Timur-dera-kul. From Saratagh I traversed the Dukdan pass to Archa-maidan (Place of the Junipers), a glacier fringed pasture ground studded with juniper groves, where Douglas Carruthers hunted for birds in 1908. From Archa-maidan I made a forced ride, in one day, down the Voru river and to Panjikent, being about 40 miles, while Carruthers varied his route along the string of lakes in the Shink valley.

In the evening, in front of a magnificent fire, the Pristav instituted Highland games among the natives, when wrestling matches and many clever acrobatic feats were performed. For setting we had a natural stage scenery of fitfully illuminated background. Iskander-kul will remain one of the pleasantest memories of our journey. In its romantic situation, the beauty of its sylvan glades, one is reminded of Scott's description of Ellen's Isle and Loch Katrine. We too, like the mysterious stranger, found a hospitality scarcely to be expected in so secluded a spot. Romantic enough were our surroundings amid these gaily dressed Orientals whose costume harmonises so well with the landscape and yet carries one back to a vanished world.

Before quitting our host's company we were treated to an exhibition of native pilgrimage, as instructive as it was amusing. In one of the side valleys of the Iskander is a cave with the remains of Khoja Isak and as this is a very holy place much resorted to by pious Muhammadans, the Pristav suggested our paying it a visit. "Come, let us enjoy ourselves by looking at the saint's skull," he said. His Yassaul-begi, weighed down by the sins of fleshly indulgence, and the Volostnoi, a haggard fanatic, as well as other dignitaries, were also keen on the expedition, so we set off. The valley of Makshevat well cultivated in its lower portion, fairly barren higher up, presented no exceptional features. The névés of great mountains were visible in the background. At the village we engaged six guides and guardians of the holy shrine. One of them, a man with a teapot in his hand, walked so fast throughout the five miles, as to be ahead of our horses all the time. The ante-room to the sanctuary was represented by a hut near a tree and spring. High above us on the face of a precipitous mountain we saw the black rift of the cave, the approach lying over a smooth desk of limestone slabs which sloped away to a cliff with a drop of several hundred feet. At one side the cliff lowered itself down to a corner where a series of easy chimneys led up to the slanting roof. Here the preparations of the Faithful began. They took off their boots and socks and washed their feet with water taken in a jar. The traverse was exceedingly slippery, all the more so as thousands of pilgrims had polished the stone, now wet with rain, to make matters worse; so we also discarded our hobnailed boots, keeping on the stockings. At first we wished to rely upon our own climbing faculties, but being irritated by the presence of so many people, we entered into the spirit of the scene and submitted to the same process with the rest. Like the Arabs of the Pyramids the guides seized the travellers in front and behind, pulling and pushing them over the ground at a rapid rate. Then, just under the mouth of the cavern, came a place the like of which I have never seen, one of those things unduplicated in the world. Near a pole hung with globe and yak-tail, swaying in the wind, a wedge-shaped balcony of stones had been built, so as to make a square and level platform on the slanting rock. Below there was a void, and our vision floated free into the distance of the lower valley. Here the pilgrims knelt, their faces turned towards the open, their souls seeking Mecca across the yawning abyss. They raised a doleful and dismal chant, that lifted itself mournfully into space. With those weird surroundings and the gloom of pregnant clouds this made a scene of lugubrious solemnity that for ever remains impressed upon my mind. The last ascent to the sepulchre was by a difficult and slippery chimney, where a fall would have had deadly meaning. One had to wriggle up like an eel with hardly anything to take hold of. In order to save time and watch the spectacle I remained behind, Albert securing the ladies with a rope. The others were helped up by the guides who, of course, knew every step and movement blindfold. All the same it was exciting, not to say uncanny, to watch two men, unroped, shoving a load like the Yassaul-begi up a steep chimney which ended in slabs and then nothingness. "Allah-i-allah, yo-ho-allah" they heaved away and shoved that portly personage to the goal of his penance. The spidery Volostnoi, who is all arms and legs, made shorter work of the climb and, once in the cave, indulged in his devotions with fanatical fervour. We did not approach until all had finished with their prayers, and I think the more intolerantly disposed would gladly have excluded us altogether.

The holy relics consist of a skeleton half buried in rubble, with some hair on the skull and a few muscles like brown ivory on his back. The so-called cave is really a chasm or fault in the mountain, closed at the top, and does not seem to go very far into the underworld. A Russian savant has written a report with an exact description of this gruesome calvary, but beyond various legends there is no historical evidence to lay hold of. Maybe the remains are those of a hermit who lived and died up here.

It was impossible not to mark the infinite satisfaction with which the Yassaul-begi regarded the accomplishment of his pilgrimage. He has the reputation of being a great Don Juan—which is possible in the Russian Duab—and the joys of the table, and perhaps, the bottle, are written large over his ample frame. So he was probably thinking of the indulgence such a feat would purchase, possibly for life and the crowning apoplexy.

Next day, at the meeting of the rivers, we parted from our genial Pristav who went into Yagnob, we ourselves descending to Varziminar.

CHAPTER XIII

TO GARM AND THE MOUNTAINS OF PETER THE GREAT

On the 27th of August, we took leave of the vale of the Zarafshan so similar in many respects to that of Hunza. It would be a delightful place to live in and second to none as a health resort for the European population of Turkestan. With a scenery sometimes verging on the sensational it offers fresh air, good water, shady gardens and many opportunities for vigorous exercise, and fair sport may even be had among the mountains.

The ravine to the Pakshif pass is comparatively short and steep. On the right slopes we saw an enormous mudspate divided into many branches. Some of these had frayed out gradually into the most delicate of welted furrows sketched on hard turf with the last and finest of sediment. Framed in the shoulders of the outgoing valley Yangi-sabak reared its massive head above the Turkestan range. The Russian map gives it a height of 20,000 feet. All loads had to be carried for the last 200 feet of vertical height separating us from the top of the pass which proved most difficult at this time of the year, so much of the snow having withdrawn from the saddle. How the horses ever got across, even barebacked, still remains a mystery to me. Very steep névé and a staircase cut into clear ice were bad enough, but then came a traverse over a rock slab, where the animals had to place their feet on a tiny ledge offering brief support for a spring of about two yards. A longer hiatus would have made the thing altogether impossible, for horses are not chamois, and many a man unaccustomed to hill-work would have refused at first. It took four hours to get the caravan over this last bit, and the battle was not fought without deep gashes or bruises for man and beast.

To the south descended a long slope to a sombre valley situated between the main divide, on which we stood, and a snowy spur, whence crawled a fine twin glacier. This, the Bodravak glacier, had retreated from the flat bottom of the typical, U-shaped valley and was now confined to its highest glacial step. Its two broad arms united a few hundred feet above the foot of the declivity, proceeding only a very short distance into the level section of the valley. The trough recently occupied showed a distinct waste-mark to about a mile from the snout. Shrinkage had even affected the Bodravak peak up to its very top. Robbed of its support the triangular ice field of the summit

125. Gorif.

had settled and slipped off in several places. The valley turns down steeply towards the south, descending to Gorif by a series of glacial steps, as far as I could make out in the failing light. Its upper portion was filled with avalanche snow, already firmly settled and traversed by considerable crevasses. At a height of about 9000 feet we rode through a full-grown birchwood, having for underscrub the leaf clusters of an umbelliferous plant which, in the dusk, gave one the impression of bracken in the glens at home.

126. The Mazar at Gorif.

Late in the evening we arrived at Gorif. The cupoles of its trees and the shafts of its tall poplars loomed through the dark, while high above, bathed in pale moonlight, rose a gaunt mountain shape seamed with ice. Wood, being plentiful, is more liberally used in architecture (Fig. 125). Wattle-work screens are sometimes used for outer walls, while the strong roofs can bear the weight of flat stones covered with earth. There are magnificent mulberry, walnut and other trees, among them a poplar measuring four yards round the girth. I was instinctively reminded of pictures of village scenery in certain parts of the Himalayas. A similarity of habitus, not easily expressed in words, made me feel as if some distant kinship was revealed, more especially in the quaint mazar (Fig. 126) decorated with ibex horns. Curiously enough the mast from which hangs the usual yaktail is not one-armed, like a gibbet, but has a regular cross-piece. This looks exactly as if the Christian emblem had been raised aloft. Horns of rams and goats, wild or domesticated, are the "regular thing" for holy graves. Right at the bottom of this cult probably lies the veneration of Eastern peoples for their most important cattle, sheep and goats, the givers of meat and wool. The horn is also the symbol of masculinity (hence an object of adoration for barren women) and subsequently of greatness, power and sanctity. One has only to think of the exalted horn, the sacrifice of rams, the horns of the altar, the horn of Alexander the Great, and hundreds of other references, to trace the emblem through Oriental life, especially Semitic. As a denizen of the awe-inspiring mountains and owing to the difficulty of pursuit, the ibex enjoys a special reputation, its horns being even found on the sepulchres of the plains. Evidently the hunter of the hills makes a sacrifice at the nearest shrine of the horns of his quarry. Needless to say these proud ornaments of the mazar at Gorif and the rude ibex figures carved in rock are connected with the same fundamental religious idea which is also linked with the wayfarer's offering to rag-trees. This can be illustrated by an extract from Lynch's book on Armenia, where he says (p. 338, vol. II), "My Circassian has told me that there exists a *ziaret*, or place of pilgrimage, in the vicinity of this cairn (i.e. near the summit of Sipan, at a height of 13,000 feet). Curious, and half doubtful, I ask him to show me the spot, which he says is close by. What is my amazement when, opening out a slight hollow of the snowy surface, we see before us a group of Mohammedan women, standing upon the ice with bare feet and ankles, and prostrating themselves before a pair of stag's horns. Indeed the antlers are so thickly covered with little bits of rag that it is impossible to say for certain to what species of animal they belonged. Stranger still is the fact that a band of women—I count twelve—should have risked their lives in this way."

Even in the Christian churches of Suanetia I have discovered horns and remains of burnt offering which betray a custom dating back to the times of Persian and Arab influence.

As witnessed by these trophies our native shikari is by no means a contemptible sportsman. Armed with his unwieldy matchlock, spiked staff in hand, he sets out to brave the hardships and dangers of the mountains, all the greater for his miserable outfit. The matchlock has a very small bore, taking a bullet not larger than a pea, and must not be considered a ridiculous weapon, clumsy as it seems. Of course it requires long preparation and cannot be used on moving objects, unless they are obliging enough to pass a certain point selected beforehand. But a hunter or sniper with a matchlock is generally a good shot. Here, as everywhere, the man who stands behind the gun forms part of the whole and, more than likely, the proportion of good stalkers and hitters-of-the-mark among mere owners of firearms is greater here than in Europe. Killing wildfowl, sitting of course, at a hundred yards is thought nothing of, and not few are the men who will lie in ambush for the redoubtable tiger.

From Gorif we wanted to explore the upper reaches of the Sorbukh river, but our negotiations for porters proved abortive. Their excuses were bad roads, ice and robbers. In reality they were either afraid of not getting paid, or of being uncashed by their officials, true to the old adage "robbery begins at home," or it was a symptom of that subtle Oriental suspicion which we can never hope to fathom. So we had to be content with a short reconnaissance instead of an investigation of the mysterious pass Kauerga marked on the ordnance map as leading over to the sources of the Yagnob. We started at 5 a.m. and, on the way, were often greeted by the white flowers of the hollyhock, the mother of our garden varieties. The road soon became impossible for horses, so we left them behind in company with several lazy dignitaries, and pushed on accompanied by a good-natured yokel. We reached the summit of a mountain about 11,500 feet high, on the left bank, obtaining a fine view of the river, as it wended its slightly sinuous course between an avenue of drought-browned ridges. For fifteen miles or more there extended to the Yagnob peaks, a vista of receding wings set in almost geometrical succession. From here one might have mapped the whole of the valley and probably such has been done judging from the big cairn which we found erected on the top. No native would ever take the trouble of building a column of victory, nor was this a place of pilgrimage. With a merciless sun we found the climb very tiring and there was no water for several hours, not even a patch of snow on the summit. The descent over hot, hard grit proved equally disagreeable and thirsty. The lower slopes were cloaked in a withered scrub of rose bushes and high weeds, that stood close together, yet each stem singly, as if nailed into the ground. Underneath this the groping feet tripped over rolling stones or were caught in the traps of marmot holes. But our troubles were rewarded by a miraculous spring welling from the torrid rock, and here we met our man awaiting us with a wooden platter of buttered cake obtained from a veiled shepherdess, whom we could see standing at the door of a hut. The welcome dish consisted of innumerable layers of baked dough, similar to puff-paste and very thin, richly interlarded with fat, like buttered toast. It was served piping hot. Its endearing taste and gliding richness made for speedy assimilation and our gratitude stood the test of a hundred yards up to the hut, where a few silver coins were pressed into a palm still bearing the seal of her handiwork.

When we came home the Karaul-begi, who ruled over the peasants of Gorif, was just holding court and mulcting a criminal into a fine of sixty tengas, because, as Mac delicately expressed it, "He has spent the night at a neighbouring village, and her father saw it." We were now in Bokharan territory, and when, next day, we descended to milder climes, we were met and royally welcomed by emissaries from the Beg of Karategin. We return to the delights of Bokhara, to clay palaces, gaudy robes, silver belts, melons, grapes, flies, hot winds, polite falsehoods, and those white turbans not a single one of which we have met in the marches of the Zarafshan. At Shingalich a house was ready for us and although it was still early in the afternoon, we had to stay here for the night. The higher the rank of a person the shorter the stages of his journey, which makes me glad to know that I am not the Amir who can only travel two miles per day. On the table stands the shining array of the dostarkhan or first offering, consisting of cold dishes, bread, eggs, fruit, sweets and tea, ready for the guest to satisfy his immediate needs. Much to the delight of the buzzing fly the dostarkhan remains on show throughout one's stay, and on leaving it is customary for one's servants to pocket what has been left over. The form of reception varies with the topography of the country. Among the high hills the aksakal, modestly clad, charges us at the most impressive gallop he can get out of his ribbed Rozinante, while one or two labourers from the field improvise the necessary adjutants, often mounted bareback. Under his roof we sit on felt mats bending our heads down to the plates on the ground, and here we frequently recognize old acquaintances from the last station by blue pencil marks left on the paper wrappers of Moscow sweets, which even prevent our servants from using the privilege of dostarkhan. Hardly have we left such a place than a courier overtakes us at full stretch, passing on with flapping saddle-bags. He carries weight, he rides a race, so that on arrival our eyes may rest tenderly upon the same little sticks of coloured sugar firmly glued to their paper cover and the enclosed scrip of printed doggerel. Up here the cheap confectionery is a decorative formality, which in our weakest moments cannot tempt us into experiments on internal chemistry. It is otherwise below, in the rich, open valleys or the plains. Already from afar we descry the glimmer of white turbans and the glittering of harness, where the silken minions wait. At their head a superb chamberlain the bearer of his master's greeting and invitation. Silver glistens on his foaming stallion's black and glossy coat. In our room we find the bounty of the land ranged in dozens of plates upon a table-cloth of snowy muslin, and fanning servants try to keep away the flies, so that we may eat and drink in peace. An odour of refinement is wafted from the steaming dishes brought out of the kitchen and tender bits of mutton are buried in mountains of seasoned rice. Huge sycamores, reflected in a silent tank, vault their canopy over the whiffs of the post-prandial cigarette.

Such are the outward and visible signs of a hospitality which, through constant delay, can be as irksome as it is pleasing and comfortable. Apart from the painful necessity of liberal tips the dream of Oriental splendour is sometimes marred by an incongruous attempt at European get-up. I have already spoken of the tables which the midnight cat on its rambles will often cause to annihilate the space between the floor and the top with a loud crash. Of a piece with these is the red cotton often spread in the guest house instead of the native carpets. Soiled and torn in a very short time the red horror is

127. Balcony Road on the Surkhab.

enough to make a rag-tree weep with envy. In small doses such spells of luxurious life are extremely agreeable, more especially to the servants for whom they mean a holiday from the daily duties of camp life. Here the ordinary mortal may enjoy the evanescent pride of a royal progress, while realising to the full, how much a crowned head can see of true life beyond the swarm of satellites. We knew the inwardness of things and the lies behind the mask, yet it must be said that everything was done in such good style and graceful manner as to make one sincerely thankful. The cultivated Bokhariot is a polished gentleman

128. Siesta.

and why should we cavil at the fact that it is the nature of all polish to be on the surface.

At Shingalich there was also presented to us a Karaul-begi, a captain of the guards, who was henceforth to be our constant companion throughout the dominions of the Amir, so that no harm befall us. Our suite grew as we went, two or three men joining on at every village, swelling into a stately cavalcade that brushed past corners of sandblown rock and trotted along broad cornice-roads over the waves of the great Surkhab. At Garm, the capital of Karategin, princely quarters awaited us; charpoys of inviting breadth, covered with silk and satin, extended their softness to our not unwilling limbs. A formal visit had to be paid to the Beg who received us surrounded by a throng of resplendent courtiers. My coat and knickers showing many honourable scars and stains reminiscent of gallant battles with mountain side or sardine tin, I was at much loss to find a dress worthy of the occasion. But at the bottom of my valise I came upon a set of garments, unused so far and fleckless. They were of pliable, light yellow deerskin, a vest and the rest, such as one wears under one's public clothes in order to keep out the cold of arctic blasts. These made a very smart, tight-fitting uniform and with the addition of a red silk sash around my waist I looked like a dashing cavalry officer. The ladies said that I was simply sweet, but somehow they betrayed some difficulty in keeping their faces from twitching. I do not care to lie myself; that is what the dragoman is for, and so through him the Beg was informed that my best general's uniform had fallen into a river and that this was only my undress suit. As I afterwards learned through Mac, there was but one unanimous voice of admiration and respect.

Garm is only a glorified hamlet, with an insignificant bazar, and very sleepy. Having replenished our stores of rice, flour and barley, we set out to cross the Liulikharvi pass over the range of Peter the Great into the Khingob valley. Our road followed the left bank of the Surkhab whose waters were yellow and gurgling thickly. A landscape of loess is boxed into that of the higher mountains, while the geological core of our immediate surroundings reveals the frequent occurrence of sandstone by red talus slopes and river deltas. How bright the sunlight is one realises when putting on one's snow-goggles, which are a great comfort. Darkening the view but imperceptibly, they enable the eye, protected against dust and glare, to appreciate the most delicate touches of the scenery. Magnificent plane trees now became a feature of the gardens on our way.

At half-past five a dense fog of loess dust set in, as we were nearing Langari-sha where a tent stood ready. It must not be imagined that loess fog is the outcome of dust whirled up by a storm; on the contrary, I have seen it thickest in the midst of summer when not a breath was astir and a genuine pea-soup obscured everything beyond a circle of ten yards. When the sun has been baking the slopes and cliffs, ascending currents of hot air lift the finest particles of loess. Floating impalpably in the atmosphere they are very gently and slowly precipitated to the ground, while new matter is continually being produced at the sides of the valley, as long as the sun shines. At night all settles. Though I believe this explanation to be correct in a general way, there are particular collateral causes to be cleared up, for the phenomenon does not happen with unerring regularity day after day, although to our ordinary observation the weather conditions seem exactly the same on every one of these days. As far as I could see the zone of greatest density is not very high, sometimes a hundred feet, sometimes more, and marked off against an upper atmosphere of comparative clearness. Evidently there is concentration in the lower stratum, due to slower precipitation. The phenomenon is not a gross one that can be touched with one's hands, for the effect is chiefly optical, next to which comes that upon the sense of smell. The tangible physiological effect is very small, perhaps a light fluff on one's eyebrows, most of which is even due to the dust of the road stirred up by the traveller. Theories of atmospherical humidity and electricity may also have to be invoked for a satisfactory solution.

The sensation in our camp was that of an autumnal afternoon in northern lowlands. Around us were stunted willows with creviced trunks, partly deprived of their bark, sending up long, smooth shoots with the tender feathers of their leafage. In front of the tent a few Sart stools softened by wadded quilts and a low table, almost a bench, covered with printed cotton. Here we are leaning over our diaries by the light of a lantern and flickering candles. On the ground coloured felt mats are spread, forming an island in the pale meadow which loses itself towards a row of blurred trees. Near by a burn with reddish sandstone water winds through the battered turf. Beyond is the fire of our servants glinting behind a veil, that muffles the low voices of talking companies and the sounds of a dotar (guitar). Men come and vanish in the fog, passing in between the silhouettes of our horses, whom we can hear neighing and snorting, rubbing their sides against a tree or turning their impatient circles around the pegs. Sometimes an astonished cow stumbles through brittle and crackling weeds on its

129. Sumptuous Quarters.

way to the village whence the odour of the evening meal creeps through a thorny hedge. In the far distance a mountain whaleback sketches an uncertain outline overhung by the lemon lamp globe of the rising moon. But looking straight up through the dry mist we see the pale blue of a fading sky dotted with a few white clouds.

Next day we had a fine impression of the Surkhab valley, where the broad gravel plain of the river with its network of channels lies full length between two mountain chains. The summit range of Alai is set far back, but the sheer and snow-streaked crest of Mount Sarikaudal rises abruptly above the foothills. Through the gap of a lateral valley this bold sierra is best seen. Above the little yellow cubes and green trees of a village lies a belt of hill slopes, forming a broad middle distance. Behind this, in contrast to the foreground, the sudden upshot of gigantic precipice, crowned with ice. But however much the eye may cherish the peaceful hamlet or try its courage on the beetling crags, it will ever return to the fascination of the middle gradient, the perfect utterance of wave lines in dry Duabic hills. How wonderfully plastic is this flesh of the mountains, how soft are these bights and arches, yet how clear and hard is their surface. Unmarred by sudden rents or gashes all shapes are perfectly blended into each other, the bosses of rising tiers being bent into one continuous flow of outline. Deep water cuts are hidden from our view and wherever we see an edge it results from the dovetailing of a trough into a rounded back. The whole looks as if cast in a mould without undercuttings or wrinkles, without rocks or forest; a revelry of curves, undisturbed by chisel or tapestry. It might be a plaster model tinted a light drab, with patches of grey-green islands of weeds and yellow squares of cornfields. Below is the pedestal of great terraces sloping gently and then breaking off in vertical bluffs to the river plain, while up above is reared a crenelated wall of forbidding rock. Such is the type of duabic rolling hills, of the swell of the land crowned with the breakers of glaciated crests. Subject to infinite variations this is a general description of the bare hills and middle slopes of the Duab, and they, like nothing else, are the bearers of the country's characteristics, embodying its physical forces and expressing its climate.

On these barren undulations water has pursued a sculptured theme, unfolding its craft of modelling and relief. In the interior valleys, among the high mountains, the active element works on large lines as shown, for instance, by the titanic grass ridges and ravines of the upper Khingob. In such places there cannot be a strict repetition of shapes, because the rocky peaks above stretch their uncouth limbs into the middle inclines. But the further we get out towards the steppes and deserts of the plains, the more we discover ridges dissected into subdivided similarities (dichotomy) or rows of equalities (monotony), whereof Figs. 196 and 199 are good examples. It is in such places that the slanting rays of the declining sun fashion plastic maps of shaded contrasts. Towards the outside of the mountains there is also an increase of sharp features and carvings, without detriment, however, to the general scheme of billowy curvature. Everywhere the heavenly waters of the Duab work under the guidance of the same climate, but the results of their efforts vary according to local situation of the surfaces. From the interior towards the circumference the soil, formed by particles of disintegration, becomes finer and softer, while the slopes are more and more deprived of permanent vegetation. The nearer we approach the upper regions, the more frequently high mountains rise above the middle hills, influencing them from above, so as to turn them into middle slopes. Outside the hills and downs are individually perfect from head to foot, whereas inside we find them as subordinated steps and stages of the superimposed architecture of naked rock. The stony frame of the alpine summits and the active energy of their streams impress the moulded slopes from above; the openness of the plains and attraction by gravity operate from below. Hence the undulations or delicate sculptures of the middle slopes and hills represent a row of gradual transitions from the rough hewn and crystalline angularity of the crests to the flat surface of the plains.

At noon, on September 3, we reached Kalai-liabi-ob where the road to the Liulikharvi pass branches off. Lunch was served by the proprietor of a rich and comfortable mansion (Fig. 17) who, among other eatable things, regaled us with flat loaves three feet in diameter. The substance and shape of bread varies infinitely in the Duab, and if one were to make a special study of it one could locate the origin of each sample within very narrow limits. We have seen it in all colours from white to black, in all thicknesses from that of paper to two inches, and in all sizes from a watch to a cartwheel.

A haze had again settled over the Surkhab, treating us to the phantasmagoria of a bay on the English coast. The mountain landscape of the opposite bank is blotted out with the exception of a cliff rising from the alluvial plain and acting the part of a chalky promontory. All the rest is lost in a misty infinity, save the river flats at our feet, mimicking the ooze and rills of a foreshore at ebbtide. To make the illusion complete the road is flanked by kitchen gardens behind low walls of stone. In spite of the midday hour a certain coolness prevails,

130. Mount Sarikaudal.—Yughan and Rose Scrub in the Foreground.

perhaps due to an interception of heat rays by the floating particles of dust.

On the way to Ganishau, our night quarters before the pass, the veil lifted and there was revealed to us the overpowering spectacle of the rampart of Peter the Great, Sagunaki on the left, Sarikaudal on the right, the former a needle-pointed pyramid of sandstone, the other a serrated limestone ridge. Separated from us by a few loessy undulations they towered fiercely above the hilly steppe where camels graze and coloured dresses speck the powdery road. One of the hamlets which we passed was quite pink, ground, houses, and all, owing to the admixture of weathered sandstone to the loess. Projections of this rock crop out on the sides of the flat-soled valley by which we ascend, so that it looks like a fortress moat.

High piles of winter fodder were already stacked upon the roofs of Ganishau, whose inhabitants amuse us by refusing tengas, the legal tender of their country, and asking for Russian kopeks instead. Unless they suspected us of circulating false coin, their request was quite unreasonable in view of the fact that the Russian government has fixed a legal exchange of 15 kopeks for each tenga. The explanation of this preference for Russian money is, that the peasants of Karategin go to Ferghana in large numbers, to hire themselves out as labourers and servants, returning home with their savings. Until the sun disappeared we sat on a knoll, gazing at the red breast of Sagunaki, while above us a company of twenty-six kites had foregathered in an evening flight.

Ganishau is the highest village and here also the last trees were left behind, clumps of slender willows, like groups of bamboo. Not even juniper is visible on these bare, northern slopes. On the other hand we find large surfaces covered with "yughan and rose scrub," one of the most typical forms of vegetation of the Duab mountains (cf. Figs. 130, 172, 181). It is a thicket from three to four feet high, consisting of yughan (*Prangos pabularia*; seen in Fig. 181, left bottom corner) and wild rose in about equal quantities mixed with a number of other plants. Yughan is an umbelliferous plant of great importance as a winter store for cattle, and forming most of the conical hayricks so characteristic of the mountain villages. During harvest time drying bundles of yughan are lying about everywhere, exhaling a pungent, acrid aroma, by which one can scent the vicinity of human settlements from afar. When the dead leaves have been wetted by rain the sour smell is especially noticeable owing to fermentation. A piece of this weed rubbed between the fingers and held under the nose will at once recall scenes and places of the Duab. From amidst this tangled bush are thrust the isolated candelabra of asafoetida (*Ferula Asafoetida* and *F. Jaeschkeana*), from six to seven feet high. At this time of the year we only see the skeletons of this gigantic umbellifer, some of them still standing upright and undamaged, like the candlesticks of the temple, others felled by wind or roaming animals, reminding one of a clearing with blighted and broken fir-trees (Figs. 171, 181). The asafoetida has enormous roots growing for five years before producing the flowering stem, which dies about two months after its appearance together with the root. This plant yields a disagreeably smelling waxlike resin, a favourite drug of the Oriental pharmacopoeia, and, in olden times, also of Occidental apothecaries. A sister is the zumbul (*F. Sumbul*) of the plains, with a stem eight feet high, while *F. Narthex*, of Tibet, grows up to ten feet.

After long stretches of yughan and rose scrub, slopes of shingle and soft earth led up to the first snow patches where we found a number of sink holes, some of them containing pools of water. Their origin must be attributed to pressure and seepage from melting snow upon a porous soil. The northern ascent to Liulikharvi is easy throughout, sliding its serpentine loops over gentle ridges, where from time to time the summit of a sub-pass surprises us with fresh constellations of the view. Eastwards Sagunaki is a grey phantom in the blinding flood of sunlight. Before us a long, red frontage, sharply fluted with pleats of silver in the purple rock. The glacier, fed by gullies of amazing steepness has shrunk back from a bed between two dams of lateral moraine (Fig. 130). We can see how this glacier, during periods of growth and recession, has built itself a raised channel above the level of its surroundings. In this way it has gained length instead of splaying out, although there is plenty of room at the sides of the valley. Thus a striking analogy is offered to welted furrows and rivers (such as the Hoang Ho) with natural embankments high above the plain. The causes are the same in all these cases: a rapid forward movement, to prevent side pressure; and a great load of detritus coupled with as little water, or ice, as possible.

We are fortunately able to time the vagaries of the northern glacier of Sarikaudal, for in Lipski's book there is a photograph taken from almost the same standpoint, and showing that it was about a mile and a half longer in July 1897. According to the Russian's photograph it reached as far as the morainic bed is visible on my picture, but already in 1897 it shows a considerable waste-mark on the insides of the lateral dikes. In its then condition it also gives one the idea that the front half was beginning to detach itself by a waist, so as to be strung off and

131. Looking Eastward from Liulikharvi Peak.

left to wither, like the severed tentacle of a cuttle-fish. Referring to our record again, taken nine years later, it seems that a squat and rubble-smothered mass of "dead" or nearly dead ice is still extending to about half a mile from the present, vigorous and "live" snout. We dare not, however, make much chronological use of this glacier for dating the retreat of the Zarafshan and others, having to reckon with the likelihood that huge masses of ice and névé may slip off bodily from the precipitous flanks of Sarikaudal, thus leading to a sudden increase of the glacier. Moreover, with its elongated channel, it will translate comparatively small causes into wide fluctuations, like the thin column of a long thermometer.

The summit of the pass remains hidden to the last behind a shifting of the divide, while in front of us rises Liulikharvi peak (14,400 ft.) which lies immediately to the east of the saddle. We examine it curiously, having placed it on to-day's programme of our joys or sorrows, whatever Allah may think fit. Its various small glaciers have burrowed deep trenches into the friable rock. The last stage is a very gentle snowfield with the dirty and sunken groove of heavy traffic. Here we lead our panting horses to the top, a plateau formed by a sort of displacement of the main crest. Held between oblique spurs the rock waste is broadly spread, seeking the watershed with hesitation.

Sending the animals some distance down the other side of the pass, we began our attack upon Liulikharvi peak. Some crushed oyster-beds were seen in the limestone and various slabs showed a faint trace of Karren fluting. In places the screes were pressed so flat by snow as to resemble a macadam surface, and many of the larger stones were weathered into a scraping roughness, like that of a file continually catching the hobnails of our boots. The looser débris were often so sharp as to cut into the leather. An easy climb of two and a half hours brought us to the top. Only just below the snow cap we encountered a rotten and disagreeable knob, about sixty feet high and almost entirely composed of pink kaolin, requiring careful management and the cutting of steps. We stood midway between the two bold steeples of this, the tail ridge of the mighty range of Peter the Great, which begins at the Pamirs with a broad maze of ice mountains and tapers to an acute angle in the fork between the Surkhab and Khingob rivers. To the east we saw Sagunaki (Fig. 131), behind a pinnacle of similar shape, called Kamch. Sagunaki is approximately 18,000 feet high and owes its extremely fine point to the overlapping of steep strata, probably sandstone. Red colours predominate on this side, while towards the west grey proclaims the rule of the limestone pinnacle of Sarikaudal. On the Surkhab side the long mountain wall is only supported by a few short buttresses and wing-like counterforts, but southwards it encompasses with branching arms the basins and glacial feeders of the Farkikush river[1].

[1] See map in *Zeitschr. d. Ges. f. Erdkunde*, Berlin, 1907.

The mountains are coloured here on a grand scale, one of them being quite purple, with patches of luscious green grass kept fresh by water trickling from large strips of white snow. If it did not look gorgeous one would call it a shame that a big, grown-up mountain of 15,000 feet should paint in so outrageous a fashion. A tarn embedded in morainic wreckage seemed filled with viscid red lead, adding glacier blood to our list which already contains glacier milk, glacier ink and glacier soup. What would have been the object of greatest interest, the file of Alai, marching from west to east, was dissolved in a trembling

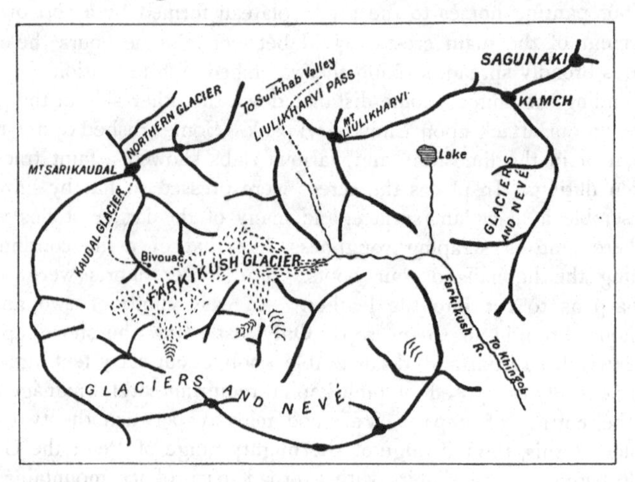

132. Plan of Sagunaki Mountains.

haze. For a long time we watched a couple of lämmergeiers playing and billing as they sailed past the slopes.

Running down a series of couloirs filled with fragments of soft slate, towards the Farkikush, we joined our horses far below the pass, and were soon engulfed in narrow gorges choked with avalanche snow, a very unwelcome contrast to the amiable openness of the Surkhab side. Knowing that water is rummaging under the white pavement makes one feel creepy down the spine, especially when broken portions of the vault unlock the roaring terrors of the underworld. It is not always possible to escape to the slimy rubble and slippery rocks of the banks, so that his conscience pricks the mountaineer, who allows his caravan to pass unroped. One hardly risks a peep into these holes where the red rush of the torrent is seen to vent its fury against the polished sandstone. Fortunately the snowy mass is firmly crammed into the gorge and sometimes so thick as to make inaudible the thunder of the glacier stream. In early summer, when the pass is first opened, the southern approach must also be very convenient, forming, then, an uninterrupted and easy road of snow, while the last travellers in late

133. Stratification and Cleavage.

autumn, have to pick their precarious way from side to side over treacherous snow bridges. Debouching into a wider part of the Farkikush valley we found a convenient camping site on a terrace covered with yughan scrub and crippled trees. Here a dozen peasants awaited us, sent up by the Amlakdar of the Khingob with offerings of bread and salt, butter, legs of mutton, eggs, omelettes, cake, and milk. Near this place, on the left bank, occurs a remarkable instance of stratification with joint planes and cleavage (Fig. 133). Fine cracks, absolutely straight like lines drawn on a sheet of paper, traverse the slabs parallel and at right angles to the strike. Into these fissures grass has grown, marking them off with greater clearness.

On the following day, having set our minds upon Sarikaudal, we rode to the upper reaches of the Farkikush. Once past the Liulikharvi ravine the valley floor became continuous, but also very dirty. We were on a regular avalanche glacier, as proved by the age of its mass, its crevasses, potholes (moulins) and moraines. It may be said to form a continuation to the neve glacier pushed out of the upper basin, but

134. Mudspates on the Snow.

not endowed with sufficient force to reach so far down. Hence we have the case of a valley glacier made directly out of snow and supplied at intervals. Longitudinal advance must be exceedingly slow or practically absent, because there is no steady shoving from behind, and a fresh block occurs in front every year from the avalanches wedged into the narrow gorge. Annual feeding maintains the level of its surface, so that increase or retreat would be paramountly expressed by changes of thickness instead of length. Throughout the whole stretch, from the lower snow arches to the fields of névé, I have never been able to tell where the avalanche glacier begins and the ice glacier ends, an almost hopeless task by reason of the enormous masses of surface rubbish. Probably this phenomenon corresponds to what A. Schuster calls the half-developed glaciers of Tibet.

Beginning at the mouth of the Liulikharvi glen, our progress becomes more and more a miniature edition of the Zarafshan glacier, for the back of the avalanche worm has its cones, with facets, thin crevasses and accumulations of detritus. Naturally the icy substance is very soft, the youngest layer being on top. Soon we notice a greater hardness in the ice peering through the mud here and there, and finding ourselves just opposite a lateral ravine from the right, we may assume that its small glacier, hidden under rubble, has penetrated here. For hours we struggle over a regular morainic bog with deep pools of mud. There are very few crevasses, many of them probably stopped by avalanches, so that the surface water is not drained off. Furthermore the avalanches carry much earth which, when melted out, forms impermeable dams and streaks by entering as a mortar between the rougher fragments of moraine. We zigzag through an indescribable labyrinth of hard snow patches, dry islands of stone, and swampy stretches, where the horses sink up to their knees. Sometimes a dike bursts and a tank of sludge is poured out in the shape of a small mudspate. It is a glacier wallowing in its own filth, a dirt glacier *par excellence.* After having passed the first right and left tributary gullies above Liulikharvi, we begin to emerge from the domain of great snowslides, the moraine becoming a little less sticky. Hearing the noise of the stream, we examine a place where the lowest trough of the valley narrows down to 150 feet and where we are in doubt if there is ice beneath us, if the main glacier squeezes through this gate, or not.

Then the prospect widens out to the desert of boulder moraine, half a mile broad, poured down by the southwestern névé basins. Grey masses of stone everywhere, with occasional lengthwise ridges and several knobs of rock, that once were sérac shoots, but which now have become peaceful islands owing to the shrinkage of the glacier. Among the rocks we saw a good deal of "pencilled" slate, so called because it is in the habit of falling asunder into thin, crooked sticks, somewhat resembling basaltic columns in miniature. Taking up a stone it will sometimes come undone like a bundle of matches. One of the most striking sights was offered by mudspates poured out over hard snow from an adjacent slope of soft screes (Fig. 134). These thick, ruddy snakes stood out like mounted specimens from a white table; they were experiments on a large scale, showing the shape and dynamics of the mud gush, unobscured by any transition from the moving material to that of the substratum. Linear growth by self-bedding could not be better illustrated.

As far as I could determine with a probability approaching certainty, the main glacier, or what should be the main glacier, is disjointed in a

135. Mount Sarikaudal with the Kaudal Glacier.

curious way. The right tributary ice stream opposite our bivouac (Fig. 132) runs some distance down into the main valley, but is not connected with the glacier coming behind it from the west and just missing it by a couple of hundred yards. Likewise the next lower glacier on the right curves into the trunk just escaping the end of its higher, westerly brother. Having built a wind screen at a spot selected for our bivouac, we went to reconnoitre our route for to-morrow's ascent, at the same time examining the Kaudal glacier (Fig. 135). We found its tongue in a remarkable state, clearly visible on account of its comparative cleanliness, for most of the surface rubble has fallen into clefts. The front part of the snout, sunken and much broken, is marked off by a step from the white field above. To the general cause of shrinkage we must add local conditions to explain the peculiar aspect of this glacier. The upper ice with the white edge rests upon a relatively level platform (probably ground-moraine) while below there is the sudden descent into a steep, isoclinal gutter. The glacier splits off in strips which, cracked and bent into hundreds of joints, settle down to the concave profile of the new bed, and, the remainder of the journey being short, a bulging snout can no longer be reconstituted. There is no terminal moraine on the slabs of rock whose isoclinal ridge meets the advancing glacier like a knife. On the outcropping side of the strata a terminal moraine has been dumped down in the shape of a cone over which the glacier, astride on the roof, sends a shorter flap. Here blocks of ice, bearing some resemblance to *penitentes*, can be seen standing on the screes, awaiting dissolution after having slowly detached themselves from the main mass.

On September 6 a bright moon saw us off from our sleeping place at 2.30 a.m. Quickly crossing a small spur between us and the Kaudal glacier we trod an excellent surface of hard snow, leading high up towards our goal (background of Fig. 135). These convenient slopes afforded a good grip to our crampons and we went at a rattling pace, favoured by bracing air and a physical condition unimpaired by the recent luxuries of the Surkhab valley. We reached the highest point (17,700 feet) at 8.15, after an hour's scramble over ledges and through gullies familiar to every climber of limestone peaks. The only difficult place was in a deep notch of the final ridge just below the summit. Here we had to surmount a knife-edge of snow, about 60 feet high, set upon that terrible northern precipice at which the wanderer looks with awe as he wends his way towards the Liulikharvi pass.

CHAPTER XIV

TUPCHEK AND THE ASCENT OF GREAT ACHIK

NEXT we were bound for the uplands of Tupchek by way of the Gardani-kaftar pass.

The range of Peter the Great is not exactly a single chain, but a complicated system of knots and ramifications between the Muksu and Surkhab on one side and the Khingob on the other. Nor does the main divide run along an uninterrupted crest (Fig. 136), for certain sections of the mountain roof are shifted against each other,

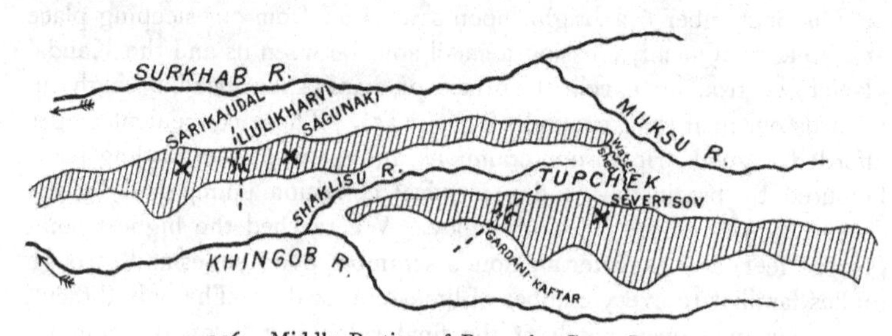

136. Middle Portion of Peter the Great Range.

being thus scarfed, or joined together like the mast and topmast of a ship. A notable instance occurs at Tupchek, where the watershed doubles across from Mount Severtsov to the Sagunaki chain, forming a hydrographical divide on the intervening valley plain of Tupchek. Hence, and as the Shaklisu valley is trackless in its lower reach, two passes have to be crossed here, one, the Gardani-kaftar, from the Khingob to Tupchek, and one of a choice (Kuliak or Yashilkul) from Tupchek to the Surkhab. On a small scale the same phenomenon is observed on the Liulikharvi pass.

The dove-grey waters of the Khingob swept past us as we rode towards Liangar which lies at the foot of the pass. Between the sandstone blocks strewing the terraces, were forests of withered thistles in which a horseman might hide himself. Liangar is famed for two things, its wonderful walnut trees standing with grave fulness of leaf on the level valley floor, and a natural grotto of white limestone shaped like a Gothic portal. Under the arch, which is about 200 feet high, we found a tiny spring surrounded by stone walls, and also many indications that it was a place of worship kept in good order. Hardly had we arrived at Liangar when there came a messenger with a letter from the Beg of Darwaz. He had done the journey from Kalaikhumb within the 24 hours, changing horses five times on the way, whereas we subsequently took three days for the same distance. In the gardens of the Khingob valley we saw a new kind of instrument for frightening birds. Thin laths of wood, decreasing in length each way from the middle were fastened to a stick so as to form a double Pontifical cross. Hung from a branch it girated and glittered in the faintest breeze.

Our ascent to Gardani-kaftar lies over a vast, gentle, endless incline whose folds and wrinkles are gathered at the pass. The torrent remains invisible in a gorge to the east (left) so that our route is not secreted in a trench, but winds openly among a hilly scenery of scarps, terraces, shallow troughs and morainic dikes set upon the massive flanks of the mountain. Along the margins of the lower fields milfoil, pink, scabious and chicory are in blossom. Then for hours through yughan and rose scrub, topped here and there by small junipers or maples and staked with the bleached limbs of asafoetida.

Mac and Albert were busy after marmots, getting off their horses whenever they saw one of the red little fellows sitting in front of his warren. Albert's score mounted up very quickly in this competition, because he shoots straight like a true Tyrolese. It is necessary to hit the animal in the head to prevent it from scurrying back into its lair. Besides their skins the alpine rodents had also to yield up their fat, in the camp cauldron, for marmot grease is an esteemed and expensive household remedy in the villages of Tyrol. At night the Kirghiz dogs were immensely grateful for the unexpected meat rations provided by the carcases. The Duab is the home of the red marmot (*Arctomis littledalei* and *A. littledalei flavinus*), whereas another species, the brown marmot (*A. centralis*) takes its place in the Ferghana mountains and the Western Tianshan. Carruthers, who has collected new species of marmots and discovered their distribution, says that the red marmot is the only representative of its genus all over the Pamirs, Bokhara, and indeed on all the mountains lying south-east of the Ferghana range and on the south side of this range itself. But immediately one passes north and east over this range the brown variety takes the place of the red. However, it is worthy of notice that on the Arpa pamir the red and brown marmots live together, as if the homes of the two just overlapped.

The notch of the pass pierces fields of névé at a height of 14,000 feet and represents the last convenient gap for pack animals, for now the chain becomes a menacing rampart of rock and ice continued into the heart of the Pamirs. We saw mighty peaks rising near us and during the descent every side valley revealed new giants on their silent guard, with glaciers crawling at their feet and rubble dumps kicked out beyond the latter reach. Although the longest, Gardani-kaftar is also the easiest pass we have crossed, the gradients and roads being easy on both sides. A few ups and downs have to be negociated, such as the climb from the Puli-sangin river (Fig. 137) to a terrace at the corner, thence a drop to the Zeriu-zamin, up again to the plateau of Little Tupchek, and again into the Shaklisu valley before one can attack the last ascent to the heights overlooking the Surkhab. The only check opposed to us was the Zeriu-zamin torrent, a boisterous child of the glaciers, unwilling to wear the fetters of a bridge, for beyond three fragments at different spots we found none. One of these we managed to repair, the result being a piece of engineering fearful and wonderful to behold, for as much water passed above it as below. Men secured by ropes carried over the loads, while the horses, pushed and pulled, struggled across by the skin of their teeth. Then we surmounted a steepish rise past a rockery where small, scraggy juniper bushes, like Japanese dwarf pines, were disposed among red blocks.

It was nearing sunset when we landed on the edge of a lonesome and treeless expanse, on the prairie of Little Tupchek. Out west a disk of pale silver was sinking fast, and we just caught a glimpse of the sharp

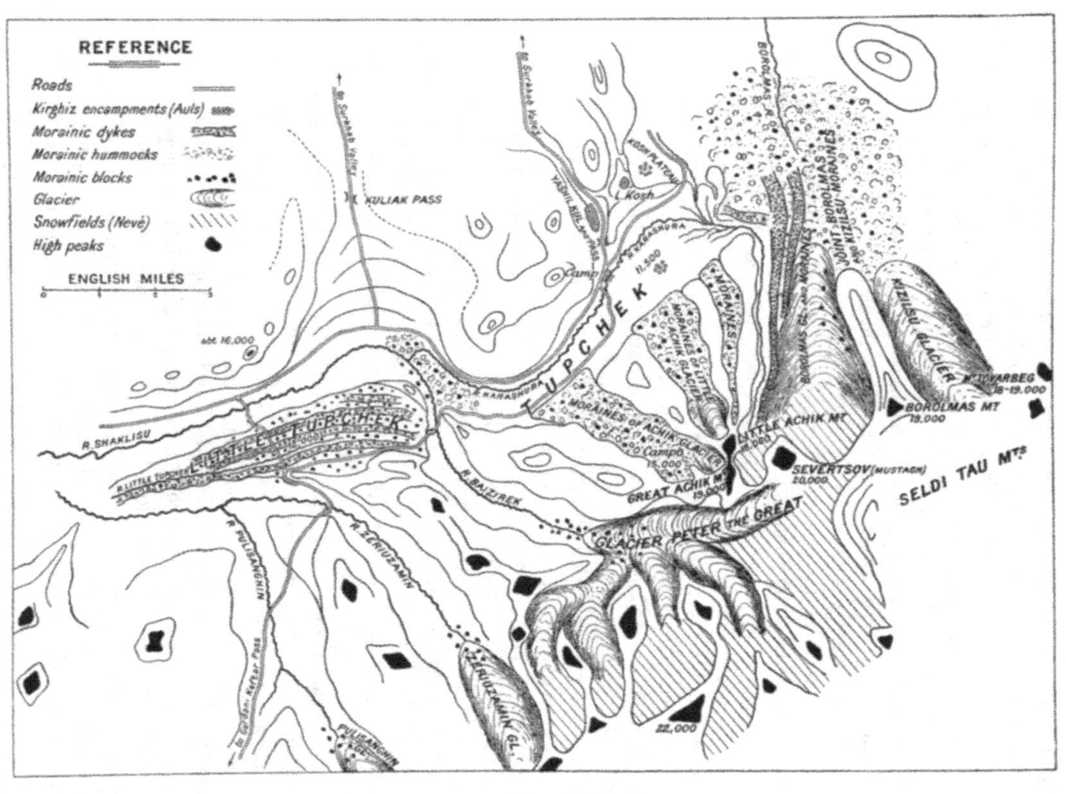

137. Plan of Tupchek.

fang of Sagunaki pointing upwards from a row of lesser teeth. As the horses were very tired we camped on the banks of the Little Tupchek stream, a tiny moorland brook in a coombe between lines of downs studded with erratic limestone boulders (Fig. 160). As I contemplated my surroundings and inhaled the crisp air, the idea struck me at once that this must be a pamir, a broad pasture land in a spacious setting of alpine crests. We passed a comfortable night behind a wind screen made of our baggage, awaking to a misty morning. Crossing several depressions between huge morainic hog-backs, we made towards the other brink overlooking Shaklisu. In the chill greyness this country of great rolling waves receding into the mist recalled the downs by the coast of a northern land. A slope of 1500 feet led down to the Shaklisu, and it was noticeable that over the edge trees began to appear again, among others a mountain ash with glaring red bark. At higher altitudes slopes become more favourable for plant-life than horizontal places, on which the snow remains longer and the wind blows harder. As we proceeded eastwards no bold mountains were visible with one exception—Mount Severtsov, gleaming like a crystal prism. At 3 p.m. we arrived in a Kirghiz aul on the Karashura plain or grazing district of Tupchek (Plan 137).

We are on a flat space wedged between the shunted western and eastern outruns of two ridges (Fig. 136). Looking towards the direction whence comes the Karashura brook, our eye dives into a wide and open prospect instead of knocking against the funnel of a valley head. The parting of the waters takes place somewhere near the eastern end of this floor, the divide (what Richthofen calls a crossing watershed) stepping over from the Borolmas mountains to the hills around Yashilkul.

The northerly margin of the Karashura pastures is formed by rounded hills, the last eminences of the Sagunaki range. Southwards a row of glacier peaks rises above a foreground of undulating slopes. These are the summits mentioned by Oshanin under the name of Tupchek. He saw them from the Surkhab valley, and the natives, upon inquiry, would, of course, call them after the important pastures at their foot. Geiger suggested that the highest should bear the name of Severtsov, one of the most deserving explorers of Turkestan. This baptism, once having been consummated, has to be upheld, and I have identified Mount Severtsov with the imposing pyramid of ice which dominates the others. Lipski calls it Mustagh, but as this simply means ice mountain, it is perhaps better that such an uncompromising determination should not figure on the map. True, the giving of personal names should be avoided, but once the thing has been done in honour of a great man, it cannot very well be cancelled.

West of Severtsov are the two Achiks (to the right, on Fig. 153); to the east follow Borolmas, Tovarbeg (Fig. 151) and a long line of

138. The Pamir of Tupchek (Karashura Plain); Looking Southwest.

other mountains, stretching away on the left bank of the Muksu in the direction of the Pamirs. This range, ending with Achik, is called Seldi-tau on the Russian map, but the topography still wants clearing up, for these summits are separated by deep valleys eaten far into the background. They are either the promontories of spurs from a main divide, or a more complicated arrangement prevails. From the Seldi-tau a great number of parallel glacier valleys debouch at right angles upon the Muksu. From west to east their names, as given by Lipski, are Borolmas, Kizilsu, Koshkul, Seldara, and Oshanin. I am here in a great quandary, having to decide between the names of the military map and those given by Lipski. The Russian explorer obtained his names from the natives, but so did the army surveyors, and a dozen future travellers would hear another dozen sets of names. "Oshanin Gl.(laciers)" is written on the 10-verst sheet across a region corresponding to the Borolmas, and as Oshanin had to remain in any case, it would have been better to leave it in its original position close to Mount Severtsov. Instead of Koshkul, Seldara and Oshanin (Lipski's) the previous names Kosh-sai, Sildi and Temirgandara, found on the map, might have served as well. They are the names of the rivers, it is true, but there is a sound rule enjoining that glaciers should be called after their streams as much as possible, especially when we have to do with native names. After all it is we, the Europeans, who wish to have a convenient and organic nomenclature for the purposes of practical topographic description. When we have a map, we travel by it, finding our way according to the drawn features and not by asking a Kirghiz where such and such a peak or glacier is. I cannot risk making confusion worse and therefore adhere to Lipski's choice, leaving the final decision to the ordnance bureau.

The lower portions of the parallel glaciers just mentioned, that is to say their sections comprised between the high peaks and the Muksu, are divided by very low ridges, and I suspect that in many cases their old moraines are merged towards their ends. Thus there would extend, as an easterly continuation to the flat plain of Tupchek (Karashura), and comprised between the outposts of Seldi-tau and the Muksu, a strip or belt of hummocky landscape of morainic bulges, alternating with low hill ridges. The right bank of the broad and massive flow of Borolmas is little higher than the greatest of the morainic humps. If this condition is repeated further eastwards, as I have reasons to assume, the situation will be very interesting from a glaciological point of view. That the Muksu valley was thickly choked with detritus from this compact array of glaciers is evident from one of Lipski's photographs, showing the walls of the Muksu canyon as consisting of rain-sculptured boulderclay similar to that of the Iskander-darya. Korzhenevski (*Yearb. Russ. Alp. Cl.* 1905) has succeeded in traversing the Muksu valley at great danger to the lives of the party. He describes the gloomy canyon as compressed between vertical walls of

139. Karashura Plain (Tupchek) and Little Achik Moraine, from Yashilkul.

loose stuff whence falling stones were continually threatening the caravan. They actually came upon a native gold prospector whose brother had just been crushed beneath a rock. Every year one or two people in search of gold are thus killed in this gruesome trap of boulderclay heaped up by many glaciers against the opposing ridge.

The Karashura plain owes its existence to the accumulation of morainic material from the three small Achik glaciers against the transversal bar intervening between their mouths and the Surkhab. While the Shaklisu-Baizirek has trenched a deep valley up to the snout of Peter the Great glacier, and the waters from Borolmas have pushed their headward edge to the Kosh plateau, the Karashura, fed by insignificant affluents, has not been able to dismantle the filled-in block of drift forming the substratum of its valley plain.

Looking in a south-westerly direction from our camp (Fig. 138) one sees the spur of glaciated mountains lying on the left bank of the Baizirek river and abutting on the platform of Little Tupchek. But the most striking aspect is offered by the morainic outpours from the three Achik ravines (Fig. 139). It is difficult to find the proper words for the beauty of these plastic formations from nature's workshop. One cannot call them either romantic or picturesque, yet there is an impressive grandeur in their moulded massiveness and flowing outline that pledges the eye to admiration. The viscous flood is petrified, and still its wavy bulge expresses movement.

It is an intense fascination for man to see things happen, to watch Niagaras, avalanches, floods, bores, eruptions and explosions, and the bigger the better. Mere results, be they never so startling, are not half so satisfactory, even when we can quite imagine how they came about. It needs an effort of the brain, and often special training, before one realises the magnitude of those processes, generally slow ones, which model the face of earth. Be the bulk uplifted never so huge, the riven hollow never so abysmal, they are results lacking actuality. However much they may appeal to our sense of beauty, our love of odd shapes or our taste for uncommon size and number, they can never claim that eager interest always on the alert for wholesale transactions and catastrophes. What we want to see is cause followed by effect at once, instantaneously, if possible ; we wish to see matter forced into shape so as to be one with force. Every hill is the adequate form of an energy,

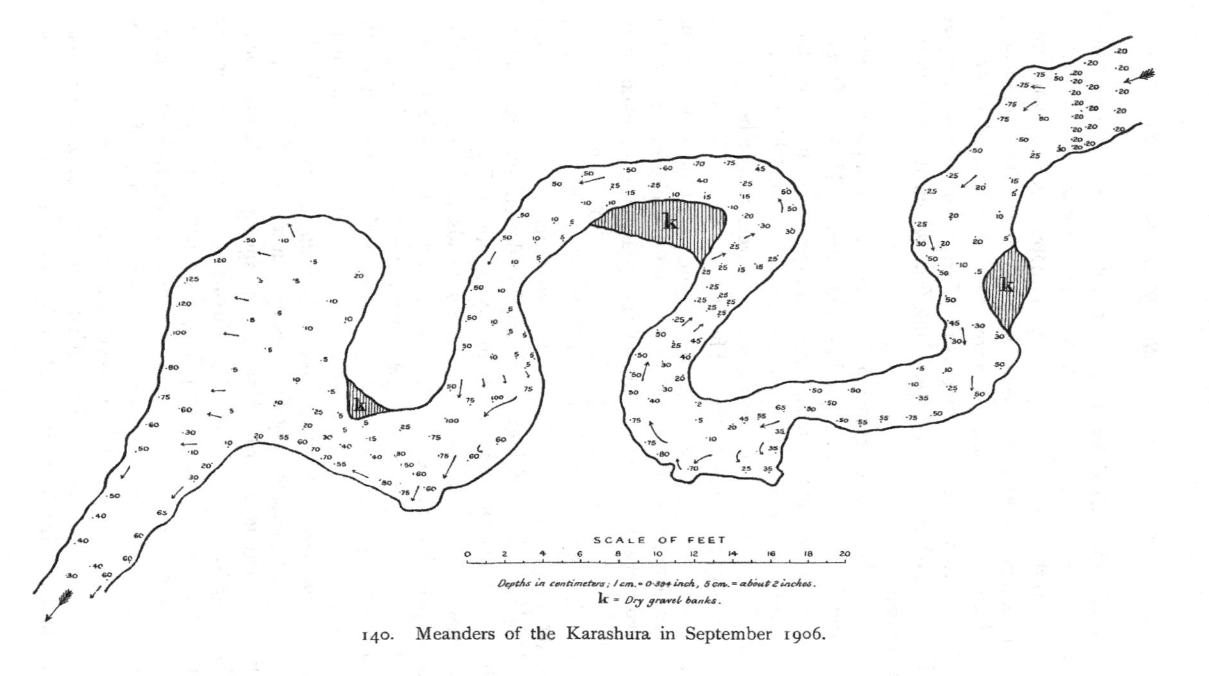

140. Meanders of the Karashura in September 1906.

but we are too short-lived for its patient truth. Hence the charm of waterfall and ocean wave, the spell of silent barkhans in the desert. And so it is with these moraines, for their shape is perfect, while their movement seems not extinct. Who, not knowing better, would deny that the sluices of the mountains were loosed but yesterday and the whole shot out in one far-spreading avalanche? Moreover near above there lies the gash in the mountain flank from which it seems torn at one fell swoop.

We had pitched our tent near a Kirghiz aul (encampment), which was to provide us with dairy produce and meat. Herds of goats, sheep, horses, cows and camels were roaming over the Karashura steppe, a grey and yellow floor over which a dry wind blew most of the time. Dust and wisps of dead vegetation were always being sifted into our cooking pots and plates. The only colour on this expanse of sere grass flashed up among the brown yurts (felt huts) where the women moved about in their red dresses and white turbans. All of them looked very robust and some of the younger ones even quite comely, like village girls of the Alps, with ruddy cheeks and clear eyes. The day after our arrival they all came with hospitable offerings, kaimak (cream), sour milk, kumis (fermented mare's milk), and two hundred cakes baked in butter. Near by flowed the Karashura, describing tortuous meanders between tufted banks (Figs. 140, 141). I have charted some of these bows and bends, showing how the current deepens its channel on the outside of a curve, while accumulation takes place on the inside. Although the filled-in valley plain is comparatively young, the winding course of the stream gives an aspect of maturity. But it is a ripeness threatened from below where gullies gnaw their way into the declivity of the upper landing. At this time of the year there is very little water, but the small flood plain as well as numerous dry rills hint that in spring the brook swells, streaking Tupchek with untold silver threads (Fig. 142). Even the flattest stretches of the Karashura steppe are not perfectly level, nor could they be so, seeing that here we have not the work of a large river evening out its bed. The surface is really composed of extremely gentle fans which, dovetailed into each other and neutralising their differences of height, have produced what might be called a peneplain of accumulation. It is the combined washplain of several glaciers, and from a favourable viewpoint we can still see the slight swellings of shallow deltas. Where the sole of the plain is lowest and nearly horizontal it is generally moist, often boggy, forming a greenish belt fitted to the basal contours of fans and slopes. This strip more or less corresponds to the flood-plain of the river which runs nearer the right (north-western) margin of the valley, having been forced there by the greater influx of rubble from the Achik side. Wherever the

141. Oxbow Curve (Meander) of Karashura. Accumulation at inside of arc; compare Fig. 140.

ground dips ever so slightly it is quite dry and tunnelled by marmots and other rodents.

Searching for the watershed and the Karashura sources, we march in a north-easterly direction. The green, swampy band narrows, the cushioned border of the streamlet changing into a gravelly beach, until finally there is nothing but a corrugated steppe of low welts and channels. The supply of the Karashura is easily traced to a valley held between the old Borolmas moraines—a cataract of hills—and the mountain slope (Figs. 137, 143). On the cone delta, projected from the opening of this glen, numerous distributary branches radiate in every direction, two of which now (September) suffice for guiding all the water there is. The left arm is the Karashura, while the right one, showing signs of some improvement by spade work, goes to the Kirghiz camp of the Kosh plateau. Hence there is bifurcation, the waters of the same stream descending over different slopes. From a minute, mathematical point of view this watershed changes with the seasons, for in spring a network of overflowing arteries unravels itself into the partisans of Khingob and Surkhab. The divide at this critical section of the range of Peter the Great therefore runs from Mount Severtsov to Great and Little Achik, down a ridge of the latter to the beginning of the left lateral moraine of Borolmas, and along the crest of this dike, then jumps across to the forks of the waters, continuing between them to a low hill spur on the opposite side of the plain, and thence to the Yashilkul pass.

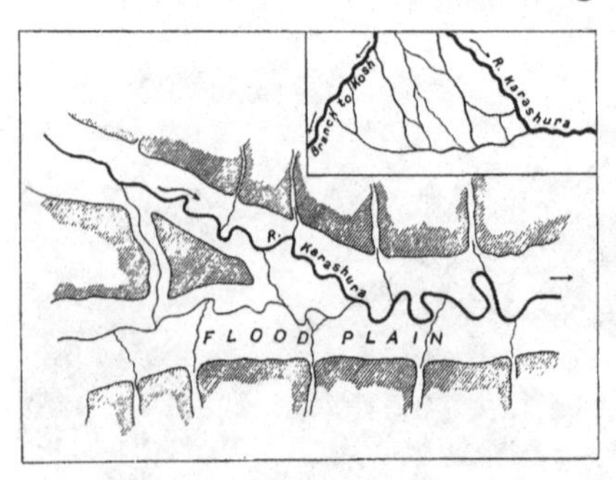

142. The Karashura in September. The heavy lines show water flowing at this season; the other beds are dry.

The Kosh plateau may have been a lake hemmed in by the old Borolmas glacier, when it was banked up against the opposite hills, and by the somewhat higher level of Tupchek, i.e. the combined washplain of the Achik glaciers. But I have never been near enough to make sure of this, and probably the Kosh platform is a residual fragment of the old Borolmas washplain. This parlour to Tupchek is a terrace between two steps, one high and steep, falling off to the valley of the Borolmas river, the other (the Kosh edge, cf. Chap. XV) forming a low embankment which supports the higher level of the Karashura plain. The lower brink is encroaching upon the Kosh plateau and will reach Tupchek before the small river (i.e. the Kosh branch of the Karashura) has succeeded in setting back the narrow Kosh edge very much further. The high lower drop is being sapped with the headward-reaching help of the strong Borolmas river which, having undercut the slopes of its own deep valley, provides the affluent with steeper gradients for its attack upon the tier, whereas the Kosh plateau is but a poor base level of erosive operations against the top storey of Tupchek.

This shows the effect of joint forces as applied to the relations of a river to its tributaries. The main trunk, by deepening its gorge, stimulates the feeders into greater energy, enabling them to accomplish more destruction than they could have wrought if they did not belong to an organised system. But, as in all things, such conditions hold good only locally and for a certain time, the increase of denudation or transport in one place being compensated by corresponding reduction in another section. The invigorated side streams carry more rock waste down their acuter declivities, thus burdening the master river at an early stage of its career, obliging it to begin deposition much sooner than it would otherwise have done. Hence, all things, such as hardness of rock and precipitation being equal, and tectonic uplifts ruled out, the theoretical day of absolute levelling of the heights should come at the same time to non-union and to organised rivers. Ten weak runlets debouching upon a plain will always retain an even gradient to the bottom of the slope; each spreading its sediment unhampered by the other; in common they will build up the lesser slope, which is to be the mountain of a still gentler one, and so on with infinite approximation to the ideal plain. On the other hand, if these ten streams join together, their trunk will make a gorge whereby the base of erosion is brought much nearer below their sources, so that they can strip their steepened inclines far more rapidly. But the reaction is sure to come, for the big river washes forward more detritus, and all through one outlet, thereby raising a vast conical plain which must ultimately grow into the gorge and up to the slopes of the ten affluents.

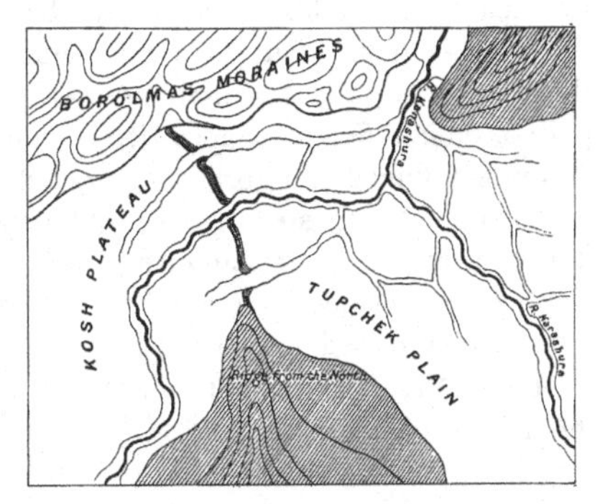

143. Bifurcation of the Karashura.

In the very long run of earthly waters their equalising work will come out level and up to time, no matter how they are combined. In truth this is only a theoretical and heuristic formula enabling us to understand these problems, for in practice the planing of continents depends upon the relation of the drainage systems to the absolute base of the ocean with its unlimited swallowing capacity, offering unfillable dumping space to great rivers which thus can never get cramped at way-head. Stating typical extremes one may say that unorganised aqueous energy—which in absolute perfection cannot exist, all active forces being directed—levels continuously from the beginning, while a catchment system dissects or destroys, levelling afterwards. The one denudates and builds concurrently, the other at different times and in different sections of its field of operations. In youthful highlands of the present day the destructive power of ramified systems is still unimpaired, their different shapes determining the character of their work. A regular fluvial tree, branching radially all round, produces the type of the funnel with steep slopes and a sudden engulfment at their foot. The herring-bone arrangement, like that of the Zarafshan above Varziminar, results in a gently inclined trough, because the middle flow is augmented very gradually.

Lake Yashil, our next object of investigation, lies embedded in the range of hills running N.E.—S.W. between Tupchek and the Surkhab. The portion S.W. of Yashilkul is well seen in Fig. 155, while beyond them a faint suspicion of the Alai snow-fields is peering through the haze. An ascent of 400 feet is all that comes between us and the top of the pass, there being, of course, a drop of 6000 feet into the Surkhab valley. That the Yashilkul once emptied towards the Karashura is convincingly shown by the old outlet on this side. The lake, whose level is about 80 feet below the pass, occupies an elongated bowl sunk transversally into the hill range. No trees fringe its banks, but a few ducks rose from the clear water curled by wind. Trickles from the last snow-patches around provide for the escape of a tiny brook, following which we descended half a mile on the Surkhab side where, however, we saw nothing but a misty void. South-west of Yashilkul two higher lakes rest in hollows of the hills. I doubt not in the least that a lobe of the great Tupchek glacier was pressed through the Yashilkul opening, whereby part of the ice found a short cut to the valley of Karategin. Similarly the old Inn glacier of the Eastern Alps sent branches into the Bavarian highlands over passes on its left bank.

Another excursion was made to the ridge of Little Achik overlooking the Borolmas glacier. Ascending by the north-easternmost and smallest of the three moraine-filled ravines we surprised three wolves which trotted off in haste over the snow-fields. These prowlers seem to abound in the neighbourhood, for at another time we saw one of them chased away from the aul by the watchful Kirghiz after an unsuccessful attempt on their sheep. The wolves are always eager to take a bite of plump mutton, and the expression can even be taken literally, for it sometimes happens that a piece is torn from the fat appendage, leaving the sheep apparently none the worse for this loss of palatable substance.

144. Kirghiz Women bearing offerings.

On the top of the ridge we found ourselves immediately in front of the pyramid of Mount Severtsov, flanked on either side by grand, glaciated corries. The great slab of summit ice with its broken lower edge seems to stick precariously to the slanting rock. Catastrophal ice-slides like that of the Altels suggest themselves, with their irregular feeding of the glacier below. At our feet the latter presented itself as the usual mixture of ice and rubble, but gradually transformed into a majestic flow of old moraine. The total length of the Borolmas, its living and dead portions considered as one individual, is about eight miles from the back of the corrie to the end at Kosh, the difference of level being roughly 2000 feet (Figs. 145, 146). It is a fine sight of outstretched and plastic regularity; a ponderous yet magnificent dragon, its sinuous body lying firmly in a bed of mountains, its warty head nosing the distant gorge. The dikes of moraine, graded in different shades according to age, wind in many curves towards the bosses and humps which overlap the plain of Kosh. The left lateral moraine (Fig. 154) is especially well preserved, stretching away in undisturbed perfection for many miles. On the slopes of Borolmas peak we saw some sharp ribs of rock nearly drowned in screes. In places the foot of the talus has subsided into graduated waves caused by the shrinkage of the glacier mass. We also visited the lower portions of the glacier, where lingered true denizens of the steppe, small bushes of ephedra and tufts of feather-grass (*Stipa pennata*). In the opening appears a fine section of the Alai panorama lorded over by the Shumkara peaks (extreme left of Fig. 145). For some distance on each side of these imposing summits the range is much lower with but faint traces of glaciation.

As long as we had only seen the triangular diadem of Mount Severtsov peeping over ridges, we were in high hopes of an attack upon the warden of Tupchek. The last 2000 feet of crowning ice seemed not too steep, admitting of a safe route between séracs and crevasses. Unfortunately our trip to the Borolmas glacier revealed a terrifying strength of lower defences, and the time of such a journey as ours is far too valuable to be thrown away on abortive attempts. Cenci and Albert brought better news regarding Great Achik, which they had reconnoitred one afternoon. It was a fine piece of scouting, deserving of high credit. Owing to our failure in the Fan group I was somewhat suspicious of any but very easy mountains, especially in view of their elevation.

Difficult rocks and the cutting of ice steps are far from tempting at altitudes above 18,000 feet, while every threat of bad weather, avalanches, or falling stones exercises a depressing influence a hundred times greater than in the Alps, where heights are lower and rescue parties can be expected within twenty-four hours. I also remembered my experiences on Mount Ararat, where we all suffered from lassitude and hard breathing at a height of only 17,000 feet. But, as far as the Duab is concerned, I have shifted upwards by 5000 feet, the level at which I would begin to fear rarefaction of the air, as such. All four of us, two ladies and two men, felt perfectly well and strong on the top of Great Achik, even going the length of smoking a cigarette. Of mountain-sickness, headache, or fatigue there was no trace.

It may now be taken as proved that the atmosphere at 20,000 feet is, in itself, incapable of disturbing the condition of a normal individual in good training. There remains, however, the mystery of local

145. The Borolmas Glacier and the Alai Range.

influences, for it appears that mountain-sickness is more prevalent in the Andes of South America than anywhere else. Everybody dreads it there, and the accounts of explorers seem to point to climatic peculiarities of air density, unless indeed we assume that they spent too long a time in high camps. Mountain nausea is a complicated ailment, inasmuch as a study of its symptoms entails a distinction between the absolute effect of diminished air pressure and subjective causes such as fatigue or a lurking indigestion. Up to a certain height, varying with locality, weather, difficulties, and individual disposition, rarefaction is only an unfavourable circumstance likely to aggravate little irregularities of the system, whereas above this level it must become the chief factor. Hence there would be transition from thin air as a concomitant cause, to thin air as a predominating cause. But even taking wide limit values the separation will always remain uncertain, for the man who was not tired at 20,000 feet may have become so by the time he got to 24,000 feet.

The specific physiological influence of a rarefied atmosphere can only be studied under the air-pump or in a balloon, while the question of mountain-sickness at very great heights can only be satisfactorily decided by the individual climber for himself according to his own experiments and experiences. Still there are a few conclusions of general import. Our case proves that for the normal mountaineer in good condition the level of mountain-sickness lies above 19,000 feet, and very likely two or three thousand feet more might have been added before distinct symptoms appeared, always provided that the climb was easy.

My opinion is that a high mountain, let us say of 25,000 feet, should be attacked as quickly as possible. Do not stay long in high camps, but make one long, last ascent, say from a sleeping place at 18,000 feet. But to spend many days at 18,000 feet and above is very exhausting, and the idea of acclimatizing oneself by a prolonged sojourn at great altitude is now, I believe, given up by many climbers in favour of a surprise assault. This at least is the feeling I have in the matter. Moreover this view is supported by a theory ascribing specific mountain-sickness to insufficient feeding of the blood with oxygen. Clearly this condition must become worse and worse during a prolonged stay, for we cannot get acclimatised to violent starvation of any kind, be it food hunger or oxygen hunger.

As to physical means, we must remember that brute force or the training of an oarsman (for races) do not weigh in the balance against a reserve of adipose tissue and latent energy. The mental state of the mountaineer is also of utmost importance. He ought to diet his soul with equanimity, directing his attention in the valley towards finding as easy a task as possible, after which an Oriental repose of mind should be cultivated. The choice of an easy mountain is therefore imperative. It is a proud thing to overcome obstacles, but it is also very disagreeable to know that those difficulties below you, the steep ice slope, the rotten rocks, or toppling séracs may endanger retreat. This anxiety of how to get down speedily when tired or overtaken by a change of weather, may sap one's confidence and energy. Ambitious struggling is futile without the most careful assembly of all favourable factors and conditions. Body and mind should be perfectly fresh and in a state of calm deliberation, founded on a conviction that nothing unforeseen can happen, that the road is clear. One must patiently search and wait for the lucky combination of practicable slopes, good weather, personal fitness, and a happy mood.

146. Head of Borolmas Glacier; Eastern Corrie.

Those giant monsters must be courted with languid ceremony, followed by quick persuasion at the psychological moment, the general advice being the same as to stalkers of live game: lie down and wait for a sure shot. One might also to great advantage cherish the pleasant illusion that one has come out to enjoy oneself, which is a miraculous guide to all the best treasures nature can yield up. Obeying these maxims we were able to climb three fine peaks without undue sacrifice of time and energy.

On September 12 we started from Tupchek, riding in about three hours to a good camping place among the youngest moraines of the Great Achik glacier, which to-day is nothing more than an ice-field, though its old moraines and hummocks descend into the Karashura plain. Here, at a height of 15,000 feet, we emerge upon a platform with red blocks disposed among a garden of turf and flowers. Here also grew a whitish lichen which favoured the tops of large rocks, making them look as if they had been sprinkled with lime. A colossal boulder sheltered our tent from a restless breeze. This rocky screen was cleft by a chimney deserving of the name, for it did good service by sucking in the smoke of our fire. As fuel we had a horseload of wormwood, on which the Kirghiz depend for their kitchens, tearing it up together with its ligneous roots. Supper consisted of dried cream, curds and bread, while rice boiled in milk and apricot jam were served for breakfast, tea being, of course, our daily drink at all meals. I never took much in the way of tinned meat on these journeys, because mutton and dairy produce are obtainable nearly everywhere. On the other hand I laid in a good store of Russian fruit preserves, a kind of jam, and not too sweet. They proved a godsend in the mountain valleys where vegetables are almost unknown. Tinned fish or meat were only kept for emergencies when cooking could not be done, or cold dishes had not been prepared beforehand. What one always needs and ought never to run out of are tea, sugar, jam, chocolate, a few condiments, and tobacco. With these, and the fat, or lean, of the land, one can always evolve a manifold bill of fare.

147. Our Achik Camp; Morainic Boulder.

The glacier, or what remains of it, forms the outrun of a wide but steepish névé gully between Great and Little Achik. Making for this couloir, at 5 a.m., we scaled the hard snow with crampons on our feet, gaining height very quickly. Only a few steps were cut somewhere near the middle, where a small icefall had to be dodged. From the saddle a broad and gentle back of snow led straight towards our goal, enabling us to maintain a fast pace. Already further down we had to stamp frequently and wriggle our toes inside the boot, so that any delay by slow work would have been distinctly nasty. A zephyr, deserving though it was of its mild name, steeled the teeth of the cold and winged our steps to the double. The very last bit was sheer ice covered with powdery snow, obliging Albert to wield his axe for a dozen steps or so.

At 9.30 a.m. we stood on the corniced arête of the summit, but not for long, as Cenci's thermometer—which she swings conscientiously every hour of the day and in every place, on horseback, while walking, at dinner and in bed—stood at fourteen ($-10°$ C.), and every one of those eighteen degrees of frost was driven into our bones with threefold force by the caressing breeze. Only a rapid glance was vouchsafed to the eastern horizon, where loomed with spectral dimness the castles of the gods, Sandal, Shilbe, and Muz-jilga, in front of the dazzling sun. With joy we discovered that the south-west ridge allowed us to vary our route for the descent. Fig. 148 shows this ridge coming towards us through the middle of the picture, the up-track lying over the sky line on the left. It was dry from top to bottom, and hurrying down to the nearest sheltered nook we undid the boots of frost-bitten sufferers to rub their toes, then basked in the warmth and generally comforted the outer as well as the inner man or woman.

Here we also began to relish the view (Fig. 161), unveiling to our eyes the gorgeous panorama of the glacier of Peter the Great, where terrific mountains soar to heights of 22,000 feet and perhaps more. Just opposite stood a noble mass of Gothic architecture, a bold combination of rock and ice (Fig. 161, middle foreground). The little glacier sunk into its front gives one furiously to think on the controversy of glacial protection *versus* glacial erosion. As I have suggested before (p. 150), this vexed question is pre-eminently one of local situation and definition, even more so than most other problems of surface activity. Much depends upon whether one is thinking of snow-fields, steep névés, or valley glaciers in all their untold variations of shape or size. In a general way this truth holds good: the larger the glacial mantle or stream, the more it protects; the more dissected or the narrower it is, the more it erodes. We cannot help seeing that the small ice-worm just mentioned must have rasped out its cradle.

In the Duab the snow-cloak of the monarchs is very ragged, hanging but loosely together and dropping tattered fringes over spiky ribs. Strips of moving ice are close to naked rocks exposed to a scorching sun in a dry atmosphere, and under a sky which is clear for a hundred summer days. At night the cold of planetary space darts through the limpid air, gripping the overheated stone, so that it shrinks and shivers with sudden spasm. No wonder then that here we see so much evidence of glacial erosion, for here the dry weathering of the desert is joined to the weighting and sliding of alpine ice in such a way, as to allow each the maximum force possible in such a union. They even hardly interfere, the desert enjoying more or less a free hand in summer and early autumn, while the snow falls during seasons of the lesser sun. The result of this fearful alliance is a power for destruction undreamt of in the Alps.

148. Great Achik.

Turning towards our bivouac, we trace the present decline and past career of the Great Achik glacier, great alas no more, beginning with its dirty little tongue licking the last residuum of former greatness, an outpour of red stone as from a monstrous pot of blood upset. Then furrows amid welts of grass like a crawling swarm of snakes or lizards, striped green and red. And finally the descending wallow of morainic bulges akin to the stiffened nodules of a guttering candle.

It being early in the day, we lingered over the feast spread to our gaze, before going down. The ridge became still easier and gradually changed into a comfortable whale-back, thickly padded with a grit of thin scales of frittered slate. This material is a kind of residual soil,

149. The Glacier of Peter the Great.

i.e. rock left in its place after disintegration. Here upright, parallel strata of friable schist have weathered into delicate fragments, and as the layers are of all the colours of the rainbow, except blue, a wonderful, striped surface results, which moreover has been smoothed by the pressure of snow (Fig. 150). Elsewhere we saw a similar occurrence where the outcrops of harder strata have formed rows of loose blocks.

So deep and soft was this rainbow stuff that it gave a nice, mealy sensation under foot, enabling us to descend twelve hundred feet at a gallop with perfect unconcern, not needing to look before we stepped. Among the rocks we found specimens of *Saussurea*, a thick-headed, thistle-like flower, almost stemless, which is in the habit of burying itself in scree up to the neck. *Saussurea* lives highest of all phanerogamous plants. *S. tridactyla* occurs in Western Tibet at an altitude of 19,000 feet. At the camping site we spend another lazy hour, drinking tea and gathering a rich harvest of autumnal flora: edelweiss, aconite, asters, buttercups, stonecrop, and saxifrage. Then we mount our horses that seek their way patiently until they feel the broad track of the steppe under their hoofs, and run with quivering nostrils towards the round huts, where the fires of home send their trembling shafts into the evening sky. In the last dim light of a far distance there fades away the streak of infinite Alai.

The reader in search of a compact review of the nature and exploration of the Pamirs proper, cannot do better than refer to Curzon's essay in the *Geographical Journal* of 1896, which will afford him a solid basis for further inquiry.

I have already explained in Chapter II that the particular Pamirs, those spelt with a capital "P," are the long and wide valleys grouped together in the region bearing their name. Topographically they are characterised by the great width of their floor and its high elevation above sea-level, while no fixed type can be attributed to the configuration of their bordering slopes. Some sections of the Pamirs are hemmed in between big and fairly steep mountains, so as to look exactly like the alpine reach of the Zarafshan (Fig. 93), only ten to fifteen times wider; others show a gentler transition and hilly ridges, like Tupchek (Figs. 138, 156); some may almost be called rolling downs (Figs. 160, 171). It is the valley plain which is the bearer of all the distinguishing features of the Pamirs, of every pamir, and the broader this plain becomes either through the wider spacing of its divides, or through their undulating gentleness, the more ideally it represents the type. So telling are the peculiarities common to them all that one feels driven to make the name of the Pamirs a general geographical term. Younghusband was the first writer to apply "pamir" as a generic term. Just as "alpine" has become the byword for rocky mountains with snow-fields or valley-glaciers, so "pamirian" must conjure up vivid associations with him who has seen or studied the originals.

In both cases the word for summer pasturages has stood sponsor, which is perhaps the best illustration of the old truth that the most valuable parts of mountains, namely grazing ground and passes, were the first to claim the attention of their users, and to impress their names upon the outside public. We can well imagine how in reply to an early traveller's query an inhabitant of Berne said, "Those are the alps whence comes our cheese"; or how the people of Kashgar answered, "Those are the pamirs where the Kirghiz keep their flocks." But though pamirs, like alps, have endowed a vast region with their name, the general meaning of pamir is naturally more restricted, for whatever else it signifies is already covered by the earlier and wider term of alp.

The Pamirs are alpine as well as pamirian, being pamirian valleys among alpine mountains. As a short definition of a pamir I suggest: "A wide valley with grass-steppe above the timber-line of an alpine district." There is more in these few words than meets the eye. The definition centres around the "steppe," the shape and location of which are determined by the other elements. Steppe is vegetation subject to a period of drought, and as there are many assemblies of such plants we must mention the fact that only grass will meet our case, or at least a very low and grassy growth.

All grass steppe in the Duab is grazing land, and all pastures are occupied by nomads or semi-nomads, and in nine cases out of ten they are Kirghiz, who are *the* nomads of Asia. The word grass-steppe therefore conjures up a picture of pasturage, flocks of sheep, horses, and camels, flowery spring and parching summer, wind, nomads, or more particularly, Kirghiz. "Wide valley" roughly indicates the shape of this steppe, while "above the timber-line" points to a great height above sea-level, treelessness, and absence of cultivation. Finally, we expect the neighbourhood of the snowy peaks and ranges of an "alpine district," in order to complete the landscape of a pamir. Although the definition is one of outward aspect, and cannot be made dependent upon genetic principles, it goes almost without saying that a pamir will usually have a glacial significance, it being, on the whole, impossible to imagine the origin of wide valleys at great height without the help of glacial factors. Most pamirs are

150. Rainbow Grit on Mount Achik.

troughs deeply choked with morainic detritus, which the rivers have not yet been able to remove, owing to an insufficient water supply. Consequently they are bound to play a rôle in every discussion of the interesting problem of the desiccation of Inner Asia, maybe furnishing some day important clues towards its solution.

Not having been on the great Pamirs myself I shall dismiss them with a short topographical summary. One usually counts eight of them, such as the Great Pamir, Little Pamir, Alichur Pamir, etc.

(cf. Chap. II and Fig. 20). They are on an average from two to four miles wide, being surmounted by ridges of growing height as one proceeds westwards from the watershed. Hence the scenery of the upper or eastern end of most Pamirs more nearly approaches a landscape of wide hills, gradually changing towards the west as the valley entrenches itself more and more between chains of bolder formation.

A few touches given by one's imagination to a great Alpine valley, such as that of the Rhone in its middle portion, would produce a pamir in its lower section. Think away all trees and human habitations and raise the flat sole by a filling of rubble so as to be about flush with the middle reaches of the tributaries. Their floor is from 11,500 to 14,000 feet above sea-level, which figures may be taken as representing the "biological limits" within which a true pamir can exist, because lower down the valleys become deep, narrow, cultivated, and bosky, while above the grass begins to yield to snow and rock.

Accordingly the mightiest of the Pamirs are stretched full length between those two levels from head to end, from the limit of vegetation to that of physical formation. Owing to this rise from west to east the majority of Pamirs can take advantage of the fact that the snow line recedes in the same direction, thus securing the greatest possible extent between levels. This climbing up of the snow line is due to lesser precipitation, so many clouds being intercepted by the western rim, and partly to the higher temperature and stronger insolation reigning in the midst of mountain masses. This contrast is especially marked between the Pamirs and the outer districts, for it is a law of climatology that all high levels (vegetation, snow line, human settlements) rise towards the interior of compact elevations of the land, because the inner bulk is more intensely heated in the sun. It must not be forgotten, however, that this solar or direct thermal effect is merely contributory, raising still more a snow line already high owing to reduced precipitation.

If rainfall were very much increased, other conditions remaining exactly the same, the snow line of the Pamirs would be lowered considerably all round, forming, if pushed far enough, a tremendous ice-cap over the great mountain pedestal. To accomplish this the solar climate of the entire region need not become colder. Frequent clouds shade the sun and rain cools the air, whereby the snow line is depressed, i.e. snow changed into water at a lower altitude; secondly, greater masses of snow are accumulated, cooling the surroundings still more. Hence the dryness of the Pamirs is chiefly responsible for their high snow line as well as for the feebleness of erosion. Pamirski Post is the driest of all the meteorological stations in the Duab, its annual average of precipitation being a third less than that of Petro-Alexandrovsk on the Oxus, among the sands of Kizilkum. It is an amazing case of extremes that meet, the one place 10,000 feet above sea-level, the other barely 300. November and December, each with less than a tenth of an inch of precipitation (in the form of snow), are the driest months of the year (save March) at Pamirski Post. At the same time winter is intensely cold, so that by this clashing of the driest and coldest season glaciation gets a very poor chance indeed.

Although the Alai valley does not bear the name of Pamir, yet its upper basin has all the necessary characteristics. The Arpa and Aksai "plateaus," i.e. comparatively shallow valleys in an uplifted and warped region of the Western Tianshan, so well described by Carruthers (*Geog. Jour.*, Nov. 1910), are genuine pamirs. Minor pamirs are Katta-karamuk, Tupchek, Sagirdasht, and others. The high, arid, wind-swept, wide and partly undulating, nomad-inhabited pasture valley of Tupchek exactly meets our definition. We found the climate very bracing and exhilarating, so much so that we felt as if, after a prolonged stay, it might even have an over-stimulating effect upon the newcomer. About its general healthiness there cannot be the least doubt.

The nature of the steppe is upheld by the mangy growth of the turf, and by various leading species of the flora. Feather-grass (*Stipa pennata*) is the same here at 12,000 feet as on the shores of Lake Aral, while wormwood (*Artemisia*) which supplies the Kirghiz with fuel, is one of the familiar herbs of the lowland steppes. But we must attach more importance to the climatic aspects of the vegetation, notably the withering effect of autumnal drought, than to the kinds of grasses and flowers, for there are close bonds of similarity between the plants of the high Alps and those of the steppe.

The more frequent wettings enjoyed by the vegetation of the Alps suppress the interpolation of a spell of drought between summer and winter, enabling the plants to refresh themselves from time to time. But a more important influence upon their organisation is exercised by the conditions ruling them during the intervals between rains, when the plants of high mountains are subjected to strong sunshine and a dry atmosphere during the day, and to sharp frost at night. These factors threaten the vegetable constitution with excessive loss of water by transpiration, against which alpine plants have therefore to guard themselves as much as those of the steppe. Hence the smallness, hardness, or hairiness of so many of our Swiss friends, many of whom have twin brothers in the barren plains of the Duab, not to speak of the pamirs, where alpine and steppe conditions are combined, so that here we find edelweiss, feather-grass, *Saxifraga*, leek, *Astragalus*, *Androsace*, *Artemisia*, *Saussurea*, *Gypsophila*, *Geranium*, forget-me-not, lousewort, etc., in various spots according to height, season, or local moisture from streams and snow-patches.

Any lingering doubt one might still entertain as to the pamirian title of Tupchek is speedily scattered by the felt domes of Kirghiz yurts endowing the scene with the last convincing touch. One might almost say: Where there are Kirghiz there is steppe, where there are Kara Kirghiz there is pamir. The great nomadic nation of Asia call themselves Kazak, whereas the name of Kirghiz by which we know them means "robbers," having been coined by the settled population. Among the mountains and in the pamirs we have to do with the branch of the Kara or Black Kirghiz, also called Wild Kirghiz by the Russians, owing to their more unruly temperament. Their language is a fairly pure Turki dialect like that of their relatives in the plains, from whom they differ very little as regards racial character and customs. Their number is estimated at anything between half a million and a million, the larger portion living on Chinese territory. Many of the Kara Kirghiz are half-nomads, the only important point distinguishing them from the lowland tribes. Our hosts of Tupchek have permanent winter villages of mud-houses in the Surkhab valley, where they also practise some tillage. This they can do the more easily, as the summer camps are relatively near, it being shorter work to dog the heels of spring through altitude than through latitude. Fields among the mountains are not so much dependent upon organised effort or complicated irrigation, and can be pretty much left alone, once the crops are sown. A few servants remain as caretakers, and if the master wishes to waken them up, his journey is not unduly long, perhaps six hours, perhaps a day or two, according to the site of his pasturage, whereas his less fortunate colleague on the Sir-darya is often separated by 500 miles and more from his winter quarters which, moreover, are entirely unsuited for agriculture. The mountain nomads wander up and down between the snow lines of spring and autumn; the Kazak of the plains go north in pursuit of the grassy season, returning south with the approach of the hard man. These movements are not such a random affair as popular fancy imagines, for each tribe or clan has its definite reservations, having followed the same route to and fro since time immemorial. The wide expanse of the steppe is as little unowned, or vacant land, as the orchard of the Sart.

In many ways the Karas enjoy much better conditions than their kinsmen of the plains, being able to lay in fodder and stores at the hibernating hamlets. Winter on the vast, open steppes is much bleaker and windier than in the mountains, where slopes screen from chilly blasts and catch the heat of the sun. This again shows how one-sided is the fear of the highlands observable in civilised nations inhabiting fertile plains with good roads, and a mild, sloppy climate. For them the mountains must be bare, cold, snowy, inaccessible, and dangerous, although in modern times they have discovered the genial beauties of Alpine winter with its dry, crisp air and sunny warmth. But the wildest mountains can be a regular Riviera when January is driving his white, demented legions across Asiatic spaces. I would rather be left to fight my way amid the wintry dangers of some glacier basin than be marooned in the flat loneliness of the storm-swept and terrible Ust Urt.

A few of the roamers of the Pamirs do not descend into the lower valleys, but spend winter at heights up to 14,000 feet, taking advantage of the dry ground, for the Kirghiz has only one great enemy, and that is snow, which makes it difficult for his cattle to find food. Cold and the onslaught of the elements he braves in his yurt, by far the best movable shelter ever invented, combining the stability and roominess of a house with the lightness of a tent. It is a detachable framework of sticks, covered with felt mats, and whoever enters one for the first time is astonished at the liberal largeness of the premises, and the cosy temperature reigning inside. It takes the women of the family only about a quarter of an hour to set up the hut, and an explorer with a large camp and many servants can do no better than buy a kibitka (the usual Russian word for it), especially if the party consists of several Europeans, there being accommodation for half a dozen men with their sleeping kit and baggage. In this circular and padded vault one can sit at a table or stand erect without feeling cramped, enjoying shade under a blazing sun, as well as protection from the most searching wind, while a small stove would turn it into a den of luxury, even at zero Fahrenheit. At all seasons I consider the yurt preferable to a native loess house, for should it ever get messy, one can take it to pieces, shake it, wash it, and pitch it on a fresh site. Our makers of travel-outfit should spare no trouble in constructing a tent on the principle of the yurt, as taking weight for weight, no device of movable dwelling gives more space, coupled with rigidity. It ought not to be difficult for western engineering to invent a much lighter pattern, but equal in strength, and of different sizes to suit the needs of travellers.

The different nature of the ground and the absence of thorny plants make the mountain pastures less acceptable to camels, while recommending them to cows. Hence, as against the plains, numbers are reversed in the Pamirs, there being very few camels—generally a Kirghiz speciality, namely a cross of the one-humped dromedary and the two-humped Bactrian—and a considerable proportion of kine. The yak, or grunting ox of the Pamirs is not found at Tupchek, his home beginning a little further east. I only saw one of these animals installed at Sagirdasht by the Beg of Darwaz, who had received it as a present from some henchman of his upper districts. Horses are kept mainly for their milk, otherwise their numbers would not be so great, exceeding a hundred times those needed for riding or transport. Kumis, the celebrated drink, is made of fermented mare's milk, and whoever has become used to its strange smell, will taste the richness of cream, relieved by the sparkling gaiety of champagne.

While we were at Tupchek it was just the season of sheep-shearing and felt-making, so that we could watch the ladies of the aul kneading and rolling the thick wool into firm carpets. All work connected with camping, loading, household duties, industry, etc., is done by the women, for a Kirghiz gentleman considers manual labour an indignity unheard and undreamt of. He visits his friends, hunts wolves, rides to his trained eagles, or plays at mutton-football in a baiga, while the intervals between these pastimes are devoted to the arduous task of repairing muscular energy with food and sleep. Although nominally Muhammadan the Kirghiz are not strict observers; their women go unveiled, and so far I have not seen a man at prayer, which does not prevent these people from being the most sympathetic race of the Duab, honest, straightforward fellows, upholding their traditions of sport and hospitality. Carruthers has made a very interesting comparison in his Aksai paper, where he says, "The experiences I have gained since amongst the nomadic Beduin have taught me the many differences which exist between these two wandering peoples. Although they are so very like in many respects, yet environment has caused important differences. Compare, for instance, the wealth of the Kirghiz, who have a fine country to pasture their flocks in, with the poverty of the Arabs, whose heritage is the barrenest parts of the world's surface. And, again, as a natural result, the Kirghiz are milder and more open-handed, less fanatical, and of a gentler disposition than the Beduin. For their life is more easy and free from care; their struggle for existence is not so severe as that of the Arabs; their herds are more numerous, and their pastures more rich, neither are they constantly harried by hostile tribes. But the Beduin's struggle for existence is relentless; he lives in constant fear of enemies, and is always in want of water. Hence he has become cruel and warlike."

The so-called nomadic instinct as an ineradicable tendency is much exaggerated, if taken to mean the lugging about of one's entire household. We have only to remember how much better a man can travel, fight, and hunt when he leaves his possessions at home. Nomadism is the result of circumstances such as the dearth of arable or irrigated land and the periodical want of food. When these drawbacks are removed human nature more or less quickly reverts to the other fundamental impulse, that of settling down in a fixed home. Even the wandering tribes are not homeless, simply having many homes instead of one, or, rather, a large home with many locations. Man certainly loves to roam and hunt, but this bent he can indulge much better from a central compound where the cumbersome household goods, the women and children are stored once and for all. Animals need movement as well as rest, and man forms no exception. Homing and roving are both strong instincts in us all, being mixed in varying degree. That nomadism and settling are relative terms is shown by the pasture system of the Alps, resembling that of the Kara Kirghiz. The Turki peoples and Semites keep up nomadism wherever obliged to do so, or wherever an opportunity offers, not made use of by other races, witness the Yirik and various scattered flocks of Semitic and Turkish "gipsies." This is, of course, partly due to inherited tendency, which makes these races more inclined to accept the offer, but no tendency, however firmly ingrained, can remain hereditary without constant exercise, nor will it resist the allurements of favourable conditions. Wherever to-day nomadism is a necessary form of life, Semites and Turks are the overwhelming majority in possession; wherever it presented an opening they have stepped in; but there are sufficient examples showing that if a really tempting chance is offered to a compact number of nomads, especially in the shape of conquered land, they can gradually overcome their proverbial restlessness.

CHAPTER XV

THE GLACIERS AND MORAINES OF TUPCHEK

HAVING described the lie of the land and recorded various observations in the preceding chapter, I shall devote this one to glacial phenomena.

Beginning with the easternmost glacier, the Borolmas, I draw the reader's attention to Fig. 137. On that more or less diagrammatic plan of Tupchek I have shown the left lateral moraines of the glacier with their sharp rectangular bend towards the west. Between the moraine and the soft northern ridge of Little Achik lies the small valley or dell of the Karashura, part of which stream subsequently descends into the Kosh plateau (cf. Chapter XIV and Fig. 143), afterwards joining the Borolmas river. A short and low ridge from Borolmas peak divides the present Borolmas and Kizilsu glaciers which in former times flowed together, leaving behind a mass of morainic undulations common to them both. For the sake of simplicity I shall apply the name of Borolmas moraines also to this joint product.

We now turn to Fig. 151, a view taken from a position just below the angle or knee mentioned above, in the midst of the morainic hills which in themselves form quite a mountain world with deep valleys and bowls, ridges and humps. The Borolmas peak is easily identified on the right. Our standpoint is in that portion of the moraine spreading out fan-like towards the Karashura and Kosh plateaus, while the main direction of the flow is almost due north. Immediately in front of us is a comparatively small moraine (with a large, square boulder on top), namely the one branching off at right angles from the great lateral moraine[1]. The enormous wall of the latter,

[1] Fig. 153, "Mount Severtsov," is a continuation to the right of Fig. 151 as regards the background, but the camera having been shifted to another place the foregrounds overlap. We can identify the boulder and follow the continuation of the small moraine to the right of it.

151. Tovarbeg (on left) and Borolmas Peaks; Borolmas Moraines and the Dry Bed in Foreground.

nearly 400 feet high, is also remarkable in so far as it forms the watershed between Khingob and Surkhab. Before occupying ourselves with further detail, let us examine Fig. 146, which takes us to one of the grand cirques or corries at the head of the glacier. On the right is Mount Severtsov; near us the great lateral moraine clearly distinguished up to its beginning under a black spur descending from the Achik ridge. Behind this spur is the dirty ice bulge of the glacier branch from the western corrie (here invisible) between Severtsov and Achik. As already described by Lipski, the union is very sluggish, only part of the ice connecting with the eastern contributory, while the greater half, probably screened off by old moraine, ends in the dark terminal face, whence issues a small stream. This, for a time, runs in a valley between older and younger lateral moraines (Fig. 154), afterwards losing itself by percolation among rubble and ice, joining other waters on their road towards the Muksu. Lack of time prevented me from finding out where exactly the Borolmas river emerges, but Lipski saw it somewhere east of Kosh. The Karashura is fed by the snows and springs on the flanks of the Achik ridge, the great lateral moraine deflecting the water into the Tupchek plain.

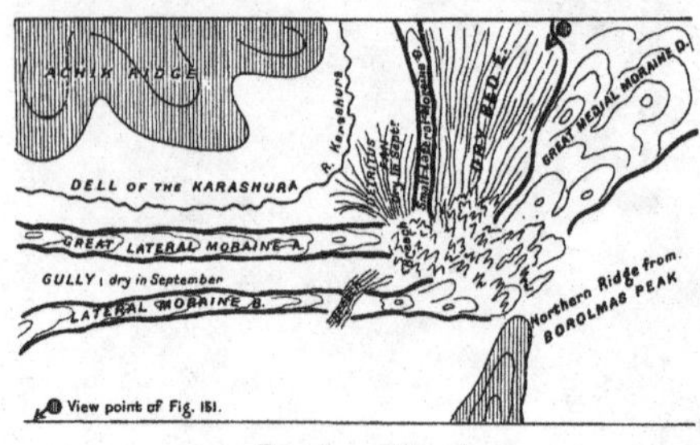

152. Borolmas Moraines.

We now return to Fig. 151, studying the state of affairs in connection with the diagram Fig. 152. In the foreground between us and the small moraine C is a flat piece of ground or dry bed E. Its even floor, composed of angular detritus and thinly overgrown with the vegetation of the steppe, lies about 150 feet higher than the adjoining reach of the Karashura. Its lower end is narrow, showing signs of forcible erosion. Undoubtedly it is the wash deposit of a glacier stream. Disturbances have taken place in the neighbourhood of the spot where the small lateral moraine C turns off sharply from the huge dam of A. The connections are severed and we behold a tumbling disorder of naked blocks, a scar in the flowing lines of the surface. What puzzles us also is the position of moraine C, for which I shall suggest an explanation later on. So much is certain that this critical spot lies near the old junction of the Borolmas and Kizilsu glaciers. Although separated to-day, as living glaciers, the lowness of the northern rib of Borolmas leaves no doubt that they jointly piled up the further continuation of morainic hills. Now, the variations in the quantity of ice and the vagaries of water in the glacial intervals, and, up to this day, must have wrought the greatest changes just in this crucial locality. The western or Borolmas glacier, being the lesser one, must always have shrunk first, and it may be that portions of the Kizilsu ice, or at least of its water, for a long time persisted in wandering to the left. That advance and retreat of a glacier are a potent factor of stream deflection we all know. A striking example is near at hand when we study Fig. 152. Between the great or outside moraine A and the younger, inside moraine B is a gully, now dry, but bearing all the traces of a river bed still in use. It empties into the Karashura dell by a breach and a cone of dejection. This gully was dry in September, but Lipski, who was here at a much earlier season, saw red water issuing from the breach and joining the Karashura. Hence there is a periodical bifurcation of glacier water to the Surkhab and the Khingob. It is the brook which comes from the open end of the western glacier branch (Fig. 146) and runs between the two moraines (Fig. 154, looking down stream; moraine A in middle of picture) before being swallowed up. In spring there is an excess of supply over percolation, thus permitting the stream to pass on to and through the breach. At one time it poured into the bed E (also compare Fig. 151), but during a high flood, or owing to ice pressure, part of its volume overflowed. This lateral escape, finding a lower base of erosion in the Karashura dell, quickly ate a gap through moraine A, leaving the bowl E high and dry. Similar occurrences may have led to the breach in moraine B a little higher up.

These happenings are, of course, not very important as throwing any special light on the great history of the glacier. They are local changes in morainic topography, repeating themselves over and over again in the course of a century. Their instructive value lies in showing the conditions reigning on glaciers of this type. Our Borolmas is dead at one end, and very much alive at the other, where its icefalls thunder. In between it is "dead and alive." The transition from ancient moraine to recent glacier is so gradual that it becomes quite impossible to define a boundary. When using the term "glacier" in this neighbourhood one cannot help thinking of the entire combined length of ice *and* moraine, no matter how old. Smothered under a burden for which it cannot find a dump, the glacier squeezes itself laboriously forward as well as it can, forming a mixture, half stones,

153. Mount Severtsov.—Lower Portion of Borolmas Moraines in Foreground.

half ice, towards the end. No wonder then that it gradually raises itself by working surface rubbish, swallowed by crevasses, into its bed of ground moraine. Consequently the topography of this labyrinth is very unstable in the middle region of the glacier, while the older morainic landscape, forming a direct continuation of the more recent dams and cones, is closely affected by all changes in the upper reach. Whatever happens to the ice configuration of a modern Alpine glacier, its waters usually meet again on an old washplain or in an intervening valley stretch, entering the distant and isolated deposit of old moraines by a fixed passage. Here, there is continuity, and therefore much community of disarrangement. If a big load of rubble slips from a cone, if an ice wall crumbles, or a tunnel collapses, new ways may be shown to the water courses. These new channels will often mean the forsaking of old beds and the making of new ones in the ancient moraines so closely bound up with the fate of the active glacier. That the changes just described habitually concentrated their effect upon the point of junction of the two glaciers (Borolmas and Kizilsu) is easily understood, for here is a weak spot where the least difference of level will weigh heavily in the watershed balance. When I saw the Borolmas

154. Old and younger Lateral Moraines of Borolmas Glacier.

glacier, the lowest ice appeared to be about level with the end of the ridge from Borolmas peak. Lipski's report seems to corroborate this observation.

Since the formation of its great lateral moraine A the Borolmas glacier (i.e. above its junction with the Kizilsu) has never stepped beyond it. Subsequent lateral moraines have been telescoped into the outer ones, while Fig. 145 shows concentric crescents of more recent stages of retreat within the innermost zone. In Fig. 154 we see how the great moraine sends a long slope far down to the left, to the Karashura dell. Inside is a later one (B), while on the extreme right is a portion of the big, crater-like funnel identified near the middle of Fig. 145. This is one of the most striking instances of a raised bed, the younger moraines standing on a much higher level. Such a condition is partly due to the complete shrinkage of ice from the lower banks, for at the summit of a glacial period lateral moraines must very often have been medial moraines between the valley glacier and the ice coat of its bordering slopes. There must have been a small glacier in the Karashura dell, or, to express it more correctly, the space, now hollow, formed part of a great névé basin. On the extreme left of Fig. 154 is a black triangle, the end of the Achik ridge, and between this and the great dike we see the lower stream of moraines set at right angles to our line of view. This is really the joint product of the old Kizilsu-Borolmas glacier, as a glance at the map will make clear.

We are now in a position to understand the apparent and otherwise inexplicable rectangular deflection of A into C. The bend is far too sudden to be a curving round of the great lateral moraine. When after the height of the last glacial epoch recession set in, the Borolmas branch had already withdrawn beyond the corner of disturbance, while the much stronger Kizilsu still reached forward. Not having to sustain any pressure upon its left side, the Kizilsu now deposited its last terminal and lateral moraines over and against those of the Borolmas. Hence the small dike C is the lateral dam of an ice lobe from the Kizilsu, and its substance may also be partly terminal moraine of a dwindling residuum of ice on the Karashura slopes. Needless to say this piling of rubble across the end of the Borolmas glacier interfered with the drainage of its streams, thus leading to many of the complications described above.

No large flow of water from the Borolmas can have been lasting towards the west, as otherwise the Karashura plain would have been deeply eroded. On the whole therefore, the dam of the Borolmas moraine, which nearly closes the Tupchek valley to the east, has had a preserving influence upon this steppe resting as it does on the fluvio-glacial substratum of a still earlier glacial period. The Kosh was the true washplain of the Borolmas whose streams are responsible for the terracing of the Kosh edge.

The three Achik moraines debouching straight upon the middle of the Karashura plain show the same gradual transition from old to recent accumulations, without the intervention of a wash deposit (cf. Fig. 139). All three being practically the same as regards the formation of their valleys and the present state of their glaciers, I shall only describe that of the Great Achik. Its smooth, bulging plasticity, as it pours forth its solid volume into the flat expanse below, is a striking sight indeed. With the exception of a deep nullah or two, and a slight washing out of the margins, the whole heap of disembowelled mountain lies undisturbed, an assembly of billows and swellings. Its upper end is formed

155. Looking Northwest from Achik Camp.

by a stage or platform marking the beginning of a period of rapid retreat. This flat place strewn with boulders of surface moraine is shown in Fig. 155, the view passing freely over the brink, with its steep gradient, to the hills on the northern side of the Karashura steppe. Lower down the moraine also contains blocks, but fewer in number, most of them having been worked into the finer material. Turning entirely round we face Great Achik and its snow fields which give birth to a stunted glacier. This miserable bit of ice is all that remains of the stately cascade that carried down and moulded the morainic cataract. To-day the area of alimentation strikes us as ludicrously small, and we wonder how glaciers can have risen from it large enough to form those enormous deposits. But it will help the imagination if we remember the Duab interglacial climate and weathering and the friability of the slates of this region.

As to the age of the three Achik moraines and those of the Borolmas, I venture to identify them with the Pakshif period. I give this opinion with due reserve, but we may, for the time being, accept this dating as a working hypothesis. They rest upon the foundation of older wash-plains, like the Pakshif crescent upon the Zarafshan terraces, and it requires no very bold speculation to assume this parallelism. As to the considerable difference in level, it must be borne in mind that the snow line at Tupchek is much higher than in the Zarafshan valley. Whether the Pakshif period should be made to correspond with the Würm period of the Alps, or not, I have thrashed out in the tenth chapter. The Würm was the last of the big glaciations of the Alps, and the Pakshif the last considerable advance of the Duab. If we accept as true the assumption that the climate here was as much drier than that of the Alps then, as it is now, that would account for the difference in length between diluvial glaciers here and in Europe. We must remember that the Pakshif moraines of the Duab are small compared to those of the last Alpine glaciation. Their huge size impresses us because they have been preserved—dried, so to speak. In the Alps the last moraines have been converted into alluvial deposits, terraces, valley floors, etc. Generally the only moraines which have retained their shape and arrangement are the frontal arcs on the Italian side. Thus, as far as landscape moraine is concerned, the size of former glaciers is best shown by the distance of old terminals from the ice of to-day.

Finally we turn our attention to the glaciers of Peter the Great, the system of about ten branches feeding the Baizirek river. We have seen them in the panorama from near the top of Great Achik. The reader, now accustomed to the types of Duab glaciers, will have noticed how these ice streams barely manage to unite. They flow between high walls of rubbish, and the open crevasses at the junctions show that there cannot be much pressure. The most important of their affluents seems to be the one under the lofty white pyramid, nearly filling out its bed and carrying very little surface dirt (Fig. 149). What I wish to urge above all is, that we have here the largest area of alimentation in the neighbourhood, a fact to be borne in mind with a view to what follows.

I am now going to describe the phenomenon of the "isolated" moraines of Little Tupchek. This might be called a classical example; classical because it is simple. Nature's methods are simple, but she has been working on the same old material of the earth's crust for

156. Edge of Little Tupchek seen from the Karashura Plain.—Compare Figs. 157 and 158.

such a long time that the deciphering of her palimpsests is often a hopeless task. One feels grateful for documents writ large. Like the stone of Rosetta they form the safe foundations of research.

First let us take up a position at the western end of the Karashura plain as indicated by the dot and arrow A on Fig. 158. From here we see the view of Fig. 156, and Fig. 157 is a tracing of this view. On the left and right are two pyramids which plainly tell the fact that they are built of solid rock. The one on the right is on the other side of the deep Baizirek valley (see Figs. 137 and 158). The left one, on our side, is the end of the western ridge of Mount Achik. Between the two eminences a long, horizontal line is seen, the rim of the plateau of Little Tupchek. This we shall describe before going further. From the Karashura plain we descend about 400 feet to the powerful glacier torrent of the upper Shaklisu here called Baizirek. After fording it with great difficulty, we climb more than 800 feet over the steep slope of gritty rubble facing us in Fig. 156. Having reached the upper edge we find ourselves on a

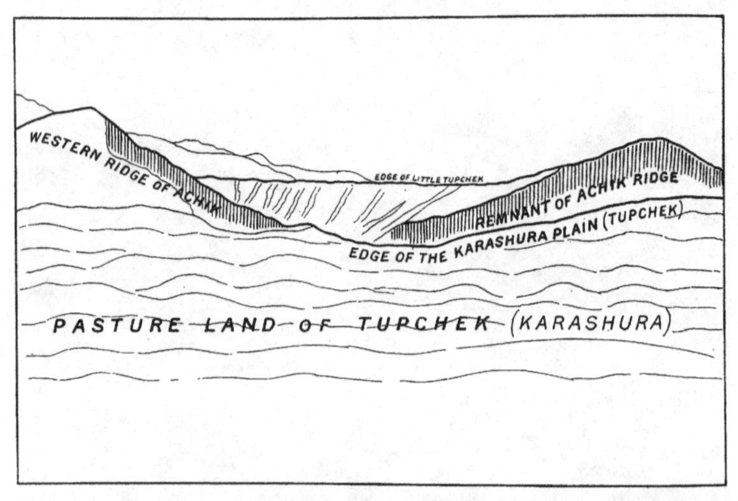

157. Tracing explaining Fig. 156, being the view from A in Fig. 158.

rolling prairie (cf. Fig. 160, our camp on Little Tupchek river, Chapter XIV), where several rows of morainic downs of great magnitude extend in a westerly direction, with a total length of about five miles. Save for a spring giving rise to the tiny brook of Little Tupchek, the plateau is very dry, having been cut off from all water supply and all energetic water action. The low hills on its northern side, a western spur of the mountains around Peter the Great glacier, can only provide a little water while the snow is melting. The higher, ice-covered summits of this ridge drain off into the Baizirek (Panorama,

Fig. 162). Moreover, most of the moisture received by Little Tupchek runs away subterraneously, as evidenced by numerous springs on the steep slope towards Baizirek. Near the edge there are plenty of sink holes and bowls with a layer of dried mud on their bottom.

That we have here a very old moraine nobody can doubt, even after the most cursory inspection. Whence it came we must find out at the edge of Little Tupchek, where, looking east (from point B in Fig. 158), we embrace the panorama of Fig. 162. In the middle the Baizirek flows straight towards us, and we easily identify several landmarks known to us from other photographs. We at once come to the unhesitating conclusion that none other than the glacier of Peter the Great can be responsible for the moraine of Little Tupchek. By their direction the long medial rows prove themselves a continuation of the Baizirek valley. But from the sole of this ravine we are separated by a drop of 800 feet. The moraine of Little Tupchek is entirely isolated, has been left "stranded high and dry" in every sense of the phrase. Between the edge and the glacier the valley is quite clear and clean; there are no big terraces, and any remnants of lateral moraines on the upper slopes have been blotted out by denudation. We feel quite sure that the river would not have spared the high accumulation on which we stand. Why did it make the sharp bend instead of demolishing the great plateau on its left bank? The solution is this. The rock peak seen on the right of Fig. 156 is really a former continuation of the Achik ridge, which the Baizirek has cleft from top to bottom. It is a remnant of this ridge and the fellow of the eminence on the left of the picture (cf. Fig. 157). I have given these two pillars of Hercules a darker shading in Fig. 158. Evidently they are counterparts, and a depression or saddle existed between them on the ridge.

Let us now reconstruct what has happened. The great glacier (say of the Zarafshan age) lay in its original valley indicated by the flow of the medial moraines on Little Tupchek, and after retreat left it filled with detritus up to the level of the plateau. Perhaps already at an early stage a tongue of ice found its way from the main body over the depression between the two hills of the Achik ridge, thus beginning to wear it down still further. It may also be that the work was only begun in earnest after the glacier had retreated beyond this spot. Some accumulation in front of the snout, some change of current, such as we have observed on the Borolmas glacier, caused the stream to fall over the depression in the ridge. Probably there was a drop on the other side, so that it cannot have taken the river long to gnaw a cleft through the rock. Once this had been done, the new course was established and fixed for ever. However soft the material behind this point may be, the stream could no longer surmount it on its left, as soon as its highest water-mark had been lowered ever so slightly below the surface of the morainic plateau. Rivers can undercut and

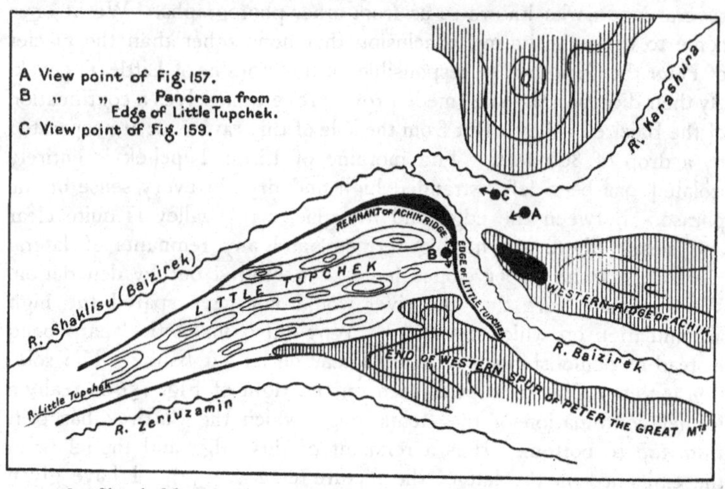

158. Sketch Map showing the Origin of the Moraines of Little Tupchek.

wash away loose stuff on their banks, but then they must be able to swing from side to side. Where two fixed points, such as gorges, projections and other hard places are near together, the river has no room to fetch a sufficient circle for the removal of all rubble from the rock slopes of the valley. By the curve of the Little Tupchek edge on the right of the panorama we can see that the Baizirek did swing a little into the material of the future platform, but further it could not go. It was irrevocably held fast by the notch in the ridge, and it may also be that the advance of the glacier during the Pakshif period took advantage of the gap, widening it sideways, whereas water only would have sawed a narrow, vertical cleft. Through the opening the stream washed all the morainic accumulations between this point and the glacier. The work was thoroughly accomplished by the great volume of water supplied by the glacier of Peter the Great. With the exception of (probably) Little Tupchek we have so far never seen any morainic hills and dikes of the Zarafshan period, all having been flushed out or levelled and converted into washplains, as witnessed by the fluvio-glacial terraces of the Zarafshan.

Thus the bulwark represented by the remnant of the Achik ridge has preserved intact the moraine of Little Tupchek, removing it from the destructive influence of the torrent. That the protective screen casts a lessening shadow, is clearly shown by the narrowing of the plateau towards the west. The further the river runs beyond the rocky knob, the more room it gets for impact upon the side, the more unhampered become its inroads upon the morainic stack which it has been obliged to circumvent. The effect must be all the more pronounced, because the solid hills on the right bank offer greater resistance.

We have now gathered a few facts enabling us to form a picture of the glacial history of the Tupchek pamirs. The Karashura plain (see left side of Panorama 162), the plateau of Little Tupchek and the boulder clay of Muksu are composed of deposits which, until further notice, we ascribe to the Zarafshan period. The Karashura floor has been levelled, but owing to the comparative smallness of the streams it has not been deeply indented by canyons. Between the Borolmas and the Great Achik moraine its surface is nearly flat. Only from where the Great Achik brook is united to the Karashura (which meanwhile had also taken up the effluent of the Little Achik ravine) the plain is more uneven and traversed by a gully. This drops down towards the Baizirek, first gradually, then by steeper gradients. Thus the Karashura filling has been partly preserved by want of water. Its eastern end, the edge of Kosh, offers a minor analogy to the edge of Little Tupchek. Thin streams from the side of the Borolmas glacier have swept a crescent-shaped slice out of the Tupchek plain. We must assume that during the Zarafshan period the western branch of the Tupchek ice mantle flowed into the Baizirek valley, joining the glacier of Peter the Great somewhere round the corner of the rock pyramid.

That Peter the Great was the larger one is self-evident from its enormous area. In Fig. 159, which is a view taken from position C in Fig. 158, we see the Baizirek river flowing away from us and disappearing round the rock pillar or remnant of the Achik ridge. About 400 feet above its right bank is a long shelf or terrace, all that is left of the Karashura morainic pile at this point. Here the two glaciers, separated by the then unbroken Achik spur, ran parallel to each other with a slight difference of level. Where the dividing crest stopped they united, gradually merging their levels. But possibly they joined much sooner burying the ridge almost entirely within their enormous bulk of ice. The small terraces low down at our feet, near the trees, are probably remains of the later, or Pakshif, activity of the glacier of Peter the Great. That this ice stream has not left large mounds like those of the Borolmas, during the Pakshif retreat, cannot astonish us, for neither has the Zarafshan glacier left a bulky morainic landscape of that time. Both these mighty glaciers must have nearly the same catchment area with a corresponding quantity of water, and both have converted into alluvial deposits their rocky freight of the Pakshif growth.

We have assumed, in the light of prevailing theories, that purely denudative and erosive happenings decoyed the river into its new bed. We do not know what was the particular combination which gave the first impulse. For we have, in this locality at least, the choice of another initiating cause. Instead of being due to gradual blocking of the way, or to an overflow of the glacier, the altered course of the river may have been started by an earthquake, which rent a fissure, however small, into the dividing ridge. And why not? True, modern feeling is not in favour of rapid upheavals and sudden ruptures; erosion is everything. Whenever a new revelation comes to mankind, we are apt to be swayed by it to the almost total exclusion of earlier theories which, through the unconscious tendency of the human brain, come to be regarded as contradictions. From catastrophes we have changed to slow, imperceptible foldings and infinitesimal stages of submersion. Glaciology and erosion have taken the place of convulsions as the modellers of post-tertiary landscape. Undoubtedly the latest theories are better and more important. They offer a truer and better balanced, if more complicated picture of evolution. I agree with them. But we

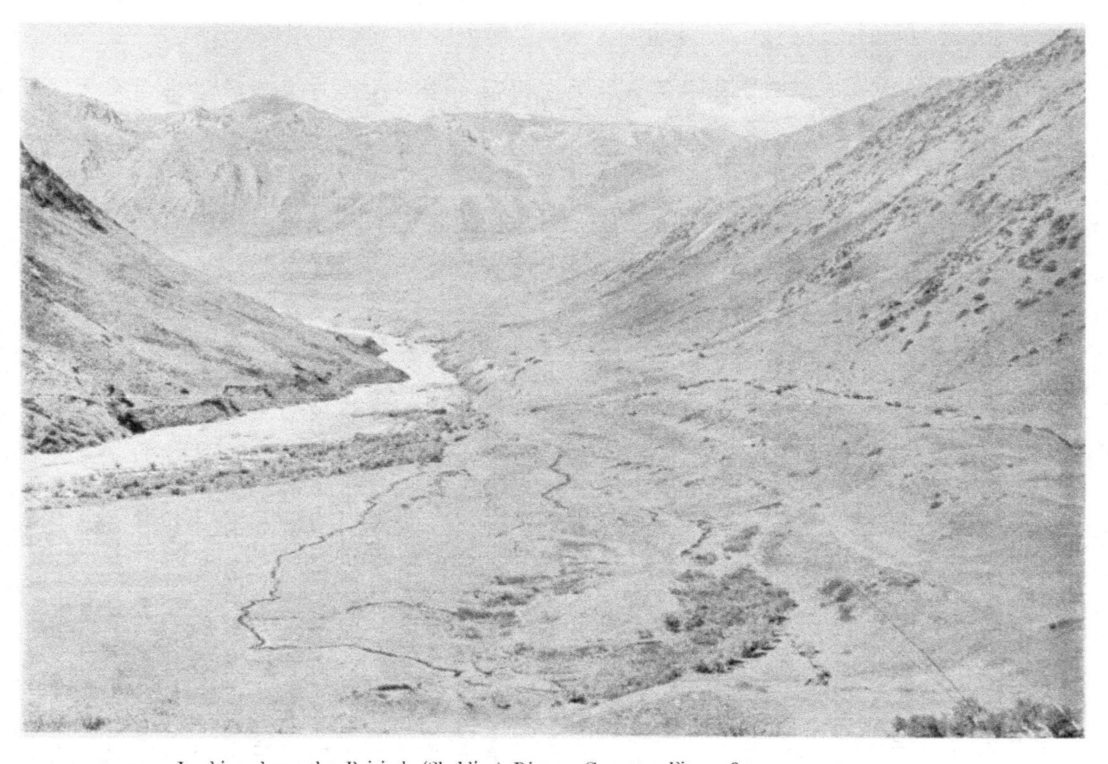

159. Looking down the Baizirck (Shaklisu) River.—Compare Fig. 158.

too easily forget the older ideas which cannot have been altogether wrong, seeing that our present knowledge has been built upon them. No new theory is a revolution; it must be looked upon as an addition, reducing the previous ones to a more limited and more reasonable circle of activity. To neglect entirely what we knew before, means committing the same mistake corrected by later views, i.e. explanation by one exclusive axiom.

I am entirely in favour of the gradual growth of all things in the universe, leaving apart the pertinent philosophical question by what standard of time or movement the term "gradual" is to be defined. But we know that slower movements are measured by higher speeds and vice versa, or, in other words, that world processes are not of a uniform rate, but of changing speeds, of differential impulsiveness. Catastrophes are as possible in geology as in daily life, they are the rarer occurrences accentuating the solid march of forces in the average. To the geologist it would not make any difference whether Lemuria (supposing it did exist) was sunk in a day or in a hundred years, for to him a century is like a flash. A catastrophe is an effect condensed into a shorter time than the changes under observation for a longer period. Until we have established the sidereal calendar of geological formations and the duration of terrestrial dynamics, our mode of expressing changes is always that of catastrophe. If a continent disappears between the history of two strata, the event is as disconcerting as the vanishing of Poland from the history of empires. Does not our mind always think in terms of catastrophe! Nor is language very much slower, even if every time we were to add a saving clause that the event took "years and years" or "millions of years." We say, "The river has changed its bed and cut through a ridge." Have we not reported a catastrophe? The fact is that we are wont to allow too short a time for past events. That is a psychological truism familiar to everyone from his own experience when recalling childhood, travels, etc., or when surprised at the duration of a return journey. Knowing this, we correct ourselves again and again, becoming more cautious every time, and too cautious in the end. In the generalising scientific mind this repeated correction of under-estimates easily leads up to the final axiom, "Nothing ever happens suddenly," which is manifestly absurd. It would be scientifically meaningless owing to the lack of an absolute of comparison.

Human thought moves pendulum wise between extremes. But as man and science get older, these swings are seen to be but the transversal amplitudes of one long inertia of uniform progression. From time to time a quickening may be observed due to the birth of new theories and new men. These are not reversions or deviations, but simply periods of increased activity at certain points; they are local

160. Camp among the Morainic Landscape of Little Tupchek.

changes in the relative grouping of energies, which do not affect the great tendency underlying the whole. Thus the catastrophes and reactions within the scientific system of mind pictures are the analogy of things themselves. Nature's doings are elaborate and complicated in our eyes, but just for this reason her occasional impulses are all the more radical.

With the word "catastrophe," brought forward as an example, we have really to do here. It is hardly ever mentioned in polite science or surface geography, being considered a historical relic left to the stratigraphical geologist to deal with at his mercy, to fossilize it, as it were. The morphologist nowadays is so fascinated by the beautiful slowness of exogenism, that he fights shy of all the more brutal possibilities of tectonic or volcanic changes. We often forget other likely explanations in our endeavour to reduce everything to "natural" and "slow" evolution. The time which a river takes cutting through a mountain range, we have unconsciously accepted as the average speed of processes of denudation and accumulation which are uppermost in our thought. Enormous upheavals or similar endogenous results are certainly reckoned with, as preceding our present era of hydrodynamic agencies, or even as running synchronously with it. On the whole however there is a marked habit of shunning breaks, rents, and tears, even should nothing speak against these effective, if violent possibilities of the telluric insides. We strain at an earthquake but readily swallow the most subtle theory of minute aggregations. All I mean to say therefore is, that we should not refuse to seek "easy" explanations when nothing speaks against them. When we know nothing, the simplest theory is just as good as the more elaborate one, especially when both can be equally true.

In the case of the Baizirek river I should say that an earthquake is a possibility. Andizhan and Karatagh have been destroyed by these disturbances and Tupchek is well within the sphere of influence. Kurban-kul near Shakh-imadan in the Alai is a lake dammed up behind a landslide caused by an earthquake. Moreover Lipski reports a strong shock at his Karashura camp. Not much would be needed. A little crack in the ridge, that is all. At the same time I must confess that I like a fluvio-glacial theory as well, or even better. It is more sporting, so to speak.

From the strictly scientific point of view we are not allowed to bar the likelihood of a tectonic lesion. No external indications enable us here to decide between the two theory-potentials of earth blasting and earth modelling. The ultimate choice then becomes a question of the aesthetics of system building.

East

161.

KARASHURA (TUPCHEK)

South

Panorama of the Glaciers of Peter the Great. Taken from a Point on the Southwest Ridge of Great Ach

GT ACHIK RIDGE BAIZIREK VALLEY

162. Panorama from the Edge of Little Tupchek; looking East.

South

Panorama of the Glaciers of Peter the Great. Taken from a Point on the Southwest Ridge of Great Ach

GT ACHIK RIDGE BAIZIREK VALLEY

162. Panorama from the Edge of Little Tupchek; looking East.

West

:hik.

163. The Battle of the Plants and Stones.—An old Mudspate on the Right.

CHAPTER XVI

TO KALAIKHUMB AND THE YAKHSU CONGLOMERATES

We left Tupchek on the 16th of September, recrossing the Gardani-kaftar to Liangar. A night was spent on the banks of the Zeriu-zamin torrent and a visit paid to the glacier snout. Judging from the size of the ice-stream and the volume of water discharged, the névé basins of this valley cannot be much inferior to those of Peter the Great. Branching off from the Khingob at the village of Minadu, we followed a southerly road leading over a set of low passes in a country of rounded and steppe-like hills. A thousand feet or so above Minadu one surveys nearly the whole of the western and snowless end of the chain of Peter the Great. Seen at a distance the landscape of its slopes appears like a plaster model carved with a sharp tracery of ribbed design. A ride of seven hours brought us to Safed-daron (Fig. 164) where we enter the zone of conglomerate rock claiming so large a surface of Eastern Bokhara. At the corner of the road near Pisteliak stands a fine old juniper, and under its shade I took the photograph of Fig. 163, showing how the rim of a scree fan draws a clear boundary line between shifting rubble and settled vegetation. To the right of this scene of tenacious struggle we can discover the lobate snout of an old mudspate whose surface is now completely overgrown with grass and scrubby weeds.

Next we came to Sagirdasht familiar to many travellers as a place of important cross-roads. It is a pass of passes, akin to a turn-table of converging rails. The kishlak lies in a shallow funnel like a pea in an oyster shell (Fig. 165). It marks the meeting of all the rivers which here run together to be engulfed by the outgoing gorge. The landscape is one of soft and wavy outline, no cornered peak being set in the large halfmoon of embracing ridge. Although the valley folds are deep enough to make mountainous swells, the whole can be called a plateau, because the horseshoe rim falls away more deeply and more rapidly towards the outside than towards the inner basin. The broad backs and flanks of these massive hills, here and there tapestried with the yellow patterns of barley fields, produce as great an impression in their way as the wildest glacier scenery. To all intents and purposes Sagirdasht may be reckoned among the pamirs of the Duab. Rarely have I seen so regular a build of the watersheds dividing, as they do,

164. At Safed-Daron.

with almost theoretical precision attributable to the homogeneous structure and horizontal stratification of the conglomerate.

In order to see the capital of Darwaz we dived from the pass of Khoburabat into the ravine of the Khumbau river. In the hamlet of Khobu we found lunch prepared in a tent pitched upon the flat roof of a mill, so that we ate our repast to the homely rumble of the millstones underneath, revolved on the turbine principle by a race of darting water. Further down we saw the fresh mudspate channel of Fig. 167, traversing a cone delta corrugated by the bulges of older tracks. The limpid pools of the Khumbau contained many fish, caught by the natives in baskets placed under little waterfalls. Seven hours after leaving Sagirdasht we found ourselves gazing upon the mighty flow of the Panj or upper Oxus. Kalaikhumb is laid out on a triangular terrace sketched by a bend of the Panj. It is not larger than a moderate village of the plains, thus illustrating the poverty of the mountainous

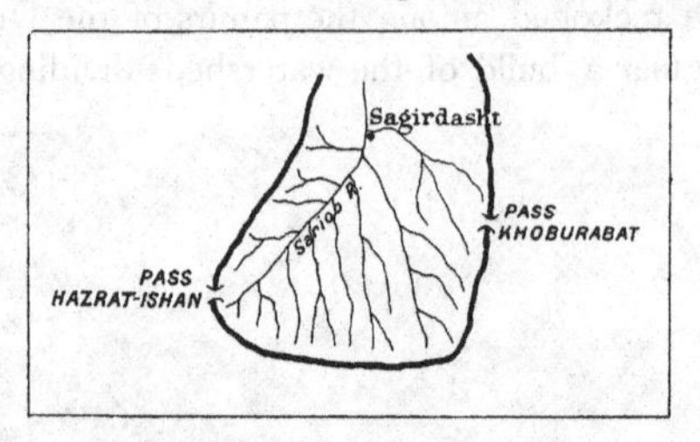

165. The Sagirdasht pamir.

province of Darwaz, furthermore revealed during a short excursion down the river. As a whole the valley presents a cheerless sight with its black slate rocks and arid scree slopes. Only here and there a cluster of trees shows where frugal peasants manage to wrest a pittance from the unwilling desert of stone. Enormous moles of rubble amassed on the borders of the fields (Fig. 169) testify to the

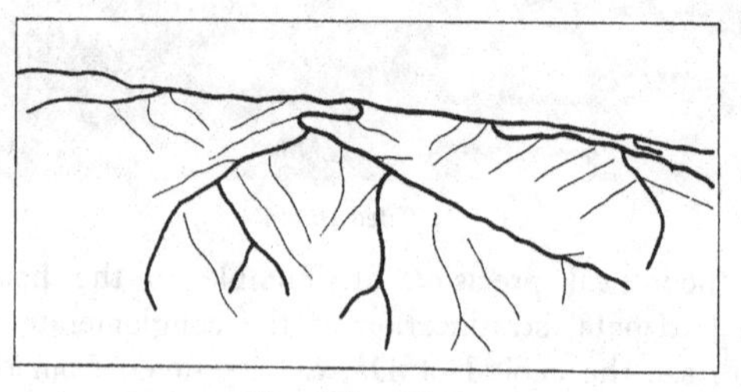

166. Ridge Dichotomy.

rarity of soil. Famine is a constant visitor to these parts where dried mulberry is a common substitute for bread, and even the pith of asafoetida is frequently eaten. At this time of year a pleasant feast is offered to the eye by the gleam of scarlet pods of peppercorn hung out to dry in rows and bundles.

The Afghan side presents nothing of special interest, unless it be the marching up and relief of the frontier posts at certain hours of the day. We paid our usual homage to the Beg and there met

167. Mudspate Track in the Khumbau Valley.

Baron Anatol Cherkasov of the Political Agency and now Russian consul at Julfa. He had just returned from a journey of inspection through Shugnan and Roshan, the Pamir provinces of the Khanate. Among other interesting things he told us of a sect calling themselves

Panj Tengi, and which he believes to have something to do with the Ismaelians. What with Kafirs, Iranians, Jusufzai, Galchas and others, the ranges of Pamir seem to hold within their icy embrace many a secret of human origins. According to v. Schultz the Pamir Tajiks are mostly Murudic Muhammadans. Their priests are called ishans and are supposed to be descendants of Ali. The head of the sect, the Aga Khan, lives at Bombay. The oldest portions of the citadel of Kalaikhumb at once recalled memories of Julius Caesar who, not content with scourging the Gauls, has laid a curse upon untold generations of schoolboys. His description of the Gallic walls seems to fit the construction of the old fortresses of Darwaz, built of alternating layers of beams and stones. Schwarz even assumes a direct ethnological connection. The only tower left standing is provided with those jutting, nose-like loopholes (machicolations) used for throwing stones or pouring molten lead and pitch upon the beleaguerers. They are the first of their kind I have noticed in the East. The courtyard (Fig. 170) is of the conventional order with its tank and square flower bed planted with aromatic herbs.

168. The Guard at Kalaikhumb.

The town boasts of a native garrison which entertains the citizens with martial music and military evolutions every morning and evening. Their appearance in a garb that strives to imitate Russian uniform has something of the ludicrous about it, being one of the many examples of perversion of taste which contact with Western civilisation so often produces in the Oriental. The barracks of the force consist of several rows of low mud huts. What makes them remarkable is the way in which they have been embellished by the tender art of gardening. Gay flower-beds fill every space, the walls are hidden under rambling leafage and rich garlands of bindweed are made to trail over lattices and arbours. It seems that soldiers are the only ones in this country who practise this effeminate cultivation for the sake of beauty alone (compare also Fig. 187) probably because they feel time weighing heavily upon their hands. Theirs at any rate are the only gardens conforming to our idea, and one must admit that charming results are obtained by simple means.

Returning to Sagirdasht on September 24, we passed on to the

169. Stones for Bread; on the Fields near Kalaikhumb.

valley of the Yakhsu. On nearing the Hazrat-ishan pass (Fig. 165) the traveller finds himself in a typical pamir landscape of great width and sweeping magnitude of line (Fig. 171). Might this not almost be a northern moorland scene with the brook meandering in its peaty track? The tallest vegetation is represented by the withered candelabra of asafoetida. Along the ridges we see perfect examples of the hollowed halfmoon markings caused by snow pressure (cf. also Figs. 155, 174), the embryonic stages of corries or glacial cirques. After crossing an intermediate valley and another pass we entered the Yakhsu valley at its upper end. Here we were in the midst of the great conglomerate region. Everywhere the openings of side valleys offer new peeps into one of the quaintest mountain formations of the world. Earlier travellers have never mentioned these sights, hence I may assume to be the discoverer of the Yakhsu Conglomerates when I first visited them in 1896. Their shapes betray much similarity to those of Colorado and the Bad Lands. This clearly stratified mass, to which

170. The Palace Yard at Kalaikhumb.

Krafft ascribes an early tertiary age, covers (as far as known) a surface of 400 square miles disposed in a strip between the rivers Vaksh and Panj. The vast accumulation of water-worn pebbles attains a thickness of many thousands of feet enabling some of the peaks to top the snow-line. As can be gathered from our photographs, the deposit has been broken up by later upheavals in spite of which some of the blocks or sections have remained in their original, undisturbed position, as shown by the perfectly horizontal bands of the layers. Fig. 172 shows a tilted

171. Nearing the Hazrat-ishan Pass. A Valley of the Pamir Type.—Asafoetida Stalks on the Right.

portion on the right bank of the Yakhsu below the village of Talbar. In the sidelight of the morning sun the crinkled chisellings of water stand out well on one of the fractured slabs.

For a few days we took up our abode at the kishlak of Pamak where Mr Pokorski conducts gold washing operations on the banks of the Safed-darya, a tributary of the Yakhsu. A native house (Fig. 184) sheltered the servants, while we preferred the better ventilation and lesser zoology of our tent, especially as Russian workmen have introduced the nimble and secretive bug hitherto a stranger to the neighbourhood. The tree in the picture is a willow and the fence is made of osiers, but as often the stems of asafoetida are used for the purpose. Imposing heaps of winter fodder consisting of yughan and lucerne are piled upon roofs and scaffolds or disposed between the forks of strong trees. These tapering stacks remind one of topes or the conical roofs of Ossete burial chambers and of temples of northern India. How far these forms betray a relationship of race I do not venture to say, nor would it be safe to speculate, if the haystack has served as a model for certain styles of architecture. Building material is a strong influencer of form. Yughan is tied into flat bundles before being stored, and slate slabs are arranged upon the Ossete tombs. Both these block shapes would naturally lead to a cone of receding steps.

After taking leave of Cherkasov, who had been a delightful companion, we penetrated into the Dandushka glen, leading into the very midst of a miraculous world. Strange shapes have here been born from the lap of earth. A maze of crests and pinnacles shoots up towards the sky, and gentle groves are surrounded by the stark bastions of gigantic castles. The smooth, towering walls are cloven by vertical rifts, and grey rubble fills the gullies fissuring the pedestal of the lower cliffs. Only the hermit juniper dares climb to where the last shred of mould clings to the rock. Mysterious like a wild fairy dream these fantastic edifices stand as if a wizard had created them to make mute the speech of man. Fortress is piled on fortress, keep stands by keep, all planned and carved as by a master hand; there are pillars as if turned out of a lathe, tapering minars, sloping roofs, vaulted domes and cupolas, surrounded by turrets and graceful spires; broad ledges are drawn along the abyss like rock-hewn roads interrupted by terraces or giddy slabs; in between are bays, balconies

172. Tilted Conglomerates on the Yakhsu near Talbar.

and niches in inexhaustible variation. Puddings, loaves, tents, coffins, turreens and other oddities of the imagination are everywhere placed among the models of a more serious architecture. Cut through this infinite labyrinth are ditches, dark and bottomless passages, so deeply sunk that hardly a ray of light enters this realm of frosty draughts. They are the corridors leading to open courts where green bushes and trees enjoy the light of day in still retreat. These are peaceful spots, and it is as if here an enchanted princess had secretly tended her bower of roses and honeysuckle, birch and maple in the most hidden recesses of the ancient and gloomy masonry. When the sun goes down, the array of gaunt palisades is outlined against the lemon tinted evening sky. Above the thin, blue veil filtering through the nether vales are thrust the shadowy gables, portals, buttresses and parapets of a spectral castle or ghostly cathedral.

These valleys owe much of their peculiar charm to the flatness of their floors where shingly beaches are spread between fuzzy copses and groups of trees. Usually among wild mountains, such as the Dolomites, the ravines of the higher regions are steeply inclined and gutter-like, so that in ascending them one walks on the bordering slope, whereas here the cliffs rise vertically out of a level plain. We revisited familiar haunts known to us from former occasions, our favourite among which is Birkendale, hidden away in the innermost folds of this stupendous maze. An ancient birch, now glorying in its amber leaves, stands near the boulder-framed brook. Under its spreading branches we have spent many days, drinking the joys of solitude. There is plenty of firewood and excellent water, while polished bathing tubs scooped out of the conglomerate are filled by the babbling burn. Near by are the tops of two high masses, Kuch-manor (10,500 ft.) a pointed pyramid, and Kuch-kalandar (10,000 ft.) crowned with a broad plateau visible on the extreme left of Fig. 181. Kuch-manor is one of the highest summits of the district, and in 1898 Albrecht von Krafft, my wife and I made many unsuccessful attempts before our efforts were rewarded.

Owing to the complicated nature of the ground and the almost invisible slits of narrow canyons, reconnoitring from a distance supplies but scant clues to the choice of a promising route. Hence we had to proceed by guesswork, trying numerous lines through an intricate

173. The Surging Rocks. Yakhsu Conglomerates.

system of gorges and gullies that might possibly lead us to the summit screened behind a succession of overhanging walls. At last we solved the puzzle by a devious crawl through sombre clefts and over dizzy ledges, without however meeting any serious difficulties. We first reached the broad platform of Kuch-kalandar, whence one descends to a low saddle in order to climb a long slope of grit to the triangle of Kuch-manor. After solution the task seemed ridiculously easy, and we were now in a position to imagine the astonishment of Columbus when he suddenly found that his egg was able to stand.

Once my wife and I explored one of the dusky side passages, a crack that seemed to split the precipice to the very core of the mountain. It was only six feet wide and three hundred deep, closed on the top by a wedge of snow partly melted into strange shapes. Portions of this roof had already fallen in, while others overhung us threateningly like the craning necks of uncanny monsters. We slunk through, squeezing gently past blocks of snow, without even daring to whisper. Finally, creeping through a man-hole, and under the spray of a waterfall, we stepped forth into a cauldron hemmed in by rocky tiers.

The upper reaches of Birkendale are filled with the jammed masses of blocks, the size of houses. Over and under these the wanderer must pick his way laboriously, and this would be, indeed, a fit place to bear the name of Devil's Gorge so dear to the tourist-trodden Alps. But why thus? To us the mountains are no longer the abode of evil spirits. Would not the "Road of Gods" sound truer and more dignified? Such are not the stepping stones of a poor, limping devil; here gods have passed. Let us seek their tracks and follow where we can. Three times within the last fifteen years we have returned to this place with which are bound up so many memories. Here Krafft received the telegram calling him to India where he died only a few years later. Here, over the embers of the camp fire, we have reviewed our lives. To me Birkendale is a grove of sacred silence where rise the pillars of a Divine gloom and where autumnal leaves tremble in the sunlight like golden flames of sacrifice.

As to their eroded shapes, the rounded forms of the softer conglomerate are easily distinguished from the jagged and fantastic outlines of the harder zone. The highest peak of the subdued category, and at the same time of the entire conglomerate region, is Hazrat-ishan (13,000 ft.), the upper parts of which are shown in Fig. 174. The main summit, on the right, is steep on the south-eastern side turned away from the spectator, because on that side project the outcrops of tilted strata. An exposure of limestone occurs on the lower dome. An interesting sight is afforded by the zigzag formation of the watershed where the heads of parallel gullies transgress against each other in opposite directions. Here one also notices the depressed crescents of embryonic corries, as well as the striking regularity and straightness of the couloirs and their sharp dividing ribs. Rarely indeed has snow such an opportunity of exerting its pressure upon so homogeneous a substance. In front of us is one of the few glaciers of which the Yakhsu conglomerates can boast. It reaches down to a level of about 11,000 feet. Two hundred feet from the upper rim a deep bergschrund cleaves it from side to side, and below this several small crevasses occur. The one terminal moraine which I had occasion to inspect, and which is a little to the right of the main ridge of the mountain, is a high mound of rubble forcing the glacier to split at this point. The most noteworthy peculiarity of this moraine lies in the fact that it consists of rounded, smooth stones instead of angular fragments. For a glacier working on conglomerate, the production of such a moraine is of course quite logical, but it strikes one forcibly as one of those quaint possibilities which the fertile imagination of nature loves to invent.

A gentle, sinuous ridge, seven miles in length, affords a splendid line of ascent from the Safed-darya valley. We used to ride to just below the glacier, completing the rest on foot. We cannot claim to have made the first ascent, for on the summit a huge cairn with a flagstaff informs us that the mountain is a mazar or sanctuary. Owing to its commanding position and accessibility it has been deemed a proper goal of pilgrimage. Native officials in their gorgeous raiment may now and again be seen wending their way to the top, bent on a visit to the holy spot. Anyone not acquainted with the real reason for such bodily exertion on the part of these languid gentlemen, must rub his eyes and ask himself if he were not dreaming—an Oriental dignitary as a mountaineer is too strange a sight. Even our fat Karaul-begi subjected himself to this ordeal, although greatly to the displeasure of his nag which had to carry him all the way up, and whoever knows the wobbly cobbles of the final slope will pity both animal and rider.

Owing to the favourable situation of Hazrat-ishan the view is beautiful and instructive. The great Transcaspian plain sends out a wedge in the direction of the Pamirs, and our mountain stands very near the middle of the sweeping crescent of Alai-Pamir-Hindukush, and just on the rim of a continuous expanse of snowy peaks. Thus we are at the apex of an arc forming the boundary line between a region of eternal snows and the lesser hills below the snow-line. Looking towards the south-west we see a country of brown, grey and yellow losing itself in the hazy distance of the plains, but turning round in the opposite direction our eye encompasses a landscape which is mostly white, there being hardly any peaks without ice or névé up to the foreground. Right and left the horns of the white halfmoon thin out into the arid steppe.

According to our Karaul-begi, to whom I must leave all responsibility for this information, Hazrat-ishan is also known by the names of Shakh-imardan, Hazrat-ali, Murta-za-ali, Haidar, Shirikhuda and Hasham-shan. Being seven in all they ought to constitute a sufficient proof of sanctity. He converted Darwaz, having left Afghanistan after a quarrel with Shakhbart-ali. He prayed on the top of the mountain, and from this point of vantage claimed for Islam the country around. The Karaul-begi also told me the following story showing the miraculous power wielded by that doughty messenger of the Prophet. Once upon a time there lived a man by name Babarafshan to whom a Jew owed a thousand tilla, which sum might be commuted by cession of a daughter or conversion to Islam. But the Jew was unable to pay and confided his trouble to Hazrat-ali who went to Babarafshan, saying, "Take me in payment of the thousand tilla." Whereupon Babarafshan climbed upon his back, and after Hazrat-ali had uttered a prayer he flew with his load to the city of Barbar. Prince of Barbar was Anka, and the country was heathen. Here Babarafshan took his willing slave to the bazar offering him for sale at the price of a thousand tilla. But nobody knew that it was Hazrat-ali. No buyer was forthcoming, and Anka hearing of the matter commanded the man to be led before him, and he asked him, "Why art thou so very expensive, what canst thou do?"

174. Hazrat-ishan.

Whereupon Hazrat-ali answered back, "What are the things thou wishest done?" Whereupon Anka said that there were three things he wished done and that he would gladly pay a thousand tilla for the work. And these were the three tasks. Firstly, there is an unruly river flowing through the city; it is a relentless destroyer and ten thousand men cannot dam it. Secondly, there is the dragon Ashdahar twelve tash in length, who devours whole towns. And thirdly, Anka desired to have Hazrat-ali brought as a prisoner, for he knew not that Hazrat-ali stood before him.

But Anka had a pahlawan capable of throwing three hundred men at once, who addressed him thus, "O Anka, thou wouldst be wasting thy money on tea; give me leave to wrestle with this man and I shall kill him." Anka gave leave, whereupon Hazrat-ali shook the pahlawan by the arm so that it came out at the shoulder, and with this arm he split his skull in twain. Then Anka ordered two thousand men to help in the damming of the river. But Hazrat-ali sent them away and prayed. There came an angel from Mecca who cleft the mountain with his sword and the river was engulfed, so that now the people complained for want of water. Hazrat-ali laid his hand upon the soil and eighteen springs issued from the earth (Hashdanhar, a district around Balkh). Afterwards he went up to the dragon allowing himself to be swallowed by the monster. Then, from inside, he ripped open its belly and cut the animal into a hundred and sixty pieces which he threw to the wind. Lastly, he asked of Anka five hundred men with ropes and chains. These he led into a forest and said, "I am Hazrat-ali, bind me and take me to Anka." And in front of Anka he tore his fetters and bade him follow Muhammad. Then the people broke their idols, and built mosques, and prayed to Allah.

The tale is certainly very enlightening, but it leaves my sense of justice unsatisfied inasmuch as I miss a definite statement at the end, that the Jew received his thousand tilla. I have my suspicions, however. It must strike one as remarkable that a saint should trouble himself about the claims of a Jew. Knowing the nature of the relations between Semites and Moslems, it now dawns upon me that poor Shylock must have been made a fool of, that Hazrat-ali, not over scrupulous in the choice of his means, used him as a pretext for the conversion of Anka's realm. I think the production of a formal receipt, after all was over, would have made his wonderful feats all the more creditable. According to modern notions the story shows a deplorable lack of finish. The saint ought to have suddenly appeared with a sack containing the tenfold sum, whereupon the Jew touched by this magnanimity would offer his conversion and give his daughter in marriage to Hazrat-ali. Finally a gorgeous baptism and wedding-feast would be arranged on the top of Hazrat-ishan.

As to the animal world in this neighbourhood, I only observed two wild mammals, marmots and a species of mountain sheep. I once stalked and wounded one of the latter, but it got away among the inextricable maze of gullies and ledges. I believe it was some species akin to *Ovis arkhal.* The habits of the marmots are instructive from the point of view of animal economy. The beginning of hibernation is determined by two different factors, the dry season and the cold season. Around the Russian colony of Safed-darya, at a height of 6000 feet, their piping is continually heard throughout the summer, but they do not show themselves any more at the entrances of their subterranean abodes after the middle of August. Two thousand feet higher their food is green much longer, and there they do not retire before the beginning or middle of September. At an altitude of about 10,000 feet only the cold will send them to sleep, because water trickling from snow-patches keeps little kitchen gardens going for them.

The same causes influence the migration of domestic sheep who, with their shepherds, infest all accessible mountain slopes, gradually climbing higher and higher until everything is eaten up. Then large flocks of fat tailed sheep are driven downwards from all directions and caravans of thousands of them throng the narrow paths leading to the lowlands. There is one glorious pasture however safe from the cattle of man. This is the mesa of Kuch-kalandar with its undulating surface of half a square mile conspicuous from Pamak, and which changes its colour from green to brown as the months succeed each other. Towards all sides steep rocks descend from it, and the broad tracks made by the mountain sheep show that here the wild denizens of the heights feel secure amid an abundance reserved for their exclusive use. Near the upper ridges we often put up the snow cock or giant rock partridge (*Tetraogallus* or *Megaloperdix*), one of the finest game birds in existence. Eagles, vultures and lämmergeiers are seen in great numbers drawing their wide circles through the air. Apparently out of curiosity, they often pass quite near to the climber in order to look at him, generally taking care to approach him from behind. All the same one has to use ball cartridge, as shot only tickles them or causes them to rock violently in the explosive wave. Only once did I succeed in bringing down a lämmergeier with my sixteen bore.

In 1898 we spent three months at the Russian colony and during our outings never felt the least anxiety with regard to the weather. When day after day we beheld a cloudless sky, we at last took it as a matter of course, and were never deceived. For a mountain district the regularity of the climate during summer is astounding. From the beginning of July to the end of October we had not a single drop of rain. In November however the Italian atmosphere changed into a Scotch one, mists came rolling down the slopes and snow began to fall. Until March these valleys are buried in snow. Then comes a period of rain and melting sloppiness lasting till May when the sun begins to battle effectually against humidity. No wonder then that during late summer and early fall, the characteristic features of the mountain steppe or desert are evident in every direction. The glacier waters of the Safed-darya, milky at noon, run through barren shingle flats. As one walks across the slopes, dry and brittle plants are crunched under foot, and at the least touch they send up puffs of a brownish yellow powder which makes one sneeze. In the stillness of the canyons the traveller is often startled by the loud rustle of drought-hardened rhubarb leaves dancing over the stones at the lightest breeze. Yet there is a good deal of pleasant vegetation. Nearly all scree fans and the river accumulations above the flood plain are covered with maple, juniper, poplar, birch, willow and thorny trees, their deep roots taking advantage of the moisture retained by alluvial deposits. The firm conglomerate is almost impervious to water, hence the rarity of springs.

Krafft has made an exhaustive study of the conglomerates which cover an elongated area with a general strike from north-east to south-west. Some parts, especially those around the Yakhsu, are very hard and distinctly stratified, while in other districts the material is uniformly soft without visible layers. Here the mountain shapes are rounded and the valleys broad, as witnessed by Sagirdasht. The stones composing the concrete are chiefly crystalline and strongly cemented together in a matrix of sand and lime. Levat's idea is that the Yakhsu conglomerates may be the fluvio-glacial product of eocene Pamir glaciers. The theory, though fascinating, has to contend against many difficulties, but it should be kept in mind and put to the test as soon as the topography of the western Pamirs and their various altitudes and levels are better known. As far as I can gather from books these conglomerates are the most extensive formation of their kind in the world. Explanation of their origin is not so easy as at first sight appears. That they are composed of river pebbles goes without saying, but the question is how they have been deposited. They cannot have accumulated on a sea bottom, because then they would show a better assortment of the particles according to size and weight. Krafft thinks that they were heaped up on a shore, there to be stratified by the surf, and so far that seems to be the most likely explanation. We may therefore imagine that several huge rivers unloaded their stony freight into a wide estuary. After the retreat of the sea came the warpings and dislocations affecting the whole of the Alai-Pamirs and to which the parallelism of the rivers (Yakhsu, Baljuan river, Surkhab) seems to stand in some relation, owing to a frequent synclinal and isoclinal arrangement of their valleys. The upturned slants near Talbar are more or less parallel to the inclined sandstone slabs of Sagunaki.

Apart from the main cleavages due to faulting, the queer shapes of this locality have been produced by erosion. It is well, however, to assume a certain tectonic blocking out of the stratified mass as giving the first lead to water action. This bizarre configuration is common to various well-known sights of the world (Colorado, Bad Lands, Garden of the Gods, Saxon Switzerland, Montserrat, volcanic deposits of South Sea Islands, Dolomites, etc.) and chiefly characterised by deep canyons with vertical walls, extremely regular shelves and copings, smooth, round masses, and pillars of monolithic majesty. Three factors, singly or together, favour the evolution of Colorado forms, as one might call them according to the fundamental and most familiar type. These are: horizontal stratification which supplies an orderly arrangement of material and guides the waterways cutting the layers at right angles; a dry climate and absence of vegetation which favour the linear attack of erosion; and a homogeneous rock, not too hard and not too soft, which everywhere offers the same resistance to the tools of nature. Some of the table lands of Colorado are less dissolved, less advanced (than the Yakhsu conglomerates) probably because the rock is harder and the climate too dry. Loess on the other hand, the canyons of which show a distinct tendency to Colorado forms, is too soft to admit of the upkeep of isolated steeples. Mountains of granite, slate and irregular and tilted limestone are not only sculptured, but also broken to pieces, whereby crooked and unruly outlines are produced. Joint planes and facets of crystallisation guide the water into various accidental directions, thus preventing it from planing large surfaces or cutting straight lines.

Firm, compact limestone and clastic rocks, like sandstone, conglomerate and grit, offer a mass of uniform strain and texture to atmospheric and dynamic agencies. In this respect the rarity of blocks is significant showing that weathering detaches separately every stone or grain from the solid concrete. The large composite blocks in the upper sections of Birkendale and elsewhere, are sporadic occurrences sometimes ascribable to the splintering effect of glacier action or, perhaps, earthquakes. Owing to the constitution of its mass comparable to the granular structure of glacier ice, the conglomerate must be endowed with a high degree of plasticity. This will enable it to yield to much folding without cracking, and even should cracks begin they will often be deflected into infinite directions between the rounded cobbles clinging together like so many billiard balls. Likewise there is a certain amount of play between the pebbles and their bedding, thus opposing compensation to the strains of changing temperatures. Weathering seems to be very slow for hardly ever does one hear a falling stone. Above the level of the river terraces and fan deltas, scree slopes are practically absent, partly owing to insufficient supply, but partly also to the roundness of the fragments which continue rolling down the steep inclines until they reach the bottom of a gully. Everywhere the mountain is armoured with a hard and flawless surface. Only here and there, on ledges, in couloirs and gentle rises we tread a dusty, gritty mortar which demands the cutting of steps when one wishes to obtain a purchase for one's foot.

Overhanging by the harder layers is not very prominent, being limited to a few slightly bulging belts. Naturally the differences of hardness in such a mass of fairly regular deposition are not so great as in the classical example of the earth pyramids, or where strata of different rock are superimposed. On the whole the conglomerate figures taper throughout, which apart from other considerations we can ascribe to the fact that whatever is more solid owing to certain admixtures, will also increase in solidity towards the base as a result of greater pressure from above. Thus we can form a general theoretical idea *why* curious Colorado shapes may somewhere originate in preference to others, but we are still far from realising the exact technique or *modus operandi* by which these fanciful objects were turned out. Who will venture as yet to give a vivid and detailed description how a Turkish tent, or the toadstools (Fig. 179), or the almost circular bastions (Fig. 2) were sketched and finished? As there can be no lack of rudimentary and intermediate forms, besides the maturer ones, a geometrical genius might be able to present us with complete graduated series of this extraordinary mountain anatomy. We may also ask ourselves if glacier ice has had a share in preparing the larger features. It is not at all impossible that some of the steps or bands may correspond to the glacial benches of the Alps. If so, the ledges at the sides of the larger valleys should, on the whole, be wider than those of the lesser valleys. I suspect that the Safed-darya reveals the lines of a trough independent of stratification, which here is, of course, the prime factor in the production of shoulders. Only there is a serious hitch, in so far as along the path of the assumed overdeepening, there occur isolated turrets continuing their outward curves round the corner away from and even opposite to the ice. Here the eye alone is powerless even if assisted by rough measurements of altitude. Only a minute determination of levels will enable us to distinguish between the various horizontal or gently sloping planes.

I have already mentioned the fact that gold is found here. The precious metal has been obtained from this region for centuries, as is shown, apart from popular tradition, by the old heaps of tailings at higher levels than those frequented by the present natives who only work near the streams. Legend associates these old diggings with Jenghiz Khan and Timur. The apparatus for treating the gravel of the placer is very simple. A set of felt strips are laid upon a flattened incline or desk of sand hemmed in with asafoetida stalks. A wooden grating is placed at the top and upon this sieve a shovel of pay-dirt is thrown and water poured over it with a ladle. The fine grit and sand are washed down, and the gold is caught by the hairy felt. All gold dust is lost, only the rougher particles being secured. Groups of five men generally club together treating perhaps two tons of crude or half a ton of assorted gravel per day. Two dig out the gravel, two prepare or assort it by shovelling it in a heap with the addition of water, throwing out boulders and biggish stones, after carefully washing off the clay; one attends to the sifting. These men are all in the hands of sweaters who advance them the necessaries of life at exorbitant interest, keeping them continually in their clutches. Consequently these poor miners are never able to attain prosperity and do as little work as they can.

The yearly gold output of Eastern Bokhara is estimated at £30,000 per year, not counting what is secretly sold across the Afghan border. But it is a mere trifle as compared to the vast potentialities of these alluvial deposits. The quantities extracted by the natives in the course of centuries have not encroached upon the main store, and they are as nothing to what European engineering might turn out here in a few years. So far the proportion of gold in the conglomerate itself has not yet been established. As the fluviatile placers result from the disintegration of the hard concrete, the gold lies already in its third bed, its original birthplace being in the surrounding crystalline ranges of Darwaz. The metal occurs exclusively in the form of thin scales, there being no coarse grains or large nuggets. The biggest piece I ever saw weighed about half an ounce. The gold is 92·7 fine. We have made extensive investigations into its distribution and I here give some of the results. Two general facts were always apparent: the richness of the gravel increased greatly with depth and the horizontal distribution is very even, for we found the same proportions everywhere. Surprises in the shape of nuggets are out of the question, but so are serious disappointments, for what one finds in one claim one is sure to be found five miles away, if only the general conditions of the locality are similar. The terraces and fan deltas were richer and contained coarser gold than the more recent alluvions in the middle of the valley, close to the river. This is of course explained by the fact that near the banks one is also nearer the bedrock where most of the gold has been concentrated owing to its weight. The bottom layer near the bedrock has never yet been explored by either natives or Europeans, and judging from what I know it must be enormously rich. The beds are clearly stratified and the corresponding layers can be more or less easily identified at distant spots. The top stratum is about 6 feet thick, containing an average of 5 grains of gold to a ton of gravel. Below this lie 20 feet of poor stuff with hardly any valuable contents. Here a line of large boulders is encountered as if to protect what rests under it, namely a blackish sand which in its upper portions yields 24 to 30 grains and from which some 6 feet lower down we were able to extract nearly a quarter of an ounce to the ton. The bedrock is at least 20 feet further down but probably very much more, so as to be 50 or 60 feet below the surface. But, of course it curves up towards the banks of the valley.

A native claims to have reached the bedrock once in the body of one of the great terraces, working obliquely downwards in the direction of the mountain side. Here he obtained a basketful of pay-dirt containing several ounces to the ton and was able to pick up bits of gleaming metal from the floor. These people burrow in the ground like moles producing what are known in America as coyote diggings. The tunnels are about 4 feet high by 2 wide and go down to considerable depths, but there being no ventilating shafts, the farthest point attainable depends on the possibility of breathing. As the strata are concave, thinning out towards the borders, they can be successively intersected at any angle. The native galleries descend steeply until the good layer is struck which the men follow until their primitive oil-lamps cease to burn. Thin, sickly looking boys carry the gravel out on their backs, in baskets. They are half naked and the few rags which manage to stick to them are drenched with muddy water. The ground water coincides with that of the good pay-dirt and is got rid of by what is technically termed a tail race. Its initial stage is some excavation near the surface from which the water is drained by a trench. As the work progresses upstream the canal is continued almost horizontally thus cutting deeper and deeper into the rising valley grade. The ditch is covered with flattish boulders and the tailings of the advancing digging are thrown over it, so that gradually it becomes a subterranean drain. From any shaft sunk near such a tail race a short tunnel is run out to connect it underground, the miners being guided by the sound of running water. In this manner the hidden conduits grow section by section. Some of them are over a mile long and all of them form a system of complicated ramifications known only to the old hands.

The native method of mining is extremely dangerous, for sometimes a gallery will collapse without warning and many lives be lost in this way. Some Western plant of gold mining machinery should prove highly successful on the Yakhsu and its tributaries. The distribution

175. Barrier at the Junction of Dandushka and Safed-darya.

of the metal is so safely assured that one can calculate to a nicety the contents of every claim, and if by now my own claim has not been jumped, even a poor scrivener like myself owns a million pounds between his boundary posts. The snow-fields and glaciers of Hazrat-ishan ensure a liberal and regular supply of water, the climate is healthy and allows of at least 300 working days in the year, the approaches are not too difficult, labourers are plentiful and provisions are easily obtainable from the thriving centres of Baljuan or Kuliab.

On the Zarafshan and at Little Tupchek we have already observed the phenomenon of passages eroded on the principle of the superimposed or inherited valley. Here we find them again, but multiplied in a remarkable manner, and represented by a row of barriers of diluvial detritus, combined with a gate of solid rock. This series of obstructions cannot fail to excite the curiosity of any one wandering along the combined Safed-darya and Dandushka valleys. The stream has pierced them all, but the geographer must tediously scramble over their flanks of cemented grit. A full view of barrier IV from the north is presented by Fig. 175. The Safed-darya river is coming

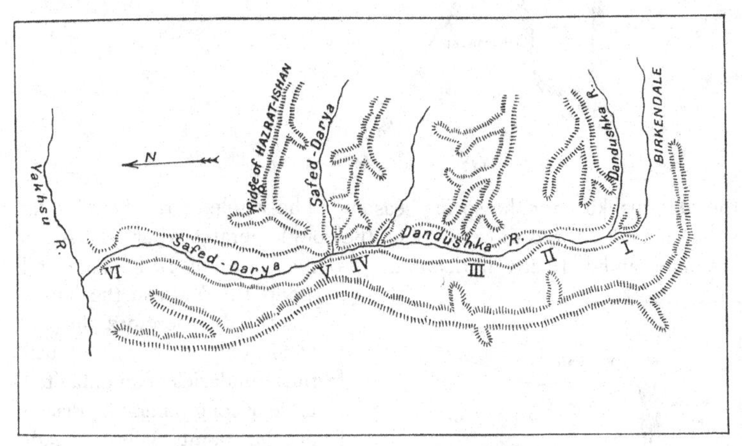

176. Plan of Dandushka Valley with Barriers.

from the left to join the Dandushka issuing from its canyon on the right. Fig. 179 shows barrier III from the south.

An idea of the topography may be formed by a glance at the plan of Fig. 176. Running from south to north there is the long and narrow valley. The upper reach belongs to the Dandushka, the lower half being claimed by the main river, the Safed-darya, tributary to the Yakhsu. For the sake of convenience I mean the entire straight line of combined valley sections when speaking of the Dandushka. The total length of this ditch is 10 miles, the height above sea-level being about 5300 feet at the lower and 7000 feet at the upper end. The gradient is 3 per cent. and very even; water is discharged into the

Yakhsu at the rate of 50 cubic feet per second (in late summer). The floor of the valley is a flat stretch of gravel bordered by the mountain declivities. On the eastern side impinge long, parallel spurs belonging to some peaks of the Hazrat-ishan type, 11,000 and 12,000 feet in height. The western bank is formed by a steep and unbroken causeway with short buttresses. From this wall not a drop of water joins

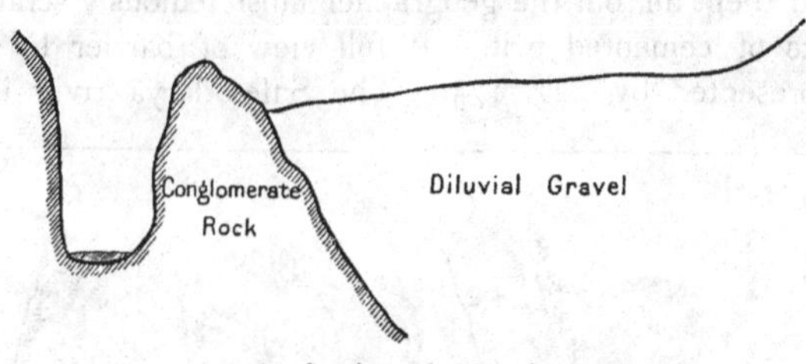

177. Section of a Barrier.

the Dandushka after the spring season. The whole of the Dandushka valley is bedded into those parts of the conglomerates where the strata are hard and perfectly horizontal. To the south of Birkendale and along the Yakhsu the layers are tilted, wherefore the extreme ends of our line touch the boundaries of an unfaulted table of conglomerate. Placed at the point where Safed-darya and Dandushka meet, the spectator sees stretching away from him towards the south an imposing avenue of fanciful fortifications or turretted battleships. If the light be favourable he can follow with his eye for five miles some of the darker bands jutting out cornice-like from the wall at his elbow, and which he still can recognize as thin black strips in the far distance.

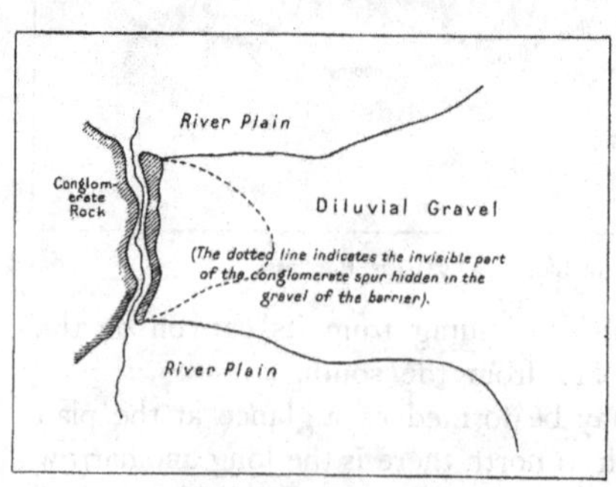

178. Plan of a Barrier.

In Fig. 176, I have indicated by a dotted line the level floor of the valley. At several places marked with Roman figures this

179. At one of the Upper Barriers of the Dandushka.

plain is narrowed in, so as to form partitions, akin to a string of sausages. These contractions are the barriers. No. III is a double bar. If one examines one of these jetties from below, it rises like a solid dam athwart the ravine. Close to the mountain side a slit gives passage to the stream. Inside this cleft, on both sides, are blank,

180. Birkendale.

smooth walls of solid conglomerate, and the same material is discovered below the water after the litter of superficial rubble has been cleared away. The average height of the barriers is 50 feet. No. I measures about 30 feet, while the uppermost terrace of V is over 100 feet above the water. The width of these dikes likewise increases as one goes downstream, the canyon also broadening in proportion to the growing

181. Row upon Row of Sculptured Conglomerate.—Barley Fields framed in Scrub; Yughan Plants in Foreground.

volume of the river. A man on horseback can just squeeze through the fissure of No. I, which is 100 feet long; IV and V taken together form a concrete canal a third of a mile in length, 12 feet across and 60 feet deep. With the single exception of the first bar, the canyon is always on the left side of the valley, close to the mountain. On entering and leaving the dark defile the river has level ground on its right, a sheer cliff on its left.

At first one may be led to think that every barrier is a solid mass throughout, but after some reflection one begins to abandon the idea of a string of tanks scooped out of the bedrock. In reality these dams are preserved strips of fluvio-glacial detritus, while the gorges are riven in the projecting ends of ridges (cf. Chap. X). A valley somewhat deeper than the present one, but otherwise practically the same, received a large supply of glacial gravel (probably during the Zarafshan period) wherewith it was filled from end to end. Owing to the greater bulk descending from the tributary glens of an important catchment area on the right, this mass of débris sloped towards the west, forcing the river to sidle down the opposite mountain flank. Excavating its bed, the Dandushka cut through some of the bulging projections and promontories embedded in the softer stuff (Figs. 177, 178). In the case of the long trenches big sections have been sliced off from the convex slope of the rock. The sides of the channel then protected the material of the terraces against being washed away. Above and below these fixed points the process of clearing out followed the usual course, and in the intermediate basins or reaches the original profile of the valley is being approximately restored. It is interesting to note how the topographical arrangement of the mountain relief has produced such marked one-sidedness. Otherwise perhaps one chance barrier or two would have been notched by stray swings or loops of the stream. But under the actual conditions, with one straight wall at right angles to the mouths of many affluents, the phenomenon has been developed systematically, so to speak, and every knob of rock projecting beyond a certain line has been lopped off with unerring certainty.

Where the barriers are near together, the outsweep has not evacuated all gravel, and the remnants of old terraces are connected by a curve. At IV a side-stream has forced itself lengthwise through

182. The "Indian Temples" in the Dandushka Valley.

a barrier. Where it crosses the main valley, as in an aqueduct, its high banks expose glacial rubble, but its entrance into the Dandushka is effected through a chasm sawed into 30 feet of rocky side-wall of the principal canyon. Thus a T-shaped incision results. On a larger scale this is repeated by the next affluent from the right, the Safed-darya, which also, before its junction, shows nothing but high terraces innocent of solid rock (Fig. 175). The top of the severed ridge is generally bared (Fig. 177), because in the early times of the process the river often flowed round it at high water during spring. All the foregoing conclusions are supported by an occurrence which can best be studied between bars I and II. There the river runs through a low stone gutter 30 feet long, having on its right a thin rim only from 2 to 4 feet high. The difference in level between this shallow trough and the surrounding flood plain is so small as to escape the eye. Thus the water seems to obey some inexplicable whim in selecting a low, rigid groove in preference to the wide expanse of soft gravel at its disposal. At flood time the whole of the valley is, of course, occupied by many branching channels. Two similar embryonic canyons are found below the sixth barrier. The bottom of the old valley is still deep down, as proved by the excavations in search of gold.

The accumulation of detritus leading to the formation of these barriers or dissected terraces is undoubtedly attributable to glacial causes. A reliable witness is the great boulder of crystalline rock which I discovered at the junction of Safed-darya and Dandushka (it lies on the extreme right of Fig. 175). Its size is about 80 cubic feet, and it rests with one of its corners on the clean floor of polished conglomerate overrun by the ripples of Safed-darya. Its foot is sunk in a saucer scooped out by swirling eddies. Over its surface are drawn many short and irregularly grouped scars, showing that it must have been brought hither in the ground moraine of a glacier from Hazrat-ishan, unless we leave open the remote possibility that it was already striated before entering into the composition of the conglomerate. Other traces of ice-action are not easily discoverable in this district, for I do not think that the nature of the concrete lends itself readily to planing by glaciers, the surface being fretted in a rough manner, owing to the breaking out of the stones. Wherever we now find a high polish it can be traced to running water.

CHAPTER XVII

THE OXUS JUNGLES, BALJUAN, KARATAGH

WE left Pamak on the 2nd of October, proceeding down the Yakhsu, where numerous kishlaks line the great diluvial terraces on either bank. The openings of side valleys on the left (south) ever disclose new forms and arrangements of conglomeratic fancy. Below Saripul the spurs of conglomerate begin to thin out, and near Sinji there is a last revelry in the shape of a comb of cylinders and slender finials[1]. Here I shall describe an excursion which my wife and I made to the province of Kuliab towards the end of October, 1898. Below Saripul we struck out for a region of wavy hills among which lies the provincial town of Muminabad. When some years later I saw the loess landscape of the Kaiserstuhl and Kenzingen in the Rhine valley I was much impressed by the close and remarkable likeness it bore to the yellow slopes of Eastern Bokhara. There is the same general aspect of rounded swellings dotted with trees singly or in files, amid dry fields and powdered vineyards (compare Fig. 185). There are the same tray-like terraces, water cuts, and dust; and there is also in proud evidence the lime-loving walnut-tree, which may be called a characteristic denizen of loess hills and dry mountain slopes. It is a noteworthy fact that it avoids the plains, perhaps because they are too moist, or because it cannot gain a footing in the thick and spongy deposits of the irrigated country. It may also be that, as a wild tree, it has long ago been cut down in the lowlands, and is not cultivated owing to the slowness of its growth. Many years ago an Armenian speculator and vandal made a raid upon these noble giants, felling hundreds of them for the sake of their burrs. By the time he had cleared a profit of several thousand pounds his depredations were

[1] Illustrated in *Zeitschrift des D. u. Oe. Alpenvereins*, 1902.

fortunately stopped by the Russian authorities, and now the walnut is strictly preserved.

At Dagana we emerged into the broad plain of the lower Yakhsu, which here traverses a basin three to four miles wide, bordered by a line of bluffs. It was now getting very cold, and a frosty mist chilled us in the mornings. A night was spent at Kuliab, which offers nothing of special interest. In the bazar we bought magnificent, juicy grapes at a halfpenny a pound and a pheasant for a penny. After Kuliab

183. Towers of Silence at the Entrance to Birkendale.

the prospect widens still more, presenting the familiar expanse of rice squares, interspersed with fens and reed-beds. On the edge of the cliffs fringing the road sat numerous birds of prey of all kinds and sizes. Like watchful sentinels they conned the fields, for owing to the mist they could not see anything when circling in the heavens. At Chubek we entered the beaches of the Amu-darya, and at Sayat touched the bank of the great river itself, here divided into two equal branches looped around the island of Urta-tugai, a wilderness of swamp and jungle, full of waterfowl, pheasant, boar, deer, and tigers. Following the right arm we finally reached Parkhar, where we spent many delightful days roaming and hunting. The landscape is real jungle, although, true to the qualities of the climate, the vegetation is not impenetrable on dry ground, water and quagmires forming the only bars to progress. To our regret the want of boats prevented us from landing on the island of Urta-tugai.

184. The House and the Haystack (Yakhsu).

The whole is a maze of steppe, bush, marshes, and channels. The plots of dry savanna form gay parks studded with coppices, rambling trees, clumps of plumed giant grass and pink tamarisks, in between which spreads a beaten floor of ruddy clay. Elsewhere are thorny forests, just high enough to weave a shady canopy over one's head. Alternating with these insets of higher ground are morasses, beds of enormous reeds, bayous, and open ponds. Further away from the river are whaleback downs, covered with extensive groves of pistachio trees. Altogether this is a most romantic and fascinating region of glades and coverts, where one has only to be careful not to lose one's way. The crisp October air, with a warm sun chasing the mists of early morn, made the place a paradise for rambles. For the hunter it must be better still in winter, when all water is at its lowest and the marshes are frozen over. Every pool swarms with duck, and in the savanna one can put up six to ten pheasants at a time. But I always had a man with the 577 express at my elbow, for tigers are very plentiful, as I plainly saw from their tracks. A cow had been killed during our stay, but my wait in a pit near the carcase proved fruitless. Russians often told me that the Duab tiger is larger and stronger than his Indian brother, but that may be a patriotic exaggeration. When we returned to the upper Yakhsu on the first of November we were received by a blinding snowstorm.

From Saripul a much kinked trail hangs over the northern watershed to Khovaling. This side of the low Zagara pass, the state of the mountain side betokens a restlessness now stiffened into quietude, but waiting to resume movement with the advent of rain or snow. Jagged fissures gape on the slope between thick lumps of clay, pasty waves, and boggy slides ready to continue their sluggish downward course after a new soaking of the soil. Such disturbances, observable in the Himalayas on a grand scale, are the outcome of drought interrupted by heavy rains or melting snow. While dry the accumulations of earth or till are able to maintain a steeper gradient, calling for readjustment under the influence of excessive wetting.

During our descent from the Zagara pass to Khovaling, there was impressed upon me a special variety of dry hillscape, which one might define as the South Alpine type. It was forcibly suggested to me when wandering through the basin of Brentonico on the eastern flanks of Monte Baldo in South Tyrol. Both these short and broad valleys are strikingly similar in their physical features, and a description of the Alpine original will do for both, thus showing the close relationship existing between these distant lands.

A gently receding amphitheatre is dissected by radiating hills and ravines. Concentrated in one point the water-courses debouch upon

185. A Village in the Hills near Muminabad.

the main valley with its level spread of fields and orchards. Between the rocky seams of the torrents, now mostly dry, are broad-backed ridges, terraced shelves or long slopes, set with jutting knolls or traversed by ribbons of outcropping cliffs. The smoothed folds of the uplands above the timber-line are carpeted with coarse, short grass, where groups of firs advance as the outposts of scraggy shrubbery striving to emulate a forest in the middle belt. Lower down is the cultivated land where larger trees such as the mulberry, Spanish chestnut and walnut, are scattered among gardens and vineyards. Plantations are laid out on terraces supported by low walls, and clumps of willows mark muddy spots where water permeates a meadow. Villages of flat-roofed houses are linked by dusty mule-paths, pursuing their erratic and relentless twists among the creases of the mountain-side. We meet pack-horses and donkeys with their loads, and there is a smell of the road distinct from that of other countries, and a special orchestra of insect noises. Thorny or small-leafed bushes, leathery, brittle or prickly weeds spring from a grey and crumbly earth by the road side.

A few minor substitutions such as juniper for fir and walnut for chestnut, and some prominence given to the loess, will make the above description fit the valley I have in view. Looking at the landscapes purely as expressed by plant life, wild as well as domestic, I even venture to say that there is complete identity of character between the valleys of the Italian Alps and those of East Bokhara.

On the way to Baljuan I noticed that many poplar trees were pitted with wormholes bored by that stately capricorn beetle the *Pachydissus Sartus.* In the early summer of 1896 the loess crannies around had yielded up to my collection new specimens of poisonous spider and scorpion, *Gylippus Rickmersi* and *Anomalobuthus Rickmersi.* I feel highly gratified that such vicious creatures should bear my name. On drawing nearer to Baljuan (Fig. 186) a change is seen in the build of houses. The preference shown for gabled roofs thatched with reeds or matting leads us to suspect some influence of the surroundings upon the habits of the people. Whether the peculiarity is limited to the provinces of Baljuan and Kuliab, or whether it marks the boundary of a larger space extending into Afghanistan, or even to India, I do not know. The moist and oppressive heat of the low-lying districts must have

186. Baljuan.

something to do with it, for reeds are equally abundant near the cube-town of Hissar (Fig. 191). Something the inhabitants may also have learnt from the nomads, for the plains of the Oxus are not far from Baljuan. Reed screens, such as the Turkmen make, are much in use, and airy sleeping sheds are erected on many houses. The place is said to suffer a great deal from malaria and the people really look very unhealthy. As at Kalaikhumb the soldiers of the garrison are good amateur gardeners and our friend in the picture (Fig. 187) has turned household vessels into flower pots. Another man had made himself an oval jardinière of clay as a fixture on the ground. Among the merchants of the bazar we met a Russian who had become completely Sartified, dress, religion (at least outwardly), wife, and all, nor did he express the slightest regret, but, on the contrary, declared himself perfectly content with his present lot.

The Beg's residence, a lofty escarpment crowned with a wooden gallery, frowns upon the city from a yellow promontory overlooking the pebble-strewn river plain. Its high wall of loess bricks bears a distant likeness to the style of architecture of the Potala at Lhassa. In one of the tea-houses a black felt mat showed a large, white pattern which, though suggesting fleur-de-lis, may simply be

a corrupted cramponee or sun-wheel (svastika), while the ansated cross would stand between the two. I do not risk any theories, however, for the comparative study of ornaments, their evolution and ethnographical or historical relationship is as insidious as etymology. Mere chance and the free flights of fancy are often the immediate progenitors of some design. Simple geometric traceries can so easily spring from the combination of elementary lines and figures, that a kind of spontaneous generation in independent centres must be allowed as possible. The svastika is known to some of the North American Indians. In connection with this item I may mention a few other similarities I have come across. The saddle-bags of Sardinia are practically identical in shape, pattern and colour with the khurjins of

187. A Soldier's Home at Baljuan.

the Duab. The woollen shawls of Fair Isle between the Orkneys and Shetlands show a great resemblance to the texture and design of Turkish and Turkestanic socks and gloves. Here possibly we have Moorish influence exercised by Spaniards cast adrift on the islands after the wreck of the Armada.

Hence onward we followed the great central highway of Bokhara along which are ranged nearly all the richest oases and cities of the kingdom. At Guzar the main road to the capital sends off a branch to Samarkand. Loess is to the fore again and near Karayar youthful valleys of extremely symmetrical formation are graven into the undisturbed deposits of late quaternary and recent times (Fig. 188). There are also many sinkholes and rudimentary shapes tending towards obtuse cones and pyramids. Everywhere the peasants are busy winnowing and under the influence of the breeze great heaps of chaff are seen to assume the aeolic features of sand and loess. From Kangurt, where the pebbles by the side of the stream are covered with saline efflorescences, we traverse to the Vaksh by the Guli-zindan pass. Its name may be derived from the thicket of roses which cloaks the north slope of the neck, while the southern declivity, only a few steps round the corner, is quite bare. Information as to distances now became fairly reliable, the tash or Bokharan mile (about five English miles; tash means stone, i.e. milestone) acquiring a more stable value. In remote districts it was elastic like a rubber tube and a mysterious economy of nature seemed to press as many of them as possible into one. While four or five tash had been liberally spread over the road in easy country, barely two were now meted out, and in the wildest mountains one tash was made to last for a whole day or even longer. But now we are in a land of peace and plenty where we jog along in comfort, where the vultures no longer wait for the dying horse, and where everyone can have as many miles as he desires.

On the right bank of the Vaksh, above Tutkaul, there is a natural quay formed by a dipping layer of limestone. It runs for a distance of several hundred feet in a perfectly straight line, and such is its smoothness that it looks as if planned by man to confine the stream. Above it are three water-marks deeply grooved into the rock and corresponding to three terraces further up the river. Changes in the gorge below this spot must have been responsible for these levels

188. A Village in the Loess Hills near Karayar.

indicating a successive lowering of the cataracts. On the left bank the path skirts a fine set of red knob-and-pockethole sandstone reminding one much of a coral reef at ebb tide. Its upper surface seems to fit the lowest beach mark. Swirling rapids passing over this shelf drilled and turned the counterparts of their curly waves. Presently we are overshadowed by the gorge of the Vaksh between Tutkaul and Norak. Here the trail dallies with the brink of precipices over which a single false step on the part of the horse would fling him and his rider into a watery pandemonium. At the spot where the river, equal in volume to the Thames at Kew, is pent into a chasm eight feet wide, the native engineers have seized their opportunity. A few beams are sufficient to span the cleft, while a hedge of interwoven branches in place of railings screens the giddy drop from the traveller's eyes. A gate on the bridge marks the frontier between the provinces of Baljuan and Hissar. In ancient times when constant wars were waged among the minor states, many an expedition must have been frustrated at this point.

On issuing from the gap white and pink outcrops proclaim the open salt beds of Norak. Needless to say a staff of excise men is in attendance, although salt, even if poor in quality, must be obtainable almost anywhere in the vast saline steppes. The Bokharan minister of finance has hardly gone the length of his Italian colleague who has the whole sea-board carefully watched for fear that illicit salt should be extracted from the Mediterranean.

The Amlakdar at Norak owned a ram, evidently a cross between an arkhal and a domestic ewe who had lent a willing ear to a monarch of the mountains. This stubborn bastard, chained to a post, had the horns and coat of the wild species coupled with a fat-tail. Our host was also very proud of a charger with pink eyes. If I can raise the capital I shall found a stud on scientific lines, for breeding a horse with black rump, brown legs, white head and red eyes which will fetch fabulous prices on the markets of the Duab. This fondness for curious horses is illustrated by a story told in chaikhanas and karavansarais. A swarthy son of Afghanistan once played a very dirty trick on the Beg of Kuliab. Appearing before him with a couple of shaven horses, which he offered as a present, he related a wonderful tale, how he had lain on the banks of the Kafirnigan river of a moonlight night and how

189. On the Zardoliu Pass. A Hazy Steppe of Loess Hills.

these perfectly naked horses had emerged from the deep to browse upon the pasture, shaking the spray from their withers with terrific noise. With great cunning he succeeded in catching them. The gift was gladly accepted and the giver rewarded with princely bounty. The remarkable animals were tended with utmost care and two months later rejoiced in a new coat of hair, but the donor had meanwhile retired to his home under the shadow of the grace of Abdurahman Khan.

The Zardoliu pass (Figs. 189, 194) gives one a very clear notion of the mountain loess. The masses of yellow earth impress one more mightily where thus one sees them piled upon the heights. It is as if a great but clumsy hand had smothered the hard sculptures of the rocks beneath a plastering of mud. Everywhere this mantle cloaks the skeleton of the country, only receding where the steep and naked gradients of alpine tracts bid it halt. Cathedral high and abyss deep it heaves over rise and coombe. But in this thickness there is, again, a saving feature, as new forms are born from the grave of buried shapes. Loess is a supple and plastic thing. In large curves it follows the underlying folds and fractures of earth. From above it allows itself to be moulded in every conceivable fashion, not only by water but also by wind, man and beast. The smallest runnel ploughs a gash into its substance; every brook lets itself down in a canyon sluice until it touches bedrock. Wherever the roads of man are drawn across its surface they sink deeper and deeper, ultimately making a narrow defile where often only one rider can pass at a time. This is used until barred by a slide or dead camel deflecting the traveller to a new line where the process begins afresh, so that sometimes two cuttings are found side by side (Fig. 25).

Alternately we saw droughty grittiness and sappy moisture as we crossed the passes from valley to valley. The watersheds being low (4—6000 ft.) the caravan route from Baljuan is exclusively dominated by loess as far as Mirshadi, the top of a pass being generally marked by the notch worn through the divide by endless rows of horsemen and camels. The spoor of caravans seeks the softest ground, hence the view from the saddle always starts with a foreground of mud-tinted banks and cuts. But in many ways the yellow undertone brings into stronger relief all other colours and formations, for whatever is not covered by loess is conspicuous, whatever has been saved from the

190. Looking up the River from the Edge of the Loess Cliffs at Diushambe.

yellow khaki uniform stands out like blots on a distempered wall. Middle distance and background offer a great variety of rock build and stratification. Dust lies a foot deep in the channel of the middle groove. The pendulum swing of the horse's feet throws forward, beats and stirs the mellow down into thick floating balloons that rise quickly, dispersing into drifting sheets. A company in a cutting fills the hollow passage with Egyptian murkiness, whence they emerge with powdery masks.

In the plains the grand trunk road is very wide, often showing six parallel ruts worn by the steady tramp of pack animals. In order to escape the dust we keep a distance of three hundred paces. When the air is still the earth-fluff remains low on the whole, settling more rapidly than one would think. A side wind is very agreeable, but a head wind makes it awful for a long file. It is the dream of my life to see a motor-car careering over a loess track at sixty miles an hour; the effect would be stupendous. But with horror one imagines the spurting slush under foot after rain, and the effect of a sudden shower upon one's clothes impregnated with dust. On the other hand there are few obstacles and we amble along at four or five miles an hour. The eye scanning the map must now accustom itself to another standard, for the distance travelled in a day seems incredible compared to the mountain marches.

Descending from the Zardoliu pass we step upon the floor of the great basin plain of Middle Bokhara, a cross-valley common to the rivers Kafirnigan and Surkhan or lower Karatagh-darya (Fig. 194). It is an oblong dish between the foot of the Hissar range and the uplift of low steppe hills on the right bank of the Oxus. The watershed between Kafirnigan and Surkhan now lies in a level country between Hissar and Karatagh. Undoubtedly the Kafirnigan once flowed west taking in the Karatagh-darya as a tributary. Later it was enabled to force a passage straight towards the south, leaving the Surkhan henceforward to pursue an independent course. Apart from possible warpings we may suspect that a blocking of the old synclinal Kafirnigan valley with glacial detritus had something to do with the change. The landscape is the same as around Samarkand, but owing to the seasonal pause in irrigation travelling is much drier now than during summer when everything is water-logged (cf. p. 61 ff.). We often brush through tunnels of tall grass waving its feathered spears over our heads. This rich transversal valley and its surroundings one might well call "Middle Bokhara" occupying, as it does, an intermediate position between north and south, as also between the eastern mountains and the western steppes. In a wider sense I apply the term to a broad strip of country extending from Baljuan to Guzar and traversed lengthwise by the trunk road. After the oases of Bokhara city, Karshi and Shakhrisiabs, this zone covers the most important region of the Khanate as regards fertility, population and trade.

The thermometer still rose to 26° Centigrade (78° Fahr.) at noon, but this was easily bearable, for the evenings were cold. The Kafirnigan was a middling stream of clear, green water, whereas I remembered it as a voluminous and yellow spate in the summers of former years. Through endless miles the grooved road unravels itself over a hazy plain; the Sarts are harvesting and on smooth threshing floors of clay the brown heaps of wheat grain lie beside the mounds of golden chaff. A cool, autumnal breeze tempers the sun. The dark tree clusters of villages form shifting wings of scenery and their foliage turns towards a motley tinge. Melons, like amber globes gleam from the meshes of their sere and rampant stalks. At a sarai luscious grapes are set before us. They are as large as prunes, their stem is already dry and the skin, covered with a faint bloom of loess, is beginning to get wrinkled, thus making the juice all the sweeter. At Diushambe our quarters were really luxurious; lofty rooms, doors to suit European stature and windows in both Western and Bokhariot style, the former being made of muslin stretched over a framework of wood. The house pond was shaded by venerable plane trees (Fig. 33) and from the courtyard a postern gate led to the edge of the loess cliff on which the palace stands and whence one looks up the Diushambe river issuing from the Hissar mountains (Fig. 190).

With the advent of the promised land higher spirits blossomed out in our two-legged passport, the Karaul-begi. That he was also a spy did not worry us, but the officials along the line of travel feared him all the more, for to them our living letter of introduction was inwardly a thorn of prickliness though outwardly a salve of boon. Before his kindly smile the bile of bitterness was ponded back, and what came out seemed as the purest, softest milk of pious welcome. Whoever has charge of the Franks holds an important office giving weight to his word in high places. Therefore all, from governor to aksakal think in their minds, "May the shaitan take the Karaul-begi, but when he speaks to the Amir about me may Allah refresh his memory, may he remember the new silken gown I gave him, and the grey filly, and the bag of silver shekels; may his words flow like honey." This explains why our guardian angel who had sallied forth with a thin horse and a pair of lean saddle bags, finished his homeward tack with two strapping and well-laden ponies. The generous givers no doubt derive some satisfaction from the conviction that ultimately he too, will have to contribute his share to the royal coffers at Bokhara. A system may appear harsh in detail, but strict observance all along the line makes for a just balance. Once we had as companion for part of the way, a mirakhur sent by the Amir on a special mission connected with complaints the Russian Pokorski had made against the local authorities. Everybody was eager to curry favour with the Mirakhur, and so he summed up: a thousand tengas from the Kushbegi of Hissar, six hundred from the Amlakdar of Diushambe, three thousand from the chief culprit, the Beg of Baljuan, a hundred at Saripul, and so on, to the few coins of a village headman. If all went well, his slender purse would wax fat in spite of the unavoidable Purgative of Court, for Allah is merciful ever.

A high official on the rampage is a terrible visitation, the higher the worse, for he and his retinue eat up everything like a swarm of locusts. But the wanderer journeying in their wake will find them a blessing in disguise, at least in one respect, for he is sure to find the road newly made and in the best possible condition. The great gun on tour is like a bear in a maize field; he devours as much as he can, leaving a visible trail behind.

It is amusing to note that if you hire servants, some of them will inevitably be used by the others to wait upon them, and this process is continued *ad infinitum*. Thus here we have five classes or differences of rank. First of all ourselves, and I think there is little doubt that Mac represents me to the natives as a European prince. Then there is class two consisting of Mac and Albert. Class three is made up of our foreman or karavanbashi who condescends to eat and drink with the aksakals of kishlaks who find it a convenient method of living for nothing, to accompany us for a day or two. The fourth class is made up by the inferior grooms who again press into service the porters and helpers engaged for certain occasions. Our own manners were subject to certain alterations, and Western chivalry to ladies had sometimes to take a back seat. Whenever we visited at a house, I had to enter the door first, because our Oriental host would have been at a loss what to do and whom to greet. Thus, in order to avoid confusion I had perforce to step in front as lord and master, taking precedence before soulless woman. Likewise I took possession wherever there was only one charpoy covered with silken stuff, especially as the ladies would be continually offering it to each other, so that I must needs come to a Solomonic decision. After all, it is quite an experience to play the Eastern tyrant.

Many people consulted me about their ailments and my medicine chest was prepared for the onslaught, every Feringi being a doctor as a matter of course, as well as millionaire and madman. The staple remedies were quinine, of which I distributed a pound during the journey, eye solutions and ointments for skin-diseases. If one wishes someone to wash a wound with boiled water it is useless to tell him so, for a nostrum he must have. Accordingly I prescribed permanganate of potash (Condy's Fluid) a farthing's worth of which will give a fine red colour to a ton of water. I told him to dissolve a tiny particle in a kettle full of the universal but despised liquid and to apply to the skin. Thus one can make him wash. A water cure, too risky for man, is only tried on beasts, and once I saw a cow submerged to her neck in an arik, where she was held down by a yoke fastened across.

Hissar (Fig. 191), the winter residence of the Kushbegi of East Bokhara (a political division including the provinces of Hissar with those to the south and east of it) has many marks of ancient greatness. The palace of the Beg is itself an imposing structure, the gateway being flanked by two huge towers quite in the style of a mediaeval castle. Our quarters here were in the house of the head of the police, a Tatar who has been in the service of the Kushbegi since childhood and who acts as his Russian interpreter. In the course of time he had become the right hand of his master and acquired great influence in his affairs. By the way he also enjoys a private reputation as horse thief, for it is said that he maintains a band of agents for stealing cattle from the Kirghiz. In the evening he introduced his spouse to my wife in whose words the visit is reported. She is a Tatar and quite a beauty of her type, with fine features and smooth skin, satinlike in its softness. The eyes are of the peculiar shape which distinguishes all the Mongol races. She was magnificently arrayed in a long khalat of cloth of gold. Her head dress was most becoming and bore semblance to the headgear of the priests of the Greek church, with the difference that in her case a veil of rich, dark blue, shot silk fell from it in graceful folds over her shoulders. Her feet were bare; on her wrists she wore thick silver

191. Hissar.

bracelets. The other women of the establishment, about five in number, were similarly clad, but in simple material. The special duenna of Rakhmetullah's pretty wife was a somewhat elderly crone whose thickly painted eyebrows gave her quite a fierce appearance. We all sat on the floor, which was comfortably carpeted, and drank tea. Rakhmetullah then opened a bottle of wine and made me drink friendship with his wife. Our conversation was carried on through the medium of Russian, my scanty knowledge of which just served for the stringing together of a few simple questions and answers. Before leaving, my hostess presented me with a gold embroidered tiubiteka or cap.

One can always test the rank and wealth of one's host by the quality of the pillau he gives you. In the humbler houses it is nothing more than mutton and rice, but between this and the highly spiced dish of high dignitaries with its variety of ingredients, there is a series of ascending grades. A native chef de cuisine of the first rank knows how to prepare fifty different kinds of pillau, whereof a kind made with quinces pleased me best. The women, on the whole, do not command a manifold menu, but some of their things are rather good, among others rice with milk, if one remembers to let them know beforehand to leave out the lake of melted butter which generally adorns the middle of the dish.

Karatagh, as we saw it before its destruction by the earthquake of October 1907 (Figs. 76, 192), presented itself as a leafy village ensconced in a bight of the Hissar mountains where the Surkhan gives up its rocky youth. It is still boisterous for a stretch and has driven the inhabitants to a device rarely seen in these parts, namely protection of the banks by a structure of branches and stones. As in all provincial capitals the gibbet occupies a place of honour in the open place where the weekly markets are held, probably because executions are a popular spectacle reserved for bazar-days. It looks innocent enough to our eyes, being in the shape of a yoke of two upright poles with a cross bar. The uninitiated will take it for a useful erection on which carpets are beaten or clothes hung, but clothes without men inside them.

Karatagh leads the grand existence of a capital only during the hot season; in winter one merely sees a few lone men wandering through empty and desolate bazars. Only in summer it holds the contents of its *alter ego* of teeming life with the Kushbegi of Hissar at the head, the governor of all Eastern Bokhara. Kushbegi means "Chief of the Eagle or Falcon," i.e. the head falconer of olden times when the keeper of the birds of the chase was a prince's highest servant. The Kushbegi at Bokhara is the Prime Minister or Grand Vizier of the Khanate. Hissar is the ancient capital of a once independent region where in the run of centuries principalities rose and fell, until swallowed up by Bokhara with the help of Russia. The Beg of Hissar was the most important of these rulers, and having surrendered of free will, he was put over the whole of the eastern and southern provinces receiving the highest title the Amir could bestow. We still knew the old man Astanakul who was a fine type of the Oriental nobleman with his high-bred courtesy of manner. He died a few months before our last visit and was now succeeded by his son, a rather bloated and uninteresting personage. In winter he resides at Hissar, leaving it behind in summer as a mere concept of supremacy and an assembly of empty houses. Almost the entire population follows him to change miasma and mosquitos for the purer mountain air of Karatagh.

192. Karatagh before the Earthquake.

The moving of the Lord Lieutenant is no small matter settled by the ordering of a furniture van. It is more like the progress of a travelling circus, though probably more interesting. An army of camels, horses and donkeys is levied. In endless succession they transport the paraphernalia of government and household, and everything connected with them. Nothing is left behind that is necessary for the acts of state or the pomp and circumstance of a dignified court. Everybody follows who claims a position in society or earns his livelihood in the shadow of the august chief. The distance is only six hours by the road, yet one asks oneself how anyone can muster enough courage to repeat this clumsy, if picturesque, transportation twice a year, and why furniture and other things are not in duplicate. The archives, of course, must go as well as the chained prisoners and rumbling cannon together with the blunderbusses, matchlocks, lances and swords decorating the guard room of the palace. Here we must also mention the stuffed tiger of uncertain outline and shaky legs and jeering face, with straw protruding from his eye sockets, a picture of commiseration, mounted on a board with wheels like a baby's lamb. Then there is the great museum overflowing with untold presents brought by the children of the land or by travelling Russians. Passing from official to intimate, let us remember the stacks and bundles and boxes of furniture, kitchen utensils, china, stable necessaries, clothes, tools, oddments and lumber. With averted face and bated breath I now mention you, O lovely Ones, the Neverseen, only chastely Guessed, you the tender flowers in His Grace's bower of roses. With your children and nurses and servants you alone make up fifty horse loads. And another dear possession, not less in weight and equally near to the heart, you the bursting coffers of the treasury. Repeat this list with dwindling quantities for the swarm of courtiers and officials everyone of whom has house, women and servants, continue it down to the soldier of the garrison who packs his bed and wife upon a donkey, and you may be able to conjure up a picture of the grand removal as it wends its way along the road between Hissar and Karatagh, dusty or muddy, as the case may be.

An inspection of our cash now begins to show a serious decline of funds. The silver-bag is getting very slender, but I hope that it will last out to Samarkand. At first it was an important piece of baggage claiming a horse to itself; later it was packed away among my clothes; now its weight is hardly noticeable in my knapsack, and before long it will find ample accommodation in my pocket.

CHAPTER XVIII

FROM KARATAGH TO SAMARKAND

As I have mentioned before (Chap. XII) a short cut from Karatagh to Samarkand leads across the Hissar or Hazrat-sultan mountains. This we did not take, wishing to save our pack animals, but with light baggage and horses lent by the Beg we made an excursion to the top of the Mura pass whence one may proceed to Saratagh by the valley on the left of Fig. 124. We followed the course of the Karatagh-darya which amid magnificent scenery paws the rocks with claws of silver and frets unwillingly at the yoke of many bridges. Near the village of Labijai a great moraine descends from a westerly ravine. On the slopes massive and noble clusters of walnut trees are strewn about in great numbers. At Hakimi we stopped for the night, our quarters being the portico of the mosque, the only shelter available. Here the view was grand. In front, framed into a picture by the pillars of the temple, towered lofty peaks holding up to the moon a shield of gleaming silver, while below the river foamed and surged in its rocky channel. Needless to say our use of the mosque gave no offence to the villagers except in so far as it may have curtailed their opportunities for gossip. It is the social centre of the place where the cronies assemble to discuss their favourite topics, where the barber is wont to ply his skill, and where any functions of like importance take place.

A short way above Hakimi the Karatagh river is joined by a left tributary torrent poured out in frisky cascades from the moraine-dammed basin of the Timur-dera-kul. I passed this lake in 1896 when on my way to Saratagh by an alternative pass to the Mura, then impassable. We again visited it in 1898. This alpine jewel of the most exquisite bluish-green tint is hemmed in by precipices on one side and by wooded slopes on the other. It is fed by two streams called the Yangilik and the Aktash. At the upper end a flat and bosky meadow runs out into a beach of white pebbles reminding one of Loch Katrine's silver strand. The cluck of the water-hen is heard among the reeds and, judging from the tracks, wild pig are plentiful in the nullahs of the juniper jungle. Unfortunately the natives have of late begun to make fields around Timur-dera-kul thus playing havoc with the peaceful groves fringing its waters. Most of the southern valleys of the Hissar range are fairly well wooded along their floors and some way up the slopes with groups or clumps of poplar, ash, birch, willow, maple, juniper, pear, hawthorn and walnut, interspersed with currant and other bushes. It is a type of "forest" peculiar to the country and one not to be likened to the unbroken timber-cloak of the Alps. These groves of Bokhara recall many Italian valleys and their pleasant, park-like aspect has always had a wonderful charm for me. True, we miss the luscious emerald of pasture lands, but I would fain exchange all the green grass of Alps and Caucasus for three months of fine weather.

The population of the Duab may be responsible for a good deal of deforestation, but this cannot have had any influence upon the climate, and the general character of tree-growth during postglacial times will always have been the same as to-day. There is much confusion and exaggeration in the views one often hears expressed about woods and rainfall (cf. Appendix). A change might just be thinkable in extreme instances, if, for example, the vast Siberian taiga or the tropical forests of Africa and South America were wiped out. This might possibly lead to a reduction of rainfall, but not sufficient in degree for preventing the rebirth of the same forests within a very short time. On the other hand areas of dense vegetation have an enormous influence upon the storage and distribution of rain water and upon the preservation of the mould, especially on inclined surfaces, whereby the problems of afforestation become chiefly connected with hilly or mountainous districts. Plant life is dependent upon two factors, climate and soil, but while it reacts little upon the former it is all the more intimately bound up with the latter. Vegetation makes its own bed and prepares its own system of irrigation, thus closely intertwining the causes and effects of its life conditions. This is less apparent on the plains where fine detritus comes to a rest, even without protection. Given sufficient warmth together with the proper quantity and distribution of moisture, every plain of the world supports vegetation however poor, while one-third of the entire land surface of earth is covered with trees.

Considering the plains or undulating hills only, and including the prehistoric woods of regions now under cultivation, we find that at least one-half of all level country in the world is immediately hospitable to forest. Here spontaneous reafforestation will speedily fill all gaps made by axe or fire, and only great changes of physical conditions (climate, drainage, submersion, impregnation with salts) would suppress sylvan formations. It is different when we come to the mountain slopes where plants, small or great, have to catch and retain the soil which in turn offers them foothold and nourishment. Wholesale destruction of the network of roots and trunks exposes the steep ground to the attacks of rain, sun, wind, torrents and avalanches which do their best in baring the bedrock. But I maintain that this difference between the plain and the mountain can only be one of time, always assuming of course, that no radical change of climate has taken place, that the general conditions of organic life have not altered during the last five thousand years or so. Thus a wood on the plains will restore itself without artificial help within a generation of tree-life, while several or even ten or twenty centuries may be necessary to reconquer a slant of naked rock, the time depending upon distance or height and the progress meanwhile made by denudation. With me it is an axiom that all ground lost through mere brute force can and will be occupied again by the forest. How, otherwise, could it have settled there in the first instance?

The original home of trees is on flat expanses, on valley floors, cone deltas or any spot in itself favourable to the accumulation of loose soil. The mountains, however bare their sides, are patiently invaded from the plains, from the bottom of declivities and from platforms or crevices serving as outposts to the advancing phalanx. The upper rim of tree-roots, assisted by a skirmishing line of lichens, mosses, weeds and bushes, builds upward step by step a layer of forest soil gaining upon the mountain, just as the Dutch have gained upon the sea by a succession of dikes. Like all climbing the process is laborious and slow, and it is only because of this slowness that we are hopeless of spontaneous reafforestation for economical reasons, such as timber supply and the prevention of damage by torrents, mudspates or avalanches. But clearly, where the artificial re-covering of a slope is contemplated, as in many localities of the Alps, the natural regeneration is implicitly admitted as possible, there being no difference whatever in the fundamental principle of man-aided and self-acting forest advance.

Since the memory of man the tree-line had already everywhere occupied its highest limits, so that the reclaiming of long and badly denudated mountain sides has not been observed during historical times. Add to this the fact that the upper trees were generally the last to be felled, unless, as in the Alps, deforestation has proceeded from both sides, from the villages below and from the pastoral summer camps above. Hence recession of the timber-line has been more frequently observed than advance. From the foregoing remarks it also seems to me that thickly wooded plains or valleys are the best and most natural basis for a gradual but sure ascent of the forest, but as the inhabited plains are the first to suffer, chiefly owing to the demands of cultivation, the existence of hill forests is sapped from below by removal of the main body and by favouring erosion of the lower rim of sloping woods. Hence length of time and the continuous influence of human life are in themselves sufficient to explain the lack of spontaneous afforestation, without having recourse to a change of climate. All trees can reach the altitudes corresponding to the latitudes of their distribution, and whatever we find in the plains or valleys is sure to climb the hills to a certain height.

In connection with this problem we must not forget that various species have different ways of associating together, of producing typical formations, and that "forest" is a relative term of vague meaning. No importance should be attached to traditional reports of the former existence of great forests in Greece or Italy, or, at least, one should not take them in the sense of Swedish, German, Russian or Alpine forests, but merely as a larger extent of the present remains of tree-formations peculiar to the region. Certain parts of Asia Minor were called Axylon, the woodless country. The woods of Nemea where Hercules killed his lion were probably similar to the scattered groves of the Duab where the bars or snow leopard roams to this day.

Owing to the scarcity of the population I do not think that the quantity of trees in the Duab has been greatly reduced during historical times, especially in view of the fact that loess is not a timber soil. True forests only thrive in the mould of moister regions, and in these the progress of spontaneous afforestation is proportional to the rapidity and exuberance of plant life. Painfully slow and ever struggling in the Alps, the growth of forests is irrepressible in the tropics where the smoothest rock is covered with rank profusion, where man has to

193. A Watery Plain near Sariassia.

defend himself against the jungle and where a defeat of vegetation is practically unknown.

Beyond Hakimi the road becomes execrable, being turned into a rack studded with sharp stones and boulders. The landscape is one of arid limestone peaks relieved by groves of very fine walnuts and elms. These mountains can best be compared to those of the Julian Alps, but being higher here they send forth larger rivers from their névés and glaciers, so that many valleys of Carniola are even drier than those of the Hissar province. Not a single bite of grass has been left along the path, and near camping places some damage has been done to trees by searchers for fire-wood. Near the top of the pass a herd of ibex scurried across a slope quite close to us, but before we had become aware of their presence it was too late, even for a flying shot. Large birds of all sorts continually circled above our heads and Albert dropped two lämmergeiers on the wing with the Mannlicher rifle. Arrived at our goal we found the view restricted and nothing much to interest us, unless it were the aspect of snow-fields closely

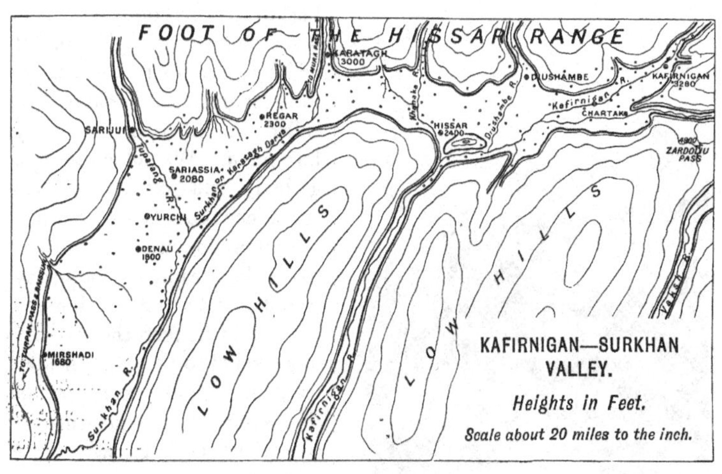

194. The Valley of Middle Bokhara.

furrowed with grooves one foot deep. Returning to Karatagh we resumed our homeward tack along the great caravan route.

As far as Denau (cf. Fig. 194) we amble through a rich, irrigated country, now mostly dry, to which tufts of steppe grass and isolated trees sometimes give the appearance of a savanna (Fig. 193). I have seen it at all seasons, oozing with sultry moisture and ringing with the hardness of a morning frost. Save for the flies which still bother the horses, the life of insects and lower animals is quite extinct; dead sheltopusiks and the empty shells of tortoises are here and there seen lying by the roadside. At Yurchi we found the forsaken remains of a mud-fortress returning to the common soil from which it had once so proudly risen. Only a thick, brick-built tower of Babylonian shape promises to hold out some time yet, offering a welcome building site meanwhile to several families of storks. The spreading nests have protected from erosion parts of the rim underneath, thus producing a kind of short earth pillar with the nest as covering stone. When I passed through this country in 1896 there had been a great flood of the Tupalang river destroying much life and property. As all the bridges had been swept away, fording the turbulent stream was a serious business. In order to help us through the rapid current which went up to the saddle, the Amlakdar of Sariassia had called together about thirty horsemen. We all crossed in a body and by the mutual assistance thus afforded reached the other side in safety. Apart from the readiness of many friendly hands in case a horse is carried off its feet, there is also much relief to the eye, shielded from the giddy swiftness of the water, by the presence of so many companions. That same year seems to have been more than ordinarily blessed with floods, for I remember that on my way from the Caspian to Bokhara the railway was interrupted by the Tejen river.

195. Giant Grass (*Lasiagrostis splendens*).

After Denau, where the Beg lives in a crenelated castle of mud (Fig. 21), the cultivated scenery gradually changes into steppe, the ground becoming more uneven and frequently disjointed by deep loess rifts (Fig. 24). Then we touch the edge of the watered plain. At Mirshadi there is a toll-house where we saw the customs official standing at his door and counting the long rows of camels preceded by their Uzbek guides on little donkeys. The loads consisted mainly

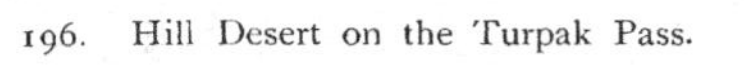

196. Hill Desert on the Turpak Pass.

of bales of printed cotton and those iron pots which form one of the most universal articles of trade demanded by the furthest nomads and the remotest dwellers in the mountains. We now climb the Turpak pass amid the surroundings of a true hill desert where rain erosion and the absence of vegetation have created sculptured forms of surprising regularity (Fig. 196). Nothing worse in this climate than a low hill, be it only a few feet above the plain or just short of the snow line; it is beyond the reach of irrigation, at the same time offering no encouragement to the condensation of wind-borne vapour.

To Sarikamish it is a thirsty half-day stage. It was cool now, but I remembered one day in July. Since leaving Sarikamish one fiery hell of stone-lined oven had followed upon the other; the white and red tiles of the mountain flank radiated a burning glare; the air, forced to a sizzling heat was glimmering before my eyes, my tongue felt like a frog in treacle and my thoughts professed an obstinate longing for the Moslem's paradise. It must be said however that weakness or lassitude did not form part of the symptoms of these milder stages of desiccation, always preferable to the torpor of steaming moisture. At last we topped the rise, looking down upon the plains veiled in a sullen haze hiding from view the green of the plantations. At our feet there was nothing but sandy or rocky soil alternating with small loess flats embedded between the ridges. Some hours still passed while we descended through grey and yellow waves where a gigantic lizard seemed the only thing alive. Suddenly there appeared a row of camels just fresh from the well, their bellies distended to a tightly stretched drum. Then, at a turning of the road, two horsemen hove in sight, sent to meet us with an earthen jug of water. From what tank it came I left to providence, and never shall I forget this hospitable forethought, one of the kindliest I ever knew. "How much further?" I asked—"One hour!"—"Well, that is nothing."—Off we went at a gallop; the horses sniffed the scent of water, and leaving behind a streamer of dust, like the tail of a noisy shooting star, our cavalcade rushed through the lowering hills to the clatter and clanking of hoofs. Shoulder to shoulder we jammed into the last cutting, then broke forth into the green oasis of Mirshadi.

A long and twisted gorge, where dribbles a salty creek, leads from the bottom of the Turpak pass to Sarikamish. Here and there noses of jutting rock have been hollowed out by wind at their base. Wind

197. Wind-shaped Hills at Sarikamish.

to some extent replaces water by carrying away the products of destruction, which action goes by the name of deflation. It drills holes into rocks, smoothens their surface and cuts small stones into polished angular shapes. Where alternating strata occur, the sand-blast undercuts the harder layers by wearing away the softer ones more rapidly. A most remarkable sight is offered by the crinkled and crumpled mounds at Sarikamish (Fig. 197) which one might call a hybrid between hills of circumdenudation and hills of accumulation. Their main body has been blocked out by erosion, but the finer features must be partly due to wind as shown by their striking resemblance to barkhans or sand dunes. The prevalence of wind-work is furthermore attested by honeycombed sandstone rocks cropping out from the surface (Fig. 198). Opposite the dune-like hills there is the curious swelling and rolling ground seen in Fig. 199. The ridge in the background, standing out clearly in the bright evening sky, is gashed with innumerable clefts, a phenomenon which we shall see repeated in the case of the Iron Gate where a low crest is also cut up by many parallel gorges.

Under present climatic conditions it is almost impossible to account for such deep and numerous cracks by erosion alone, and it seems more likely that tectonic disturbances broke up the landscape. I agree with Suess who says that one can exaggerate what might be called extreme gradualism. After all "catastrophe" is as relative a term as "slow change," "uniformity," or "stability." Some people may call sudden what others call gradual; it all depends upon the comparison one has in mind. The ice-age is a catastrophe as regards the revolution of fauna and flora; it is a slow change compared to an earthquake. The disappearance of a continent in the course of a few years or even days would presuppose no greater effort and no greater change on the face of the world, than a wrinkle which we produce on our forehead by a frown. A contraction of the brow is more than the upheaval of Himalaya. Do not let us forget that there could be no "slow" if there was no "quick" in things. If we had no climax or snap we would not know minute preparation. Both are in the plan of nature; each is meaningless by itself. Inertia must be preceded and followed by inversion. If the "accidentalists" had to be reminded of the accumulation of small effects, the "gradualists"

198. Honeycombed Sandstone at Sarikamish; the Result of Disintegration and Wind Action.

must remember that there is such a thing as unstable equilibrium, and that, as a rule, there comes a sudden return to stability. By calling an event sudden we mean that the speed of the upset was but a fraction of the time spent in the accumulation of topheaviness or overstrain.

Our tea at Sarikamish had a brackish taste until a messenger supplied us with sweet, if turbid, water from a well two miles away. Near our karavansarai we noticed an encampment of nomadising Uzbeks. Their primitive kibitkas were like gipsy tents and covered with striped carpet-mats. From here a ride of a few hours takes the traveller to Baissun. We proceed along a valley trough filled with loess between a surf of grey, low hills, bounded in the distance by the high, red wall of the Chuldair mountains. Around us an earthy labyrinth covered with crackling, grey-green herbs. Sometimes we espy a depression filled with the green trees of a kishlak, like an emerald lake sunk far below our level. Then Baissun appears, a long streak of vegetation thinning out into the direction of the Amu-darya.

At night the Beg invited us to his private room adorned with a dozen different designs of wall-paper and stuffed with a motley collection of saddles, belts, robes, chests, guns, pistols, bottles, mirrors and books. There were also gorgeous golden watches made of tin in fastidious leather cases made of pasteboard, given as presents by European travellers childlike and bland. Owing to the absence of windows and the evening gloom, this chamber, dimly lit by a candle, gave one the impression of a subterranean treasury that might have belonged to the forty thieves. Judging from the jumble of presents one sees in the store-rooms of various officials, every visitor racks his brains in order to devise some original gift. Hence we find almost every conceivable article of Western manufacture even including gramophones and cinematographs, the latter with scenes that would cause apoplexy to a pig. I advise a return to solid and useful things. To the begs one might well give boxes of choice sweets and chocolates, but such as will not easily melt in the heat or are properly packed in oiled paper. Most of the minor officials will always be grateful for a decent watch, especially if one has been careful to order dials with Persian figures. Peasants always appreciate knives, axes, tools,

199. Waves of Earth; Hills at Sarikamish.

fish-hooks and other products of European steel industry because they can estimate the practical value of such things. But I would fight shy of fancy goods, toys, cheap baubles, ornaments, and the like. All women, high and low, are fond of strong scents, perfumed soap and needles.

The Beg of Baissun, who is a mulla and learned man, read us passages from various books. From time to time he expectorated against an unpapered portion of the wall, trusting to the absorbent qualities of the loess. On leaving his presence we were taken to see the prisoners, this apparently being regarded as part of our entertainment. They were three in number, and there was something suggestive of a wild beast show in seeing them dragged out, chained to each other by the neck, from the depths of a den which seemed absolutely devoid of light. Two were guilty of theft and one of murder, and they were awaiting their sentence from the Amir. Owing to Russian interference the modern prison at Bokhara offers nothing in the way of horrible sensation, though some of the provincial dungeons still recall the sufferings of Stoddard and Konolly. Next day we walked about in the neighbourhood of the town and then spent a quiet afternoon in our room, a long bright apartment with seven "French" windows opening into the garden and admitting a flood of sunny brightness and fresh autumnal air. On the floor was a large mat with lozenges and stripes of blue, red and white woven into the flat, stringy texture. Silver-headed nails and glistening chains caught the sunlight on the carved doors. Three small iron bedsteads with cherry coloured quilts and silken bolsters made the room look like the dormitory of a pleasant school. The light from without also toyed with the china tea service and red and golden fruit on the white table cloth. Hence the view wandered through the window-doors into the recesses of vine bowers and young poplar plantations, where in spite of withered leaves breathed an atmosphere of vigorous joy.

We left Baissun on the 17th of October, being first treated to half a mile of contorted pavement. The low Tirakli pass leads through a maze of fantastic red sandstone (Fig. 200). This rock forms a large feature in the Alai ranges and Lipski reports a mass of great thickness and bold shapes near Yangi-kishlak to the N.E. of Baissun, which must belong to the same connected outcrop with a general strike from N.E. to S.W. The woolsack shapes are due

200. Sandstone on the Tirakli Pass near Baissun.

to a scaly deterioration of the stone called desquamation, chiefly due to changes of temperature. It seems to me that these forms are already partly existing within the rock where they have been roughly sketched out, so to speak, by certain strains of the substance. Thus the work of heat and cold, water and wind is often met half-way by these preformations. Granite betrays a similar tendency as witnessed by the blocks of Kemkutan. The pressure of mountain folding is certainly one of the causes, as may also be a cruder kind of crystallisation. In the concrete of conglomerates such minor distortions of the homogeneous mass do not occur, which accounts for the different style of their architecture, where only long cracks due to settling and upheavals act as first guides to running water. On top of some of the roundbacked bulges the natives had stuck little elliptical store-houses looking exactly like conning towers on the deck of a torpedo-boat or submarine. In 1898 we had made a short excursion into one of the ravines descending from the summit plateau of Chuldair above us, and I can still see poor Krafft at work in one of the gullies where he found interesting fossils after having shattered the landscape and turned over with his hammer the leaves of an antediluvian herbary.

The descent from the Tirakli pass to Derbent is down the shallow and gently sloping floor of a wide tray remarkable in so far as it has been eroded out of almost vertical strata. Possibly it is just on the top of a very steep anticline. A low ridge of upright coxcombs forms the left divide, while lower down, towards the right, are seen many parallel outcrops of the same kind. Transversal erosion has shaped these into what looks like half-buried horizontal spindles, and I have never seen any surface formation more deserving of the epithet of "rolling." The eye at every moment expects rotation from these tops with their equatorial belt of rocky teeth.

Between Derbent (which is 150 miles from Samarkand) and Shurab there is a fine display of geological colouring produced by polychromatic marls. Here and there the higher mountains throw out undulating ridges towards the plains. Through one of these crests lies the famous defile known as the Iron Gate, a gorge of reddish, jurassic limestone, nearly a mile long and at times only a few paces in width. Its mighty cliffs, hewn and carved in bizarre fashion, tower in places to a height of 600 feet, and are rendered the more impressive and picturesque by reason of the windings of the canyon. In winter or early spring a stream flows through the chasm, which is then impassable, and the travellers circumvent it by a path over the mountain. "Derbent," the name of the village at one end, means "defile." A town of the same name is situated on the shores of the Caspian sea where a narrow pass, also known as the Iron Gate, lies between the sea and the Caucasus mountains. To-day the natives call the gorge Buzgala-khana or house of goats, showing that mankind has become more prosaic not only in

201. The Iron Gate.

the West. What now is nothing better than a hiding place where peasants sheltered their herds in time of war, was, in the heroic past, the gate of the land whose iron portals shut themselves against the advancing host. Already Hiuen Tsang, the Chinese explorer of the seventh century, knows of the fame of the Iron Gate. According to the chronicles a heavy door hung in the defile, but in vain we looked for the spot where the hinges might have been.

In 1898 we had gone from the Iron Gate to Sharshauz by a short cut *via* Akrabat, Karakhoval, Kaltaminar, Yartepe and Yakasarai.

We then met four Russian doctors, two men and two ladies, who formed part of a cordon of quarantine drawn in a wide circle round the village of Anzob in Yagnob. Plague was said to have broken out and the government had at once taken energetic measures for its localisation. The epidemic however proved not to be the plague. It died out after having decimated the population, and so, after all, the effect remained the same.

Beyond the Iron Gate the road to Guzar by Tengi-khoram drags its weary miles through a monotonous landscape. We travel a little more slowly now as we are in the month of Ramazan when everybody is expected to fast, and we cannot ask our men, and especially the old Karaul-begi, to be shaken on horseback all day without food or drink from sunrise to sunset. Does fasting turn the thoughts to things Divine? I am inclined to believe the contrary, for our people were talking of nothing but food all the way, and probably thinking of it still more. At night they sat till a late hour feasting and talking. Thus Uraza is fast and festival in one, a penitence in daylight, an orgy by lamplight. Guzar lies at the fringe of the great steppe extending northwards. Leaving one track to pursue its line towards Karshi, we turn to the right so as to reach Samarkand by way of Sharshauz, Kitab and the Takhta-karacha pass. Out of the arid plain rises a dark dome of foliage welling forth from the low, yellow quare of the sarai, as a big bunch would from a flower box. That is Yakasarai (Fig. 28) where one also finds one of those bazars serving as periodical commercial centres in rural districts. On market days they are a scene of bustling life, but otherwise the gaping emptiness of the booths in the dusty, sun-baked steppe gives one a feeling of accentuated loneliness. Presently we enter the oasis of Shakhrisiabs watered by the Kashka-darya and holding the sister cities of Shar (or Sharshauz) and Kitab. Though only of medium size this oasis is one of the richest in the land and its grapes are famous for their sweetness. Sharshauz is about as large as Karshi, while some claim for it to rank immediately after the capital to which, moreover, it bears a great likeness. It has the remains of some very fine old buildings ornamented with glazed tiles, and some of the edifices have actually been renovated.

We dive into the semi-darkness of a cavernous bazar where pale merchants squat in uncertain twilight behind their wares. We catch glimpses of green tobacco heaped on white paper, of spices, piles of cotton goods, smelly meat on hooks, busy tailors and silversmiths, and fruit shops with pistachios, quinces on strings and dishes of kishmish (raisins). The ceilings are everywhere full of pendant grapes hung up to dry, and furtive rays of sunlight stealing through loopholes at the back of the stalls, turn the clustered berries into a glow of transparent red. After sundown we take a lantern and leave our guest-house to see the people of Sharshauz enjoying themselves penitently during Lent.

A solitary group at first, we grope along dark lanes where we are gradually joined by other dangling stars drawn towards some unseen goal. Then suddenly, we step from murkiness into the garishly illuminated hall of the bazar, so deadly still on ordinary nights. During Uraza the rule of early closing is suspended, for usually nobody is allowed to show himself in the streets of a city after sunset, a very convenient regulation which the native police have invented for the simplification of their duties. Everybody caught abroad after sundown is liable to arrest as a vagrant found loitering. Now all is noise and light and gaiety. In spite of the outer cold the roofed-in bazar is warm with the heat of the crowd and the steam of countless candles. The sounding passages are filled with the hum of thronging men pressing hither and thither; cookshops are busy and high life simmers in the tea booths; peepshows attract the simpler minds, while roulette-tables and amiable boys with long hair cater for the gilded youth. Sometimes a whole street is made into a festive hall by spreading felt upon the ground where hundreds are seated singing and drinking tea. We look down upon a garden of white turbans swaying in the breeze of merrymaking, and the flare of the candles is reflected from glistening biscuit-tins which line the shops. In the centre of this turmoil a guard of soldiers sits enthroned, commanded by an officer decorated with a silver star. Stumbling over rows of slippers we fight our way into a chaikhana with its raised platforms covered with carpets, its battalions of cups and tea-pots ranged along the wall, its samovar of shining brass and the pungent smell of the chilim floating between the greasy rafters of the ceiling.

By a short ladder we clamber to a separate room where bachas dancing to infernal music enthrall a raptured audience. The bubbling pipe goes from mouth to mouth and everybody sips kok-chai. The infusion of this tea, imported from India, is actually green. It has a very astringent taste and is taken without sugar, from Chinese cups without handles. Watching a party of revellers one soon notices certain movements and tricks of manner as conventional as those of average society at home. The way a man balances his cup, how he asks the waiter for more by snapping his index against the rim, how he sends the empty cup bowling across the carpet, all these are evidently some of the secrets of good form, partly characteristic of the sedate gentleman, and partly of the young blood out for a spree. The dancing boys are continually being offered a drink from the cup which, if graciously accepted, throws the honoured guest into transports of delight superficially hidden by polite indifference. A similar custom obtains in villages of Tyrol where one often notices how a peasant asks the waitress to accept a sip from his glass of beer. Taking things all round, I do not see much difference between the men of Sharshauz and those crowding the Criterion Bar or promenading about Leicester Square. What strikes us as a serious flaw, however, is the total absence of women from all public life and enjoyment. A sorry country is Bokhara. In this respect I prefer the conditions of Ladak with its agreeable polyandry so graphically described by David Fraser in his *Marches of Hindustan*, and the institution of temporary marriage established at Meshed, for the benefit of pious pilgrims whose arduous round of castigation and prayer can be lightened by legitimate bonds of four weeks' duration under ecclesiastical patronage. After all, what distinction is there between polyflirtation in Europe and polygamy in Asia?

On the 22nd of October we crossed the Takhta-karacha. A winding road leads to the frontier on the top of the pass where a black and yellow post proclaims the dominions of the Tsar. Here we encounter the rounded granite rocks with which we have become acquainted on Kemkutan, for this col belongs to the same ridge. It was the last ride with our horses and we could not but feel a sorrowful regret at our prospective parting. As if to do us a last service they swallow the miles under their feet, and the domes and minarets of Samarkand begin to rise out of the ground. We trot through the bazar and then into the Russian town where we may already consider ourselves at home, for where railway and telegraph stretch their tentacles there travelling ends, there we need only pay and wait for the most distant parts of earth to come to us. Here we shall enter the train speeding into the distant plain, that infinite plain which is the battle-field and grave-yard of countless races, where the softly drifting desert has buried ancient cities and new ones have been stamped from the ground by the grey regiments of one unbroken empire, where the reindeer wanders over the moss of the tundras, where the steppe of the Kirghiz spreads its vastness and the forest of Siberia unrolls its sombre cloak, that wide, illimitable plain which reaches in one stretch from the glacier-bearing mountains to the roar of the northern ice.

APPENDIX[1]

REVIEW

WHAT pre-eminently characterises the Duab mountain landscape is the regularity of the phenomena of erosion and the preservation of all the forms of young deposits. Nobody can fail to observe the huge terraces of the Zarafshan, nor the deep canyons cut through them by lateral streams. The lack of atmospheric moisture and the resulting absence of an evenly spread covering of grass or forest is responsible for this state of things. During summer and autumn the great rivers consist of almost nothing but glacier water, a fact which, by the way, sufficiently betrays the great extent of glaciation. The amount contributed by rain or springs during the drier half of the year is very small indeed. Therefore these river systems are made up of comparatively few lines running through barren matter. The symbol is a tree with long, gaunt ramifications, the ends of the stronger branches being glaciers, the leaves representing snow patches. With this let us contrast the northern slope of Alpine Europe, in which case the river tree must be compared to a succulent growth of innumerable capillaries filling up and draining the spaces between the trunk and branches, and between all the branches. The main channels appear embedded in a network of veins permeating the substance through which the rivers are traced. The Duabic river, on the other hand, sends out bare and lanky shoots interlaced only towards the top, like the thorny plants of the steppe. A formula by which we can express these different conditions is one giving the permanent points of confluence on a given surface. We shall then find that the catchment area of the Rhine has more brooks, streams and points of confluence than that of the Oxus. In the case of the Rhine these points are not only more numerous but also more evenly distributed ; the river net has closer meshes. When consulting maps, a large and uniform scale should be chosen, for it is the detail which brings out the contrast in a striking manner. We see, for instance, how the tributaries of the middle Oxus pass like through trains from the foot of the mountains, hardly deigning to collect additions on the way. The large European rivers grow in volume to their very ends ; those

[1] Only metrical measures and Celsius centigrades are used in the Appendix. For exact meaning of terms refer to Glossary. For references to authors see bibliographical section.

of the Duab increase to a certain point, afterwards diminishing and often running dry before a lake is reached.

All mountain-born rivers on great continental slopes are the same at their extreme ends, at the mouth and at the highest watershed, the mouth, in all cases, being a discharge into a reservoir of salt water. Going upstream we observe that the greatest dissimilarity between the Rhine and the Amu-darya begins immediately past the mouth, continuing for a long time above it. The German stream receives additions up to the very last; the lower Amu is surrounded by thousands of square miles without a single permanent runnel, the number of points of confluence being absolutely nil. After we have met the first affluent, and entered the foothills of Bokhara, and traversed the mountain gates, the contrast becomes less glaring. But in the Duab the intervals between rivulets and torrents decrease very slowly until the neighbourhood of the snow line is reached. Along the edge of snow and ice where water leaves the solid state, the two rivers are alike again. The difference begins as soon as we descend. Of the Duab we may say that the mountains fertilise the desert plains and that the desert climbs the mountains, creeping into the highest valleys. Both pass and evade each other, but where they intermingle we see the fruit in the gardens of Samarkand and Ferghana.

In a humid, temperate climate the water remains longer on the surface between the permanent streams, much of it being evaporated through the intermediary of plant life. Under the influence of this constant moistening, acting like a wet pad, combined with the work done by the roots and acids of vegetation, chemical disintegration plays a preponderant part. The surfaces rot, so to speak. The dynamic energy of rainfall is stored or converted, and only in case of great excess over the immediate storing capacity of the basin (forest, moss, lakes, caves) it gives rise to turbulent floods.

How different is the action of a discharge of atmospheric water among the foothills or subalpine districts of the duabic type. The collected raindrops run quickly off the slopes in separate grooves, leaving on the grit of homogeneous texture those sharp scars representing intermittent watercourses. Soon after the passing of the cloud everything is dry again, and the smaller particles of sunburnt lime set into a hard cement. All water is rapidly confined to a small number of permanent channels, the banks of which it rarely oversteps because the beds are deep owing to the narrow line of mechanical effect, as testified by the familiar canyon. The bare rocks suffer mainly from sudden changes of temperature, being blasted and chipped, as proved by the enormous prevalence of screes. The air being dry and therefore clear absorbs very little of the sun's energy, allowing its rays to beat upon the stones with full force. At night the process is suddenly reversed, the stones now sending their heat rays into space, unhampered by humid air. The result is a rapid cooling, causing violent strains in the mineral structure. Olufsen speaks of detonations as loud as gun shots when the rocks of the Pamirs begin to crack after sunset. Likewise African travellers have frequently described how the stones of the desert burst with loud reports, often flinging wide the chips and scales blistered from their surface. It is wonderful what a difference water vapour makes although still invisible to our eye. A certain amount of humidity in the air is equal to the thickest furcoat, as shown by the little need there is in England for this kind of garment. In the Duab however it can be equally useful in winter and summer, for retaining the heat of the body and keeping out the sting of the sun.

As compared to the Alps, the present climatic conditions of the Duab are expressed by a changed relation between the various surface agents of destruction, transport and deposition. Weathering is much stronger, transportation by water at the same time being weaker, shorter and confined to narrower lines. Thereby accumulation takes place much nearer the origin of the broken fragments, further inside the mountains and at a higher level, as shown by the size of screes, terraces, moraines and mudspates in the valleys and on the slopes. Hence the frequency of impassable river gorges circumvented by passes, and of canyons engrafted on bedrock buried in detritus. The crests of the Pamirs are embedded in a glut of glacial gravel. We observe a great preponderance of what Girardin (Haute Maurienne, *La Géographie*, 1905) calls "*phenomènes actuels*," namely processes of the more violent and visible kind, such as falls of rock, landslides, ice avalanches, mudspates, breaking out or drying up of small lakes, earthquakes, avalanche transport, river floods, etc. Glacier-fed rivers, insignificant in the morning, swell to an amazing volume quite suddenly on a clear hot day, so that the traveller experiences some difficulty in escaping from the unexpected onrush. Hedin mentions such a case in the Chinese Pamirs. Huntington gives a graphic account of how once his party crossed a streamlet without noticing it, and how, after a spell of rain, it grew into a tremendous river.

In his paper on Himalayan erosion Dr Neve draws attention to a particular factor explaining many kindred phenomena. Large deposits, such as boulder clay or slope grit, accumulated under the sway of a dry climate (or period of dry years) can rest at a steep angle, whereas under humid conditions they must splay out in a flatter gradient. If such a deposit gets soaked by a more than ordinary rainfall, by a deviated stream or melting snow, it will begin to slide or collapse, in order to assume the angle adapted to its new state. The result is seen in the shape of landslides and mudspates. Some of the slides described by Neve remind me of what I saw on the slopes of the Zagara pass (Chap. XVII). He says "One landslide has been in progress for many years at Doyen on the Gilgit road. The slope is 40°, but the whole hillside for some 2000 feet is cracked and crinkled, and is very gradually settling down towards the river." The same writer is also struck by the vivid colouring of earth, to which I have frequently drawn attention. "After some weeks of travel in Ladak and

Baltistan one gets accustomed to the sight of tilted strata of various tints, stripes of pink and brown in wavy lines, with green slate, light pink granites, and grey limestones harmonized by a soft blue atmosphere,......".

Geologically speaking the Duab mountains have preserved many traces of youth, as shown by the important part played by limestone, sandstone and conglomerate in the architecture of great peaks. In the Alps, owing to stronger transport, the granite core has been deprived much more of its sedimentary mantle, whereby all limestone peaks are degraded to lesser rank among the outer ridges. To this must be added a late continuance of tectonic movements as witnessed by the earthquake radius covering Vierny, Andizhan and Karatagh. In Chinese Turkestan Huntington has even found lava streams overlying glacial deposits. According to Krafft the foldings of Darwaz are post-eocene coupled with warpings and faultings of a still more recent age. Undoubtedly the Duab mountains give one an impression of greater general steepness, looking more precipitous and threatening than the Alps. This may be ascribed to the combined effect of greater absolute height and lesser relative glaciation. The corries of the ice age have been desiccated to a greater extent, and the belt of soft glacial valley forms has been more roughened by weathering and erosion. Thus the sharp ridges and prisms hollowed out by corries, the pyramids of the Bernese Oberland and the aiguilles of Chamounix rising above their snow-fields are, in the Duab, continued downwards to a lower level, towering fearsomely above the deeply sunken beds of glaciers and rivers.

The dryness of the atmosphere, the absence of a covering soil and vegetation give free play to the contrasts of temperature and to the erosive force of water or wind. To the characteristic traits which this state impresses upon the face of the mountains, the desert at their foot stands in direct relation, being a necessary adjunct, moreover one capable of aggravating climatic factors by the growing accumulation of hot sand. Deserts can spread with amazing rapidity wherever conditions tend in their favour or man's watchfulness relaxes. To keep the desert within bounds is difficult enough, to gain on it is the work of generations, an inch for every yard the enemy has advanced through carelessness or destruction. A great desert, by the power of its atmosphere, has the tendency of pushing its boundaries to the uttermost limit, to the main watershed, forcing back towards it the rain belt, timber line and snow level. By sending forth every available excess of running water, distributing it with strict economy, man may advance a few steps, but this border of oases is as nothing to the endless waste beyond. Underlying the oasis there is always the condition of the desert, the irrigated paradise being but engrafted upon a soil under the sway of a climate inimical to vegetation. The oasis swims in the desert like a ship on the sea. To soak the whole of the desert is as hopeless for human effort as the filling in of oceans. The lonely aspect of sea and desert, their ever encroaching malignity or relentless fury, the dangers of their wide expanse inspire mankind with fear. The least gratitude towards them is probably felt by the fishermen and gardeners along their shores. Sea and desert are indispensable factors in the household of nature, but their uses are not fully apparent to immediate observation. Between them the high mountains form a link productive of that peculiar type of landscape where yellow drought and alpine snows are interlaced in curious fashion.

I hope that from my book the reader has gained an impression of the character of the Duab as a geographical type. In order to understand fully its relative importance among the features of Asia he should also study the neighbouring districts, whereby he can make the Duab stand out as an intermediate type between the extremes of Central Asia proper (Lop basin and Tibet) and the Western Caucasus. It is hardly possible, nor even permissible, to give a short and concise definition of the term "duabic," just as little as one can properly outline a man's character by such generalities as choleric, energetic, and brave. We cannot say in a sweeping fashion, for instance, that denudation and accumulation are quicker or slower than in the Alps, as so much depends on quality, quantity and exact location within the vast region. These processes are intermingled in various ways, changing hands many times and in varying degree along the line from the high summits to the bottom of the sea. Moreover the work of climate is complicated by geological influences and there is nothing more effective than a rotation of forces. The greatest and most rapid changes are produced by combined or alternate action of tectonic disturbances (upheaval and subsidence), atmospheric denudation and disintegration, erosion, transport and accumulation by water, wind and ice. Every one kind of energy is somehow apt to spend itself in the increasing quantities of its own products, or the reduction of its levels, or the growing weight of its upliftings, if uniformly pursued for a long time. Action meets its own reaction at an ever increasing rate, so that about half way through the task progress becomes so painfully slow, as to be almost tantamount to a standstill. The ideal end, such as the absolute level of denudation, is removed to infinity. Reaction retards so much that nothing but a revolution or reform will give new life by some sort of reversion or change of tactics, by the setting up of a fresh level of energy or the finding of another angle of impact. The greatest effects are produced by a rotation of forces, by a change of climate and level, by a small tectonic catastrophe here and there, by water, wind and ice taking alternate leads in the continuation of the destructive and constructive work of their contemporaries or predecessors, attacking their stagnations from a new base, or by new devices, carrying the fragments either to the same destination or building up in fresh places, lowering where they had raised or vice versa, depositing where they had planed and hollowed. The atmosphere has little effect on smooth, waterworn pebbles; wind cannot wear away vegetation; water is strongly resisted by clay; upheaval is limited by the bulk upheaved and erosion comes to a standstill long before the absolute plain is within sight. On the other hand frost and heat are poison to the rocks freshly unfolded from the womb of earth; water drowns the desert, and wind disperses dried up clay; furrowing glaciers find good purchase upon slopes rendered friable by air-rot and cross-scarred by water threads, while the rivers load themselves with the stone flour of the glacial mills.

To the student desirous of unravelling the influence of landscape upon the nature and history of man, the Duab offers an enormous field of investigation. Bound up with the atmospheric and morphological conditions is the life of the native mountaineer. The inhabitant of the Zarafshan valley has roads over terraces or naked and precipitous slopes which the Caucasian finds obstructed with undergrowth and fallen trees. Above the tree line the Sart's horse must strike his shoe into a hard substance, while that of the Suanetian walks on soft or muddy earth or struggles with the long grass of beautiful meadows. Here the crops of the villagers mostly depend upon an artificial water supply, there they are endangered by a possible excess of rain. How architecture has been evolved by surroundings and racial temperament, is apparent to the dullest eye as it wanders from the Swiss chalet to the slate-covered fortress of Suanetia and to the cubes of the Duab. The topographic nucleus of the Pamirs is full of meaning with regard to the distribution of the peoples, forming a prominent landmark from which to take bearings of their wanderings and vicissitudes. Just as the heart of a mountain range keeps within its innermost folds the oldest rocks, so it retains the oldest remains of bygone races. It is as if the shreds of floating humanity had clung to the jagged cliffs, or as if their remotest and stillest bights had preserved the rarest sediments left by the surging waves of migration.

The Galchas occupy the higher, interior mountain valleys with the exception of the true Pamirs which are practically steppe and which have always been easily accessible. Below and outside of the Galchas are found the majority of Tajiks, on the slopes of the Alai, on the middle Zarafshan, in Karategin, etc. Thus, as we pass out towards the plains the population becomes more and more mixed, and finally quite Mongolian or Turki, so that the purest Aryans are highest, the purest Mongols lowest, while in between the blends occupy the land of plenty. As yet it is impossible to obtain exact figures for Galchas, Tajiks and Uzbeko-Sarts owing to their not being sufficiently held apart in census returns. Nor are the ethnological questions quite cleared up, so that much remains to be done by the explorer of mankind. The Uzbeko-Sarts are undoubtedly the most numerous of the settled tribes, fairly monopolising the villages and cities of the irrigated plains. If one remembers that the steppe of the lowlands and of the Pamirs is inhabited by Turkomans and Kirghiz, it will be seen that one cannot go far wrong in taking the formation of a country as a guide to the distribution of its peoples. The great migrations have flowed around the base of the mountains, although traders and pilgrims will force their way over the most horrible passes, as shown by the Karakoram. Hence in some instances, we may also assume a slower kind of migration such as between India and Tibet, by a very gradual process of filtering through alpine obstacles.

Few shapes of surface are so expressive of the interdependence of physical conditions as the Crescent of the Khanate of Bokhara, best seen on a map with coloured political frontiers. Outlined by the Oxus this broad strip is folded around the central mass of the Alai-Pamirs. Long valleys, due to stronger erosion on the side of the rain-bearing winds, converge towards the Pamirs, representing the palm of a hand with fingers spread towards the west, or the top of a huge fan projected into the plains. Flowing round the foot of the mountainous expanse, the Amu-darya connects the ends of the large tributaries, linking each radius to the common circumference. Thus the Bokhara half-moon is a sector evolved from the central mass which impresses its influence upon the surrounding spaces of earth, determining the configuration of the ground, the arrangement of the river system, and the distribution of life. In the course of history the sweeping curve of the Oxus has served as a political frontier, and it still is so to-day, while in the remote past as well as in modern diplomacy the plateau-like elevation of the Pamirs has held the imagination of travellers, soldiers and statesmen. Thus between a circumference and a middle of exceptional geographical interest the semicircular area lies deployed as between the source and the boundary of life. Midway between these lies the trade route of Middle Bokhara which forms an inner concentric circle connecting those fertile districts where the rivers issue from their gorges to unite with the yellow loess, where the many factors of geographical interaction strike a happy mean between the icy mountains and the arid plains.

CLIMATE

For this paragraph I can only claim some general observations as my own, Professor von Ficker having given me leave to use his Academy report *in extenso*. All meteorological figures are taken from Ficker's monograph and Hann's great standard work on climatology. To these I must refer the student desirous of acquiring detailed knowledge of the climatic conditions of the Duab and neighbouring regions. The reports of Schwarz and Olufsen, and the diaries of Frau Dr Sild (Cenci von Ficker) have also been referred to.

As a whole the Duab is under the sway of a uniform continentality, but it will be well to remember that it stretches over many degrees of latitude, a fact to be borne in mind when comparing the temperatures of various places, even after having allowed for altitude. Roughly one may reckon a fall of 1° for every degree of higher latitude (1° in the plains, 1·4° in the mountains). Likewise the figures given below (culled from Hann) can only convey a very general impression, enabling the reader to make comparisons with more familiar places covering a similar range of latitude.

Temperature and Precipitation of a few familiar places.

Temperature.

Place and Height[1]	January	July	Year	Range	Extremes Minimum	Extremes Maximum
Cairo, 30 m.	12·3°	28·6°	21·2°	15·8°	1·9°	42·7°
Alexandria, 32 m.	14·1	25·6	20·3	11·9	7·3	37·4
Naples, 150 m.	8·6	24·6	16·2	16·0	—	—
Rome, 50 m.	6·7	24·8	15·4	18·1	—	—
Innsbruck, 600 m.	−3·3	17·8	7·9	21·1	−17·0	31·3
Munich, 530 m.	−2·6	17·2	7·3	19·8	−18·5	30·7
Berne, 570 m.	−2·3	17·6	7·8	19·9	−14·2	27·9
Sils Maria, 1810 m.	−8·1	11·2	1·5	19·3	−22·0	22·5
Säntis, 2500 m.	−8·8	5·0	−2·6	13·8	−23·0	15·5
Sonnblick, 3105 m.	−13·3	0·9	−6·5	15·0	−30·1	9·6
Rigikulm, 1785 m.	−4·5	9·9	2·0	14·4	−18·8	20·5

"Range" stands for the mean annual range which is the difference between the means of the hottest and coldest month. "Extremes" stands for the extreme annual range which is shown by the means of the highest and lowest temperatures of each year during a series of years.

[1] Height of observatory (cf. Naples).

Precipitation in mm.

Hungary	590	Middle Italy	840
Berne	930	Southern Italy	800
Tirol	1150	Sicily	600
Säntis	2510	Malta	550
Sils Maria	970	Alexandria	220
Italian Alps	1210	Port Said	82
Po Valley	810	Cairo	32

For the purposes of description and comparison Ficker divides into eight groups the 17 meteorological stations of Russian Turkestan. As I am giving only a very cursory review, I restrict myself to the five most important sections named below and belonging to the Duab proper. Occasionally I shall refer to the Issik-kul region (Przhevalsk and Narinsk) forming a transition to the moister climate of the Tianshan, and to Ferghana (Margelan and Namangan) representing an intermediate stage of Duabic types. Nor shall I give details of every single station, although the averages are, of course, derived from the complete list.

(1) **Steppe.** Kazalinsk, 63 m.; Petro-Alexandrovsk, 85 m.; Turkestan, 215 m.; mean height of whole group 121 m.

(2) **Western Rim** (i.e. of the mountains), Samarkand, 719 m.; also Tashkent, Khojent, Jizak; mean height 477 m.

(3) **Middle Oxus.** Kerki, 245 m.; Termez, 310 m.; mean height 277 m.

(4) **Panj** (i.e. Upper Oxus or Pamir Fringe). Khorog, 2105 m.

(5) The Pamirs: Pamirski Post, 3640 m.; Irkeshtam, 2840 m.; mean height 3195 m. Owing to certain divergences of Irkeshtam I shall, on the whole, accept Pamirski Post as typical of the Inner Pamirs.

A good method for comparing different localities is that of reducing their temperatures to sea level, i.e. of assuming that a place was let down vertically to a height of 0 m. A mean thermometric gradient of 0·5° has been chosen, viz. a difference of half a degree for every 100 m. of altitude.

Annual mean temperature reduced to sea level.

Steppe	11·6°
Western Rim	16·4
Middle Oxus	18·4
Panj	18·9
Pamirs	16·8

Hence the Middle Oxus and Panj are the warmest groups (relatively) the Steppe the coldest, while the Pamirs are warmer than the Western Rim, a phenomenon chiefly due to the greater heating of interior mountain masses, and partly to latitude. But the Middle Oxus is also absolutely the hottest portion of the Duab, its mean annual temperature (not reduced) being 17·0° against 11·0° of the Steppe. The monthly means of January and July reduced to sea level are also suggestive:

	January	July
Petro-Alexandrovsk	−5·0°	29·0°
Samarkand	3·1	29·2
Kerki	3·0	30·2
Khorog	2·1	32·5
Pamirski Post	−0·2	32·1

I have chosen Petro-Alexandrovsk as being farther south and therefore more easily comparable. Kazalinsk (difference 4° latitude) is only slighter cooler in July (26·5°), although much colder in January (−11·4°). It has an extreme desert climate. The actual mean temperatures are as under:

	January	July	Year	Range
Petro-Alexandrovsk	−5·4°	28·6°	12·6°	34·0°
Samarkand	0·5	25·6	13·2	26·1
Kerki	1·8	29·0	16·5	27·2
Khorog	−8·4	22·0	8·4	30·4
Pamirski Post	−18·4	13·9	−1·1	32·3

Throughout the Duab January and July are without exception the coldest and hottest months. The annual mean is often similar to that of some Italian places (p. 489), but this figure alone does not reveal anything of the particular climatic character of regions so compared. We must therefore also supply the monthly means (all given in Ficker, l.c.) I have picked out those for January and July because they furnish us with the mean annual range which is one of the most graphic elements of climatic comparison. The greater the mean annual range the more pronounced the continentality. Rome and Naples have a warmer winter and cooler summer than Petro-Alexandrovsk and Kerki, the annual range going down to almost one half. Alexandria is mild owing to the influence of the Mediterranean. Cairo has a small range in spite of its great aridity, but this blurring of the range must be ascribed to the approach of tropical conditions whereby the solar difference between summer and winter is toned down. Hence the rise of continentality is best revealed under temperate latitudes, and by comparing places on nearly the same latitude. In other words, nearer the equator a degree of range has a higher value than nearer the poles. A comparison between Sils Maria and Khorog (difference 300 m. and 9° lat.) is also interesting. Samarkand has a much milder climate than the steppes of the plains and the high steppes of the Pamirs.

The cold of the wintry steppe is mainly due to its low and open situation. On the flat plains the lower strata of air are cooled by contact with the ground which radiates its heat into space. This may be illustrated by Aulie-ata (620 m.) on the northern slope of the mountains overlooking the great steppes. Even allowing for a slight difference in latitude, it has a warmer winter than Kazalinsk (560 m. lower), and a warmer January than Turkestan. Winter along the Western Rim is far milder than in the Steppe. The Pamirs present an analogy, having an extreme steppe climate at an altitude of 3500 m. above Petro-Alexandrovsk. In this connection a look at the figures of high Alpine places proves instructive. The Engadine (Sils Maria) shows a greater annual range than the Säntis and Sonnblick peaks. Isolated summits, owing to their lesser mass and surface surrounded by a wide atmosphere, are relatively mild, possessing a smaller range than valleys or plateaus where large surfaces of soil and slope react upon the air by contact and radiation. Bringing Sils Maria up to 3600 m. would give us Jan. – 17·1°, July 2·2° which reveals the excessive range of the Pamirs as being chiefly attributable to a much hotter summer, explained by latitude and continental plus interior mountain heating. In winter, on the other hand, the enormous dryness of the Pamirs stimulates radiation, thereby neutralising latitude.

October is warmer than April at most of the Duab stations (exc. Khorog), while at Pamirski Post these two months are nearly equally warm. Little snow falls in the plains; soil and air are therefore quickly heated in April, whereas in the mountains (Khorog is the only really alpine station) solar heat is used up for melting the snow of last winter. Hardly any snow falls at Pamirski Post, the range of temperature thereby again resembling that of the Steppe. Thus spring is warm in the lowlands, cold in the mountain valleys, the opposite holding good of autumn.

A figure strongly illuminating the excessive continental type is that of the (mean) extreme annual range, i.e. the difference between the highest and lowest temperature of the year. In this case the mean is drawn from the absolute maxima and minima recorded during ten years:

Kazalinsk	66·6°	(difference between 39·3° and −27·3°)
Petro-Alexandrovsk	61·2	(„ „ 40·9 „ −20·3)
Samarkand	54·2	(„ „ 37·6 „ −16·6)
Pamirski Post	67·4	(„ „ 25·4 „ −42·0)

The absolute extreme is shown by the highest and lowest temperatures ever observed:

Kazalinsk	42·1° (June 1902);	−30·6° (Dec. 1903);	Amplitude	72·7°
Petro-Alexandrovsk	43·4 (July 1893);	−28·4 (Feb. 1886);	„	71·8
Kerki	42·8 (June 1899);	−21·7 (Jan. 1900);	„	64·5
Samarkand	39·5 (June 1899);	−20·9 (Jan. 1897);	„	60·4
Pamirski Post	28·0 (July 1901);	−46·7 (Jan. 1894);	„	74·7

Here also the northern Steppe and the Pamirs are easy firsts.

Similar diurnal extremes and ranges can be obtained by comparing the temperatures occurring at various hours of the day and night. In the lowlands and foothills and at Khorog the amplitudes are on the whole least in winter (mostly January) and strongest in September, when the heat is still great under daylight, but radiation cools the earth violently under the clear sky of lengthening nights. On the Pamir steppes however this oscillation is greatest in December, and least during June and July. There being practically no snow, radiation into and out of the ground must be enormous.

It remains to be mentioned that the Panj valleys (in the Pamir Fringe of high alpine ridges) are very hot and sultry in summer, as shown by the records at Khorog and the unanimous testimony of travellers, and also as evidenced by vegetation and agriculture. Thus there is very slight difference between the summer means of Samarkand and Khorog in spite of the great difference in elevation. A good way of illustrating such relations is that of the thermometric gradient which gives the fall of temperature for every 100 m. between two stations. Its world mean is somewhere near 0·55°.

	Difference in mm.	Summer	Winter	Year
Khorog to Pamirski Post	1535	0·53°	0·74°	0·61°
Samarkand to Khorog	1386	0·24	0·60	0·35
Samarkand to Pamirski Post	2921	0·40	0·67	0·49

It will be seen that the gradient is exceedingly small between Samarkand and Khorog. Compared to Samarkand the Pamirs are too cold in winter (gradient too high) and too hot in summer (gradient too low). Khorog is approached to Samarkand, that is to say it is much milder than the Pamir steppes as demonstrated by the sudden jerk of the gradients above Khorog.

The Panj district has a mountain climate (in this case an alpine climate modified by continentality) while the Pamirs form a massive pedestal furrowed by relatively shallow valleys. In the Alps the relation between the gradients of summer and winter is reversed. In the Duab the descent of the thermometer with increasing height is slow in summer and rapid in winter, or, in other words, the difference between the low country and the mountain is greater in winter than in summer. In Europe, on the other hand, winter equalises the contrast between low and high places. During July the Alpine gradient is about 0·60°, during January about 0·40°. The very long gradients (Samarkand to Khorog, etc.) give a specially low value in summer owing to the relative rise (less rapid fall) of temperature towards the interior of a mountainous complex. From Ferghana to the Pamirs I have obtained a very equable mean gradient (July 0·50, Summer 0·50, Winter 0·48, Year 0·48), betraying a compromise between Alpine and Duabic conditions. Quite different again, as may be expected, are the gradients traced from Ferghana and from the northern steppes (Aulie-ata, etc.) towards the Issik-kul group (Western Tianshan) with its moister climate.

	Summer	Winter	Year
Aulie-ata to Issik-kul	0·52°	0·44°	0·40°
Ferghana to Issik-kul	0·70	0·64	0·60
Namangan to Narinsk	0·76	0·93	0·67
Mean for Western Tianshan	0·61	0·56	0·49

Among the temperature elements passing mention may be made of the duration and frequency of frost. These data can also be used to demonstrate continentality, but so far the Russian observations are not sufficiently homogeneous in this respect. Stations on the same latitude ought to disclose a growing insistence, landwards, of temperatures below freezing point.

Temperature and humidity of the air, greatly dependent upon each other, are the most important elements of climatology. Upon them precipitation depends. I only give a few figures on humidity as those on rainfall will supply a more graphic picture to most readers. The annual mean of relative humidity for the Duab is 58 %. The driest places are Petro-Alexandrovsk 57 %, Jizak 57 %, Samarkand 60 %, Pamirs 49 %, the latter being the driest of all and considerably below Petro-Alexandrovsk. The maximum of relative humidity occurs everywhere in January. At Pamirski Post September has the driest atmosphere, while the lower places produce their monthly minima in July, sometimes including June and August with the same percentage. On the Pamirs Olufsen has observed absolute minima of 5 % in July, 2 % Aug., 4 % Sept., 3 % Oct. Taken at heights above 13,000 feet these figures show increased aridity at altitudes above Pamirski Post. In the Steppe the minimum is 7 %.

Precipitation or rainfall (which is used in the same sense, i.e. including snow, dew, hail, etc.) increases with height where the moisture-laden air meets a lower temperature, but above certain levels a decrease sets in again, causing the high Pamir steppes to be even less rainy than the desert plains. This zone of high precipitation one may call the rain belt or belt of snow clouds. The sums of annual precipitation for the five groups are as follows:

Steppe	130 mm.
Western Rim	334
Middle Oxus	160
Panj	228
Pamirs	62

Monthly means of precipitation, in millimetres.

	Jan.	Feb.	Mar.	Apr.	May	June	July	Aug.	Sept.	Oct.	Nov.	Dec.
Petro-Alexandrovsk	11·6	8·9	24·6	13·3	4·7	6·6	1·0	0·9	1·3	5·3	8·6	10·3
Samarkand	35·7	24·3	65·2	76·8	35·5	6·7	5·2	1·3	2·3	25·2	31·0	28·5
Kerki	30·7	16·2	32·2	24·7	8·0	2·0	0·1	0·0	0·0	6·2	18·9	20·5
Khorog	28·2	20·8	27·8	21·4	28·6	14·6	5·7	0·2	0·2	13·2	43·0	24·8
Pamirski Post	7·8	2·5	1·6	3·6	8·4	15·4	8·0	4·4	3·9	2·5	2·1	2·1

The Duab below the 1000 m. level is a region of spring rains where the greatest precipitation is due in March or April, summer being almost quite dry. The Middle Oxus is absolutely dry during August and September. Almost everywhere the minimum occurs in August. Conditions are however very different on the high Pamirs. Here early summer is richest in moisture. The Panj mountains provide a notable contrast with their dry summer and a snowfall of 43 mm. in November; there is more precipitation in winter than in spring. During the early part of the year the humidity of westerly and southwesterly winds is condensed at lower levels, leading to winter and spring rains in the plains and foothills. Thus deprived of their moisture these winds reach the Pamirs in a fairly dry condition. In summer these winds are drier (owing to great heat in the plains) and must rise higher in order to condense their vapour. The bulk of the moisture thus carried forward is captured by the outer mountains, beginning with the foothills up to and including the last bulwark of the high rim or Pamir Fringe (cf. Chap. II). A fair amount of snow falls on the Alai, in the Zarafshan and Panj valleys, etc., the Pamirs remaining quite bare, save for a faint sprinkling now and again. This lesson is also taught by the glaciers born from the vapour catching walls of Transalai, Peter the Great, Darwaz. The longest periods of drought in the Duab are shown by stating the greatest number of successive months with an unbroken absence of rainfall: Petro-Alexandrovsk 5, Kerki 6, Samarkand 5, Khorog 4, Pamirski Post 3.

Cloudiness or the state of the sky is, of course, intimately connected with rainfall and evaporation. The number of clear days is largest along the whole course of the Amu-darya, including the Panj. Termez (near Kerki) has 191 cloudless days, Kerki 169 (43), Petro-Alexandrovsk 167 (52), Khorog 151 (48), Samarkand 146 (66), Pamirski Post 115 (55). Totally clouded days are placed behind these numbers in brackets. Nebulosity caused by loess dust is a characteristic phenomenon of the middle mountains (Alai valley, Karategin, Khingob, Middle Bokhara). In the plains dust fog is not of such regular occurrence as higher up, where convection is promoted by ascending currents along the slopes.

As to winds, N. and N.E. form an overwhelming majority in the Steppe; N.W. at Kerki; S.E. and E. prevail at Samarkand; W. at Khorog; S. and S.W. at Pamirski Post. The lowlands show a preference for northerly currents, the higher levels for southerly and westerly ones, while in between winds are fairly well distributed over all quarters. I seem to have felt a typical Föhn wind descending from Kemkutan one autumn day at Samarkand. It caused the same bodily and mental depression so familiar to me from my residence at Innsbruck.

Concerning evaporation, the available data are very meagre, and I shall speak of them when treating the problem of desiccation. Owing to their scientific interest observations of this factor would prove valuable, chiefly because such figures will enable us to compare them with calculations based on rainfall, river volume and lake level. With the exception of the snowy mountains there is probably no district in the Duab where evaporation is not in excess of precipitation. In itself this fact is not so characteristic as at first sight appears, not necessarily being a direct expression of the physical conditions of the region. So much depends upon the "where" and "how" of excess of evaporation in conjunction with the formation and distribution of morphological features. If there were no glaciers and valleys, the water would not be concentrated, but evaporated over an even spread. Thanks to the mountains there are rivers presenting a lesser surface to sun and air, whereby much of the moisture is enabled to collect in the Sea of Aral. Tapping these arteries the peasant produces an artificial distribution not properly accomplished by nature through the medium of rainclouds. The evaporative power of an atmosphere depends upon solar energy (latitude), continentality, altitude and wind. The low position of the timber line on and near the Pamirs may to a great measure be ascribed to windiness which stimulates evaporation.

According to the latest views (Brückner, Meer und Regen, *Naturw. Wochenschr.* 1905) we should dismiss the idea that the greater portion of rainfall on land is furnished by the sea. Hence we must assume a greater turnover of interior or continental circulation, and concentric areas of evaporation and return arranged around a centre of condensation such as the Alai-Pamirs. A certain percentage of the water fallen near the snow line will return within a short radius; another portion traverses a longer radius before coming back again to be condensed, while some will be lost to neighbouring regions, the deficiency having to be made up by another district or the distant sea. Much of what falls above the snow line is predestined for the longest journey, and the higher the longer. Not only does the glacier water run out far into the plains, but the snow particles on the highest rim of a névé basin are also the last to melt at the glacier snout, so that the topmost snowflake stands a chance of performing the longest descent. Aridity is due to difficulties of condensation and facilities of evaporation. These are mainly caused by a combination of latitude and topography. But the atmospheric (i.e. climatic) conditions may so react upon the surface, as to increase its tendency in the direction of still greater dryness by an extension of the desert.

These few remarks and tables provide the reader with a short survey of the figures by which the extremes of the Duab climate can be set forth in terms of scientific meteorology. Climate here obeys the bidding of a pronounced continentality coupled with extremes of morphology and altitude. It is worth mentioning that the general continentality of our region at first decreases eastwards, as shown by lessening annual and diurnal ranges on the way from level and steppe towards elevation and mountain valleys. But beyond the Pamir Fringe the contrasts of temperature are accentuated again as we emerge upon the high steppe. The Pamirs have practically only two seasons, a long winter and a short summer. These, owing to aridity all the year round, are mainly distinguished by cold and heat. In a more or less pronounced degree the same holds good of the plains, and of the Duab as a whole, the transitional stages of spring and autumn being exceedingly short. In localities with four distinct seasons the temperature of the middle spring and autumn months are usually both very near the annual mean: Cambridge 8·0, 9·2, 9·2; Lyons 12·2, 11·8, 11·7; Paris 10·3, 10·3, 10·3; Warsaw 7·1, 7·8, 7·3; Frankfort o/M 9·7, 9·6, 9·6; Nuremberg 8·0, 8·0, 8·0; Munich 7·4, 7·8, 7·3; Vienna 9·4, 9·8, 9·2. To these compare: Petro-Alexandrovsk 14·1, 11·6, 12·6; Kazalinsk 9·1, 8·0, 8·3; Turkestan 13·4, 10·8, 12·1; Pamirski Post 0·2, 0·0, – 1·1; Port Said 18·5, 24·1, 20·5; Alexandria 18·4, 23·3, 20·3; Assuan 27·5, 29·3, 25·0. Of very oceanic temperate climates, on the other hand, it might be said that they also have two seasons only, namely neither summer nor winter, but a spring of gentle, moist warmth and an autumn of sloppy coolness. The sea climate has a cool spring and a warm autumn, while the continental type often betrays itself by October being colder than April. Thus in the low levels and steppes of the Duab a warm spring is opposed to a cold fall, owing to the absence of a snowy blanket whereby the vernal sun is enabled to heat the dry soil very rapidly. This rule is modified wherever a thick layer of snow is accumulated on the ground during winter. It causes the solar energy to be absorbed by the effort of melting the snow, thereby depressing the spring temperature (relatively to autumn) of continental regions, approaching them, in this respect, to the conditions of a more oceanic climate.

Whoever has visited the Alps in winter, especially such favoured places as Davos or Cortina, will have observed characteristic extremes. The sky is generally cloudless, and the sun shines hotly through a dry and transparent atmosphere. But in the shade and at night the bitter cold is all the more noticeable. Compared to England the Alps have a winter climate of severe contrast of temperature and humidity. The Alai mountains again have a severer and windier climate than the Alps, partaking, as they do, of extreme continental conditions of the vast Tibetan, Kashgarian and Aralocaspian plains and of the Pamirs. But compared to these steppes and deserts the Alai, and still more so the Alps, must be called "mild" in summer as well as in winter.

Unfortunately space forbids my entering into details concerning Central Asia (Lop and Tibet), where the continental character is still more pronounced, still drier, still more implacable. A few figures quoted from Hann must suffice:

		Jan.	July	Year	Range	Rain
Kashgar	1230 m.	– 5·8°	27·5°	12·4°	33·3°	46 mm.
Yarkand	1255	– 6·0	27·6	12·3	33·6	13 „
Uliasutai	1635	–26·3	18·3	– 1·1	44·6	very dry
Lukchun depression	– 20	–10·5	32·5	—	43·0	infinitesimal

As to the circulation of atmospheric water, innermost Asia is probably dependent on its own turnover of drifting vapour. What holds good of the Duab as the peripheral or subdued surroundings of extremest Central Asia, also applies to High Bokhara (Shugnan, Roshan, Panj, Pamir Fringe) as the rim of the Inner Pamirs. In both cases the outer zone is less pronounced in its continental character.

Without its mountains the Duab would be an unbroken stretch of desert. As Schwarz rightly observes, the peasant of the oases pays no attention to weather, his weal and woe depending entirely upon the snows of the Alai. Elsewhere "weather," *ceteris paribus*, is more important for vegetation and (non-irrigated) agriculture than "climate," the single occurrence of a dangerous frost at a late date, a short delay of rain often deciding the fate of the season's crop. A desert is frequently compatible with a fair amount of rainfall, only that it comes at the wrong time. According to Schwarz only about 3 °/₀ of the surface of Turkestan is under cultivation, while 40 °/₀ may be reckoned as pasture land and the rest as desert. For the Duab alone the position is hardly much better, and I think that 4 °/₀ of cultivated area is a fair maximum, leaving 35 °/₀ for grazing and 61 °/₀ for desert, rock and glaciers.

I may add some remarks anent the effect of climate upon the human body. In dry climates the blood, containing less water, stimulates the nervous system whereby excitation and sleeplessness may be caused. At first this result is also felt by healthy persons freshly arrived in an arid or high mountain climate. They experience a certain restlessness or nervous energy. In spite of greater relative humidity a mountain atmosphere (like that of the Alps) is in this respect equal to the drought of lesser elevations, owing to increased evaporation under diminished pressure. The plants of the Alps and steppe show the same relation (Chaps. III, p. 50; XIV, p. 379). Humidity tones down the nervous functions, increases the exhalation of carbonic acid and retards the circulation of the blood. Dryness accelerates the pulse, desiccates the skin and makes one more liable to feel chilly (Hann, I, p. 50). Hence dry air at great heights, such as in the Pamirs, must produce these physiological effects more rapidly and in a still more marked manner (cf. Chaps. VI, p. 139; XIV, p. 378). As a popular illustration I refer to the heating of a house or room. The drier the room the more one must heat in order to create a comfortable sensation. Overheating therefore drives one into a vicious circle, for the air is dried too much, which again reacts upon the skin making it crave for a still higher temperature. The primitive English fireplace is generally sufficient in the moist climate of the British Isles. It also saves one from becoming spoilt. Americans are very chilly people and slaves to overheating.

FOREST AND CLIMATE.

That forest must react upon the climate or, to be more definite, upon rainfall, goes without saying. The question is, how much? What the geographers and meteorologists ask themselves is, whether this influence is great enough to alter the climatic character of a region, whether re-afforestation will cause a considerable increase of rain. That the wooded portion of a climatic district reveals slight variations of the climatic elements, that it affects radiation and air temperature and that it breaks the force of the wind, nobody will deny, although it must always remain difficult to apportion the exact amount of these secondary deviations. Nor do we speak here of the enormous importance of vegetation with regard to the preservation of the soil and the distribution of running water. As we do not know if certain regions, now bare, were really thickly wooded, we have no means of comparison, unless we except the Central Provinces of India, where large tracts have been re-afforested and where some authorities take an increase of rainfall as proven. It is not unlikely. But the difference cannot be one between nothing and much, but between much and more, it being unthinkable that trees will grow up in a country where rain is altogether inadequate in quantity and distribution. They would have to cause an increase during summer, during the vegetative season, or even to create precipitation afresh. As very large forests cannot be raised by irrigation, the problem is hardly solvable by experiment. As the trees will grow anyhow, if rain is sufficient, the inquiry is mainly of theoretical interest. Practically then, our object could only be one of forestry, namely that of growing a forest in order to call forth still better conditions for forests as regards soil and drainage. It cannot be assumed that the benign influence of densely wooded (artificially) areas in a dry region will spread far over wheat fields around the rim of the forest, while groves interlarded with open spaces for agriculture would be neutralised in their possible effect upon the atmosphere of the district as a whole.

At present opinion is still very much divided. Supan does not believe in any influence worth speaking of, while Eckardt even strongly denies it. Blanford and Fernow claim a distinct effect in tropical climates. Harrington, who also believes in it, draws our attention to winds blowing from and into the forest, as in the analogous case of land and sea breezes. All agree, of course, that vegetation and more particularly forests, are mediators between the climate, to which they chiefly owe their existence, and the surface of the land on which they live. Much of the action upon the air will depend upon the situation of the woods, whether on the coast, in the interior, or on mountain slopes, much also upon their density and the species of trees. Most of the typical trees of dry climates never form large associations (walnut, elm, juniper, poplar, plane, olive, hawthorn, etc.), while those protected against excessive transpiration could hardly be expected to recharge the air with moisture, or to cool it down to condensation point.

One thing which has struck me of late is the connection between forests and interior vapour circulation. They allow less water to reach the sea, hence they must send back more into the air. As the major part of land precipitation is not derived from the sea, forests should be responsible for a larger interior turnover between evaporation and precipitation. The Nile basin, full of forests and swamps, has a rainfall of 3700 cb. km., but the river only discharges 100 cb. km. or $\frac{1}{37}$. The Oxus basin receives 100 cb. km. in rain and snow, discharging about 30 cb. km. into Lake Aral, or a little less than one-third. Even allowing for the enormous length of the Nile's desert track and assuming a gross underestimate of the Oxus precipitation, the difference remains very great. Plants do not require an excess of water beyond a minimum, because they only use it for keeping up the circulation of the sap, and not for building. Land vegetation requires a turnover of water; water plants need a circulation of air and minerals. Hence the annihilation of a vast forest area may lower the precipitation, though not necessarily to such an extent as to prevent the natural re-growth of the trees. But where the present quantity and seasonal distribution of moisture is just sufficient, complete deforestation might cause a loss of water to inland circulation, and thereby stop natural tree growth. In such cases (probably rare, if provable) the forests must be the preservers of an older climate (of temperature and wind direction) which had invited spontaneous vegetation. Here then devastation would result in what should look like a local change of climate. Thus forests may preserve forest conditions (frequency of rain and keeping of soil), but the reconstruction of these conditions by the artificial restoration of forests is quite another matter. On the whole we must look upon the forest as a result of climate, while admitting certain reactions. But if we were to magnify these secondary effects they would lead us into a vicious circle of cause and effect.

SNOW LINE

There are several definitions of the snow line such as the orographic, the climatic, and the temporary. The latter merely signifies the actual limit of the mountain snows at any given time, while the snow level in general is determined by the greatest altitude to which the white mantle ever recedes during the year. Thus the snow line is the lower limit of permanent snow which never bares the surfaces upon which it lies. We can shortly dismiss the climatic snow line as a very mathematical and theoretical concept expressing that elevation where the crystalline precipitation of the year, collected on a horizontal plane, would barely melt. The orographical snow line, which might also be called the accidental or local one (though it must often coincide with the real or average one), is represented by a line connecting all the lowest snow patches wherever found, a great number of which owe their low position to the shade afforded by rocks or gullies, that is to say to orographical favour. Others are very high up, in the sun on southern slopes. Hence this curve is very jumpy, often showing extremes of 500 m. and more on the same mountain. The edge of a snow-field which the traveller meets is, to begin with, nothing but a temporary orographical snow line, a particular occurrence. Only very many observations throughout the summer, and on all sides, will enable him to determine the general or actual snow line which, for practical purposes, we obtain by connecting the margins of the large snow-fields towards the end of summer. From a distance it will appear as a more or less horizontal boundary. According to Paschinger who identifies the climatic with the true or actual snow line, it is the altitude level above which, during the hottest part of the season, the snow-covered surface is far in excess of the dry ground. Owing to the differences between north and south exposure one often expresses the snow level by limit values such as 2500 to 2800 m., but I here usually give the mean between the two. From the above it will be seen that the fixing of the snow line is by no means an easy task in a country, where we have no detailed maps of the high mountains, and where travellers only pass through at intervals, and chiefly during July and August. In the Alps the temporary snow level coincides with the absolute one some time in August, but this may quite likely happen a few weeks later in the Duab, owing to the smallness of precipitation in September. That rainfall is a more important factor than cold is shown by northern Siberia where hills 600 m. high bear no permanent snow, in spite of a mean annual temperature of $-17°$ (nearly zero Fahrenheit). Data of the Duab snow line, including my own observations, being very scanty as yet, it is impossible to use them as a basis for exact calculations. On the average it may be put at something between 4200 and

4400 m., rising from about 4000 m. in the Zarafshan districts (Western Rim) to 4700 in the Pamirs. In the Chatkal and Talaski Alatau, Machacek has measured it very thoroughly, putting it at about 3700 m. (3700—3900 m. and 3450—3600 m.). Thus a fluctuation of 1000 m. is observable within a comparatively small area.

The temperature at the permanent snow line is a problem to which I have devoted some study, but without arriving at a very satisfactory result. I mention it here in a tentative manner, with a view to further investigation. Hann tells us that the temperature at the snow line (i.e. at that fixed altitude corresponding to the permanent snow level) is the higher the greater the precipitation and the smaller the annual range. In other words, in a moist climate the accumulations are able to reach further down to meet the warmer layers of the atmosphere. Evidently Hann refers to the annual mean, and in this respect my calculations agree with the theory. Using different gradients to those in the Alps (as also later, different ones for July and the year) I find that the Duab snow line, taking a high value of 4400 m., is 3° colder in the annual mean (about $-6°$) than that of the Eastern Alps (about $-3°$). My results do not seem to agree with Paschinger who applies the above rule to July also. After mentioning the fact that in the early days the snow level was believed to coincide with the latitude and altitude isotherm of 0°, and after referring to the wide variations of snow line temperatures all over the world, owing to the different thickness of the layers to be melted, he continues: "Where the snowfall is so heavy as in the Western Caucasus great quantities of heat are needed for melting the snow which therefore reaches far down. In dry countries however solar energy can dissolve the thin covering into higher altitudes where the temperature is already below freezing point. In the Alps the July temperature at snow level is about 4°, whereas in the mountains of Central Asia it is from 6° to 8°. This fact, together with other considerations, makes me doubt the supposed extreme dryness of these regions." I, on the contrary, believe that we had better accept as true the extreme aridity of Central Asia and the somewhat lesser extremes of the Duab, and endeavour to find out why the Duab snow line has such a high temperature in July. Paschinger may possibly be mistaken in believing that the July temperature (as well as the annual mean) ought to be colder in continental regions, although his argument is plausible enough to make us wish for better information about the Duab snow line. As far as I can work it out the July temperature is about 3° higher than in the Alps, and even assuming a very steep gradient (0·60°), it cannot be lower. In this I am confirmed by the lesser annual mean in accordance with theory. It may be that the Duab snow line is nearly covered by the belt of high precipitation and high snowclouds, so that here the melting process meets with a sudden increase of thickness. There may also be topographic reasons, the snow line being so high as to coincide with the rim of very large, thick, and icy snow-fields, and basins fed by their higher portions. Finally we might remember that the extremes of climate are possibly expressed also by the snow line temperatures participating in the range between warm and cold. Be this as it may, it leaves us with a strong desire for more exact records of the Duab snow level.

Error.

A slip has occurred on p. 35, top line, "The Eastern Alps are slightly drier...." This is misleading. The snow line is, of course, highest in the great interior mountains (see Scientific Glossary), in the mighty constellations of Mont Blanc, Monte Rosa and the Bernese Oberland. On the whole it descends towards the east, although there are several ups and downs in between.

GLACIATION.

Judging from my own observations and the reports of other travellers, the number of Duab glaciers is far greater than one imagined a few years ago. Although the snow line is high, the Alai-Pamirs must find compensation in their enormous altitude ensuring them a gain of glaciated surface, so that the total amount of ice may prove equal to that of the Alps which cover an almost equal area. But this glacier ice is probably much more dissected into independent branches. In this, as in all other questions of past and present glaciology, the paucity of available data still stands in the way of safe conclusions and exact comparisons. The larger valleys are well filled with detritus, the lateral ones being cut down to accordant grade; hanging valleys, waterfalls and steps, large surfaces of polished rock or roches moutonnées are rare; small interior lakes are fairly numerous; foothill lakes like those of Switzerland are entirely absent. Old moraines are frequent objects of interest through their size and shape, their position apparently pointing to two distinct periods. One set reaches down to a level of about 1500 m., the other averages about 2300 m. In the latter must be included the Achik moraines of Tupchek which owe the great height of their ends to the smallness of isolated feeding basins. According to Berg some diluvial glaciers of the Isfara district (Alai) descended to 2200 m. Although it is not yet permissible to synchronise the stages of Duab glaciation with those of the Alps, we may at least assume that the higher level belongs to a period which I have dated by the Pakshif moraine (cf. Chaps. VII, X, XV). This Pakshif period seems to correspond to the glacial records discovered by Machacek in the Chatkal and Talaski Alatau. He is however unwilling to express an opinion on other glaciations of that district. The huge masses of boulder clay choking the Iskander and Muk valleys probably belong to the Zarafshan period. From Lipski's photographs and still more from Korzhenevski's report we can infer that the sides of the Muksu canyon consist of young morainic stuff, as testified by the steep angle at which these deposits overhang the river.

Owing to the differences of snow line, ranging from 4000 to 4700 m., we must always be prepared to encounter considerable variations of glacier ends. Added to those caused by topographic conditions, these variations will necessitate a greater number of measurements, before we can obtain a clear insight into the classifications of the Duabic ice age.

As to the neighbouring regions it seems that two glacial periods have been established with certainty. Huntington, who has observed many moraines and terraces in the Tianshan, Prinz who has worked in the Kulja Nanshan, and Friederichsen agree that two periods are quite sure. According to Friederichsen a greater number is not improbable, though some of these intervals may merely be due to interrupted recession. Gröber considers two glaciations as probable in north-western Mongolia.

As to the climatic causes of the old Chatkal and Talaski glaciations (which practically fall within the Duab) Machacek thinks that humidity was very little above the present, but that owing to a somewhat lower annual temperature there was less evaporation and a lesser melting of glacier ends. Few glaciologists are likely to agree with this view, against which Merzbacher has already protested, and which I also prefer to see modified a little. Physicists are not yet agreed as to a fall of temperature (change of solar energy) being necessarily the cause of a glacial epoch. But we shall here occupy ourselves less with the prime cause than with the probable conditions reigning during the Duab glaciations. That cold alone cannot perform the task of a powerful glacier advance is amply shown by northern Siberia. A combination is needed of moisture, temperature and small annual range (comparatively cool summer and comparatively warm winter). In a sea climate an increase of cold is of course sufficient, but in the Duab a fall of the mean temperature can hardly lead to a considerable growth of ice streams. The belt of crystalline precipitation would be lowered, it is true, but it must be remembered that the lower snow does not contribute very much to the volume of a glacier. Nor does greater cold necessarily increase the precipitation above the level of the summer snowfalls. Much depends upon topography. On enormous slopes gently rising to great heights the mere depression of the snowfall may cause energetic glacier growth, while in dissected mountains the advance of the glaciers may be small compared to the depression of the permanent snow level. Hence it is very important to consider at what stage of a glaciation a change occurred. The descent and length of a glacier will show a different proportion of growth according to the point at which it just happens to be, whether in a shallow basin or among deeply eroded slopes, whether still near the tops of the mountains or already beyond their foot.

On the whole humidity is the most important factor, and a reduction of the annual range more important than a lowering of the annual mean. I believe therefore that during the main Duab glaciations the rainfall must have been greater than now. The present aridity of the Pamirs seems to indicate a pretty complete exhaustion of vapour, so that a fall of temperature cannot give a strong impetus to the glaciers. Easton (*Periodizität; Peterm. Mit.* 1905) thinks that the European coastal climate is more sensitive to the fluctuations of solar energy than the continental climate, which means that under extreme interior conditions a change in the sun's force will not produce results of a magnitude equal to that in the Alps or Skandinavia. It may quite well be that owing to this reason we shall perhaps, in the future, discover that the Duab glaciations show a lesser number of well marked subdivisions, such as are traceable in Europe. Hann says that colder summers are not a reliable indicator of ice ages, and that the problem is exceedingly complicated owing to the overruling share of precipitation. Frankland and Becker (quoted by Hann) tell us that the formation of glaciers is a true process of distillation requiring both heat and cold. The product of distillation would be diminished, not augmented by a lowering of temperature. It is not impossible that the temperature at sea level was higher during the glacial period than to-day. Ficker is of opinion that the Duab climate during a glacial advance must have been milder, i.e. more oceanic. That would make summer cooler and winter warmer, coupled with an increase of rainfall at all seasons. The Pamirs, now a district of summer rains, will have had a good deal of winter precipitation. Summer being colder, there was less melting of the glaciers. We must beware of mixing up cause and effect. Summer was cooler, because there was more rain and a greater mean cloudiness, not the other way round. This question mainly rests upon the problem of the desiccation of Asia. If it can be proved that considerable sheets of water existed in Aralocaspia and elsewhere, then the glacial climate of the Duab must have shown approximation to that of Europe, and the more, the larger the lakes were. As our present geological knowledge points to the likelihood of such inland seas, and as cold alone appears to be rather ineffectual in a dry climate, one cannot go far wrong in ascribing the old Duab glaciers to a greater humidity coupled with the as yet unknown general cause of glacial recrudescence and periodicity. We have to reckon with two factors in the Duab, the local one and the world-wide one. The local influence leads us to the conclusion that at the beginning of climatic reckoning (i.e. end of tertiary period) the Duab climate was less divergent from the European or Alpine one than it is now, that they approach in likeness as we go back in time.

Above all we must remember that a fall of temperature will reduce evaporation. Hence it is exceedingly important to know whence comes most of the vapour condensed in the Duab, whether from the outside (or sea) or from interior circulation, whether external or internal turnover is greater, whether the glaciers and running water are mostly sent around an inner atmospheric track, or whether they form the balance between the imports and exports of vapour. If the turnover is mainly an interior one, then a uniform increase of cold will depress the snow level, glacier ends and rain belts a little, but the glaciers could not gain much in bulk as there would probably be no increase of precipitation above the present snow line. If more vapour is forced in from outside, or if inland seas are created, then the balance of moisture as a whole is augmented and the glaciers are vigorously fed. The difficulty of the problem lies in knowing exactly how the give and take of vapour is worked between neighbouring regions, for what is gained by one the other must lose. It seems unlikely that a continental area will be very much affected in its glaciation by a mere change of annual temperature, unless coupled with a different disposal of land and water, or a change of the great wind tracks. We can quite well imagine that the ice age was possibly due to a world-wide uniform fall of temperature (solar or cosmic), but the effect must vary enormously with the morphological features of the land. Cold alone is a poor instrument for raising the humidity of continents, whereas an increase of humidity is in itself sufficient to invigorate glaciation without the collaboration of cold (i.e. extra cold from outside, as apart from the cooler summer produced by cloudiness). In the Duab the question of moisture supply is the leading one, and we must seek for reasons, topographic and meteorological, capable of explaining a greater humidity of the whole Duab atmosphere. After that the universal glacial cause, whatever it may be, takes rank as the possible regulator determining the apparent synchronism of glacial events all over the earth.

In comparing the various glaciations of the same district and those of different parts, the height of the snow line is usually taken as a very direct standard, so that given a similar orography, the snow line would determine the size of the glaciers. I wonder now if this rule is as reliable as one thinks, if like ground above two snow lines need necessarily hold the same amount of névé and ice in different parts of the world. Assuming a simple case let us say that a cone or a basin of exactly the same size and shape is set upon a snow line of 2700 m. in the Alps and upon a 4000 m. snow line in the Alai. Is it then a foregone conclusion that upon these equal surfaces rests the same weight of névé? I should say not, because the snow level is not a direct expression of the quantity of precipitation and its preservation in the shape of névé, but also of the range of temperature. To this difficulty of comparing equal orographical conditions above a common base of snow line, must be added that of different surface formation. It is owing to these doubts and complications that the comparison of glacial periods in different climates and localities is much obscured. An important point is the absolute height of the mountains and its relation to the snow level. Thus the towering height of the Alai-Pamirs offers a good deal of compensation for the lower position of the Alpine snow line. A loss again must be entered against the Duab mountains owing to the greater isolation of their summits. We should also keep in mind that the maximum rainfall is bound to certain levels, above which there is a decrease. Hence shifting of this belt as a whole will have different results according to the topography which it intersects. If it is shifted by cold alone, that which is added below may have been taken away above.

A different combination of height and morphology will cause a different ratio of advance, that is to say an equal increase of cold or moisture need not produce the same increase of glaciation or depression of glacier ends below the snow line. A vast plateau just above the snow line, and furrowed with shallow grooves, favours the accumulation of ice and the formation of massive glaciers far more than a concourse of precipitous mountains, however high, where the steep ground is more dissected, exposing larger surfaces of snow to insolation. The more numerous and deeply notched the valleys are, the more they cut below the snow line, the greater will be the number of small separate glaciers prevented from joining each other by an unfavourable position of the valley junctions. When two glaciers unite they reduce the wasting surface of their common bulk, and both together are able to reach a lower level than each singly. Thus a junction just prevented, or just effected, must make a considerable difference in the size of the glacial phenomena (moraines, etc.) of a certain locality. The way in which a snow line is adjusted to the structure of the mountains, to the mean gradients, to the level of most numerous valley junctions, basins and plateaus, must cause a different rate of progress or retreat when certain critical points are passed. Hence a slight rise or depression of the snow line near such a critical structural horizon may result in a growth or dwindling of glaciers and moraines, apparently quite out of proportion to the change of snow level. A good example of the influence of topography is offered by the regular and gently graded herring-bone valleys with steep ravines. The Zarafshan glacier shows how an ice stream can be very long without attaining huge thickness, so that if our attention is riveted upon the longitudinal fluctuations of the main glacier, an erroneous impression may be conveyed of the magnitude of the change. Every one of the small but numerous tributaries adds its mite of nourishment just at the right time, and just enough to keep up the life of the main trunk. Such an arrangement causes the middle and lower portions of the glacier to be nothing but the ends of its branches welded together at the last moment. Every depression of the snow line will cause many new lateral glaciers to descend to the main valley where, joining their ends, they will add enormously to the length of the trunk. When retreat sets in, a long row of affluents becomes detached, leaving a large section of "dead glacier" behind, to waste away as a whole up to the point where lateral feeders still unite. Clearly then the estimate of the comparative values of mountain glaciations demands a close study of topography on the basis of excellent maps. These problems are furthermore complicated by the progressive work of the glaciers themselves during the time of their strongest activity, and by tectonic disturbances. Ludwig points out how the old glaciers "dug their own graves." They lay on massive heights shaped by tertiary erosion (peneplain theory; lately denied by Distel and Martonne) or on the plateau-like warpings of the earth's crust, so that compact surfaces were above the snow line. By erosion these gentle slopes were lowered and divided, whereby the glaciers descended more steeply, reaching their melting point much sooner in deeply sunken beds.

Undoubtedly upheavals are often instrumental in stimulating glaciation by raising the land towards the snow line, but in treating a wide area these post-tertiary warpings had better be left aside, unless well established locally. In limited spaces they can be made responsible for anomalous deviations from the behaviour of the region as a whole. I profess little belief in those theories which allot to tectonic lifting the main responsibility for the ice age of large earth spaces. One would have to apply the theory to the whole world by showing geologically, that recurring earth spasms always led to upheavals at the weakest spots, that is to say where foldings already existed. Post-tertiary returns would have to be proved, and North America remain a hard nut. Or we must entirely abandon every thought of synchronism, so that various big glacial periods are not connected at all.

On the strength of the foregoing observations it will be seen that, even assuming a uniform cosmic cause of universal glaciation, the graphic curve showing the march of phenomena and their relative intensities must have been different in various regions. At corresponding times the proportions of the glacial stages compared to each other may have been quite different in the Alps and the Duab. The Duab Würm measured by the Duab Riss may, for instance, be much smaller or greater than the same Alpine period compared to its predecessors. Some intermediate recrudescences and recessions may be obscured, while others may have attained a greater importance (relatively to their own past and future) than in the Alps, so that the picture of the rise and fall of morainic and erosive phenomena in the two mountain ranges will show two curves of a different ratio between succeeding amplitudes. But chronologically the chief maxima and minima must be made to correspond, if we wish to know if the ups and downs of the ice age were controlled by an outside or cosmic cause. I agree with Hess in the closing words of his book, "The cause of the glacial periods is a riddle. Its solution is best furthered by a better knowledge and closer investigation of the climatic conditions of glaciated countries and the traces of former glaciations." To every student of these problems and to every glacial explorer I recommend the Duab as a region of the highest importance where many interesting questions may be solved and new ones put.

DESICCATION.

The problem of desiccation which has occupied many scholars, is also intimately bound up with climate and based on the same fundamental meteorological elements as the problem of glaciation. But from the beginning I warn the reader against identifying the desiccation of Inner Asia with the general deglaciation of interglacial and postglacial periods.

Here again we see how important a part humidity plays in the physical economy of the world. Air temperature and humidity are inseparable factors, often vicarious. Solar heat, modified by the shape and position of earth, by the distribution of land and water, by continentality and altitude, produces an intricate system of winds, evaporation and condensation. The whole forms an organism of climatic circulation. Anything affecting the whole, will affect its parts in various ways, and topographic changes in one part will influence its surroundings. Hence the climatic symptoms of a portion of terrestrial anatomy may owe their change of character either to a general cause, or to local changes, and these local changes may be within or in the immediate neighbourhood. Practically we can take it for granted that only a tremendous change in the condition of the sun would be capable of upsetting the present arrangement of oceanic and atmospheric currents by general temperature alone. Seeing moreover, that the cooling of the earth and its geological progress were far more rapid than any cooling of the sun, we may look upon solar heat as a constant factor subject only to comparatively slight fluctuations (sunspots, cosmic spaces?). Viewed in this light the phenomena of the ice age were the local symptoms of a general cooling attacking those parts of the terrestrial organism, which were predisposed to glacial swelling by the presence of humidity. That the general fall of temperature (which so far we must accept as the best available explanation of synchronism) was complicated with upheaval, subsidence, denudation, inroads of the sea, sedimentation, etc., goes without saying, but the value of their respective share depends upon the relative duration of the general and particular causes and processes. We do not know how much of the final effect must be apportioned between the different coordinated or successive causes which ruled in the past, to what extent local physiography may have retarded, outstepped, or lagged behind the universal or neighbouring influences acting upon the march of a certain phenomenon, such as glaciation or desiccation. It is the great question of the geological calendar. The unravelling of this tangle, the calculation of the exact values (in time, weight and volume) of morphological processes is the task of the science of the future.

We may say therefore that the history of climate is the history of topography, and that consequently the question of topography as cause of climate, should always precede that of climate as a cause of topography. Hence there must be a tacit understanding that, in ascribing some physical change to a climatic one, we think of probable geological or morphological changes in the neighbourhood as the nearest agent, while the effect of a universal or astronomical influence has also to be determined according to the geographical conditions prevailing at the time of its advent. The morphologist is nearly always on the safe side when looking upon climate purely as a function, as a reflex of, and hyphen between, surfaces.

As climate expresses the more subtle qualities of a region, its spiritual character as one might say, its detailed description is best confined to the smaller geographic individuals, and its history to a short period of geological time. We can only connect a very vague idea (or too many ideas) with a "climate of earth" just as the "character of mankind" would say little or nothing. Laws should be as general as possible, but the things to which they are applied, as particular as possible. Climate being symptomatic of an individual district, its vagaries are of vital interest only as long as this individual remains the same in its salient geographical features. If it is changed entirely, the importance of its climatic metamorphosis pales into insignificance before the greater geological event; the constant variable of topography, as a coordinate of climate, becomes the function of an irrational factor. For these reasons the historical age of climate is best dated from the close of the tertiary epoch, when the climatic conditions and provinces of to-day were already established in their main outlines.

The distribution of land humidity is determined by the position of the reservoirs or seas, and the condensers or mountains. Hence the climatic history of the Duab begins with the existence of the vast lakes of Aralocaspia, Western Siberia and Mongolia, and the ranges of Alai, Tianshan, Pamir and Himalaya. Both water-filled depressions and tectonic elevations were probably in their prime at the beginning of the great ice age.

It may not be amiss to ask for a definition of desiccation. Desiccation of a given place means that this specified location holds less water at a certain moment than it did before. In geography we have to add the qualification of time ("at corresponding intervals") because the yearly cycle with its seasonal fluctuations enters into all these comparisons. Hence, in order not to obscure the statement or search for cause, the concept of desiccation must be kept free from all unconscious suggestion of cause, especially a meteorological or climatic one. Afterwards we can narrow it down. If the volume of water in a tumbler diminishes, then there was desiccation of the tumbler (the Duab) as a whole, irrespective of its upper or lower portion, or remaining air space (desert, or mountains, or atmosphere) and irrespective of cause, whether evaporation, drinking, percolation through a hole, spilling, or filling up with other matter. Geographical desiccation is found by measuring the water held within definite spaces (lake, river section, gravel bed, bog, cube of air, glacier, district, continent) in different years or centuries. Strictly therefore the desiccation of Central

Asia or the Duab means, that a distinct topographic area, including the atmosphere above it, has lost water as a whole, water which must have gone somewhere else outside this region, unless there was a reduction of original supply from the ocean. It is well to insist upon this point, it being always possible that only a different distribution of an undiminished total took place, that one part was favoured at the expense of another, that the glaciers dwindled and lakes rose, or vice versa; that the desert became drier while oases flourished; that the annual sum of absolute humidity increased without benefiting rainfall or vegetation. We may be speaking of the Duab as a whole, but thinking of a particular part or symptom only, of the plain, steppe, snow line, glaciers, lakes, rivers, population, cities, irrigation, sedimentation, rainfall, absolute or relative humidity, etc. The question may be put in a hydrographic, glacial, biological, meteorological, economic or social sense, as applied to the whole or its parts. A strict definition and limitation of the problem is therefore necessary in each case. I must refer those desirous of studying these matters thoroughly to the writings of Kropotkin, Stein, Huntington, Schwarz, Brückner, Machacek, Fraser, Berg, Hedin, Ficker, and many others.

The element of time demands a precise statement. Some scholars confine themselves to the general post-tertiary reduction of the lakes, others go into historical detail of the last 2000 years, while a few even reckon with modern events. Here, as regards chronology, I hold a strong opinion, namely that happenings of the present time, and even of the last century, are quite immaterial in this problem. We begin at such an early date that 500 or 1000 years form about the smallest unit of subdivision. The abandonment of the dead cities of the Lop basin (Tarim) is only just admissible as evidence, bringing the problem up to an exceedingly modern time, to the geological "to-day," in fact. Moreover the fate of the Lop cities may not be the adequate expression of a slow process, but a magnified projection through sudden interruption of the recording oasis (see later). Anything observed during the past century, or covered by the latest exact scientific records, such as the meteorological observations or the rise of Lake Aral during the last thirty years, does not count at all, unfortunately. These recent facts may be nothing but smaller or greater fluctuations of the general downward, or perhaps even a new upward, grade, nothing but oscillations of a curve of which we do not know the turning point. As to the future I can be brief. We know nothing of the future, and the safest assumption is always that things will continue as they have gone before during similar intervals of time, that the tendency of the post-tertiary period will still proceed for untold centuries, that of the last thousand years for perhaps another thousand, that of the last ten for a short time. As Kropotkin rightly observes the desiccation of Eurasia is a geological fact. We start from the last notable geological condition, from the existence of the Aralocaspian lake, a portion of which was situated within the Duab, and which belonged to it in the same sense as the Atlantic belongs to Europe (or America, etc.). How this inland sea originated is beyond the scope of our investigation. That upheaval and run-off are not chiefly responsible for the reduction of the great lake or lakes of Central Asia and Turkestan we may take for granted, though sedimentation (fluvial and aeolic) must have acted as a contributory cause, by making the water shallow and forcing it over larger surfaces of evaporation. It has not kept pace however with the retirement of Lake Aral from the Uzboi channel. We may therefore seek an atmospheric, i.e. climatic, reason.

Evidently then the Duab has lost water because the lake was very much larger, and because it is unlikely that the missing balance is still within the district in the shape of glaciers or suspended vapour. Since the inland sea was formed, a constant loss has been going on in the shape of a balance of vapour overstepping the boundaries of neighbouring regions, where it was converted into glaciers or rainfall draining into the ocean. This passive balance, carried away by a slight overplus of prevailing winds could probably only be shown by a millenary average. It must have grown automatically with the progress of desiccation, whereby increasing land surfaces were subjected to solar heating, and increasing quantities of ascending vapour lifted over the watersheds. Whether this mean millenary loss has already passed its maximum or not, we cannot know, but it can hardly go on indefinitely as long as the watersheds are not lowered considerably, for some water must always be caught and sent back by the Duab mountains. From what we know of natural processes they continue, until some sort of balance is attained between demand and supply, between the force working in one direction and the obstacles it sets up against itself by its own results, or simply by exhaustion of the object (e.g. water). But although the proportion of water lost to water retained may have changed, there is no reason to suppose that the loss has been stopped altogether. Most likely it was greatest, absolutely and perhaps also relatively, when the lake was about half its original size, when there was more hot air than before and more water than to-day. Now there is so little water that possibly a larger relative share is reclaimed by the interior circulation. There is also this to be considered, that the quantity of outside or imported vapour (oceanic and neighbouring) retained by the Duab cannot have decreased much since the time when the Sea of Aral had become so small, as to make no appreciable difference to the facilities of condensation, so that the constant amount gains in proportionate importance. However this may be, there is no reason to suppose that the process has come to an end. It may have slowed down considerably, but as the continental conditions are still the same, we may suspect that they are still working in the direction of a complete reduction of old lake stores, which is tantamount to a reduction of the rivers, as they must fail to make up for evaporation. These arguments make me look upon the desiccation of continental Asia as the return, by devious ways, of the water which the ocean had poured into the Aralocaspian depression. Seen in this light the general desiccation has nothing to do with a change of climate as initiating cause. On the contrary, climate was changed by the geological appearance of the inland sea, and had to reassert its continental supremacy by a long struggle against the wet invasion. I accept the hypothesis of marine transgression as the simpler one, instead of ascribing the Aralocaspian sea to a previous moister climate—Lopnor will then have been greater owing to higher precipitation carried inward from the peripheral region. We must assume the existence of the continentality of climate which was disturbed by the creation of the lake. The desiccation of the Duab is therefore the repulsion of a geological oceanic interference by the prevailing and constant condition of continentality, which still had a margin to spare (N.B. after the lake had been shut off again from the Black Sea by tectonic warping). It is not the effect of a change of climate, as commonly understood, but of the perseverance and insistence of a climatic tendency unaltered in itself, cumulating its symptoms of aridity. This steady continental reclamation must be looked upon as a straight line pursuing its course independently of periodic fluctuations. Its goal is final and irremediable; no cosmic cycle we can plausibly think of is able to reverse it. Only a geological disruption of Asia has the power of breaking continentality. We must not confuse it with the coming and going of glacial maxima or lesser oscillations, such as the Brückner cycles. Their waves are not linked to the march of continentality and retiring "oceanitis" by a common cause, although manifold reactions and interferences, weakenings and intensifications are the necessary outcome of simultaneous activity. The undulations of the cyclical diagram run alongside a stubborn tendency which can only be aggravated not softened, by the lowering of the mountains and the filling in of the plains.

Nothing lends itself more easily to a mixing up of cause and effect than the relations between topography and climate. This is illustrated by our habit of saying that the rivers fill the oceans, whereas in reality the oceans fill the rivers, namely that portion of them which permanently debouches into the sea, and which constitutes the type of the oceanic river, as contrasted with the landlocked river. Hence we may truthfully say that oceanic rivers are the result of a climate owing its surplus of precipitation directly to the sea. The rest is water which never leaves the land. Thanks to the discrepancy between rainfall and river discharge all over the world, we cannot say that the oceans supply the rainfall, but must state that they influence it. The oceanic vapour is a quantity injected into the land circulations where it acts in an initiatory manner, by raising the absolute and relative humidity to condensation point. This also enables us to understand the considerable fluctuations of rainfall in continental (interior) districts. In the Amu and Sir basins very wet years show double the precipitation of very dry ones, and in the Aral steppes the maximum can be nearly three times as great as the minimum. As the rainfall is small anyhow, an occasional increase of vapour from surrounding oceanic provinces must make a tremendous difference.

In places without escape to the sea the suspended vapour, mean precipitation, rivers, glaciers and the reserves of lakes or swamps form the total stock of interior circulation. Any gain must necessarily come from outside, and losses go there. Here indeed it may be said that the rivers make the lakes, their terminal bulbs, for the rivers are made by the climate and the climate is made by land and sea. Next to exactly measured precipitation, the rivers are the best thermometers of climate in the long run, being the links between the visible ends of distillation. As it is unthinkable that under present (post-Aralocaspian) geological conditions the humidity of a continental area can be raised so much as to refill the vast inland basin, it is safest to assume that the Asiatic mediterranean was a cause, not a result, of climatic change, that the ocean came in and fought the influence of the land. Only should this be denied by geologists, must we dive into the somewhat appalling problem of tertiary climate. One would have to show that no connection existed with the Black Sea and Mediterranean. Humidity is the decisive factor and desiccation must mean that the water capital of the Duab has shrunk. An access of moisture must enlarge the glaciers, rivers and lakes, because it increases rainfall, while reducing evaporation and the annual range. World-wide changes of temperature can hardly revolutionise the distribution of humidity over earth, though, without upsetting the general balance very seriously, they may accentuate or tone down the contrasts between land and sea climates. Local changes of temperature are only possible by topographic alterations. Uniform cold is not a good instrument for augmenting the humidity of the Duab. It will only effect certain shiftings in the running and invested water capital. If the mean annual temperature were to fall to-day, more moisture would be converted into glaciers and the deficiency show itself in a reduction of the Sea of Aral. When the glaciers melt again, the rivers will swell and the lakes rise. Thus glacier, river and lake need not be affected in the same sense, but in their quantitative relations to each other. Only a simultaneous recrudescence of snow, ice, rivers and lakes will betray an undeniable import of vapour. Owing to the nearness of the sea, the Alps, and still more so Skandinavia, can draw upon an unlimited (potential, that is to say) supply of moisture. If the world gets colder, these glaciers grow, causing greater cold upon the mountains, thereby increasing the difference of temperature between heights on land and the sea, whereby condensation is favoured, while evaporation remains more stable in the open and distant ocean. This might continue until some sort of atmospheric balance is restored or the general temperature rises again. But the Duab has no such store to draw upon and must do all business within itself. This selfcontainedness is the background upon which I see my picture of Duabic desiccation. I think it is supported by the present views of climatological science. It has still to be upheld or disproved by strict morphological evidence showing what were the sizes of glaciers, rivers, lakes and the nature of vegetation at exactly corresponding times. Unfortunately this synchronisation of diluvial events is saddled with a probable error far in excess of the necessary close approximation. A point suggested by the seasonal behaviour of the streams must also be reckoned with in the fluctuations of longer ages. In summer the rivers swell owing to the melting of the glaciers, while evaporation in the rivers and lakes is highest. In winter the rivers are smaller, but evaporation is also less. Hence the oscillation of the level of a medium or small lake need not be as decisive as we are led to infer from the size of the rivers. Since the advent of man irrigation adds another element obscuring the evidence of lake levels (cf. p. 72 ff.).

As I have said, a change of temperature will only cause a different arrangement of the forms of water in the Duab. But these glacial and interglacial episodes cannot have arrested the relentless march of post-Aralocaspian recontinentalisation. The rivers may have been bigger at certain times, especially just after a glacial climax, but this need not alter the fact that they have steadily lost since the beginning of our climatic era. In the Duab and Central Asia periodical or glacial ups and downs of climatic equilibrium should be distinguished from the constant decline of continental humidity. This uninterrupted gradient of desiccation was accentuated in some localities and by certain symptoms during interglacial periods. Sometimes the glaciers were greater, sometimes the rivers, sometimes rainfall. A postglacial swelling of the rivers lasting to the return of a normal[1] temperature would probably be long enough to cover human history, so that the dead cities of Lop may possibly be the victims of a humidity illusion created by the retreat of one of the minor glacial advances (Pakshif?). Interesting riddles are given by the fitting together of desiccation and glaciation. We believe that the main glaciations of the Alps grew weaker as they succeeded each other, partly owing to lowering and dissection of the land by denudation. These factors must equally apply to the Duab, with the addition of desiccation. This would explain what I have suspected before, namely that the Alai-Pamirs have diverged more and more from the Alps, that the descent of Duab glaciations was far more rapid. The absolute quantity of each recrudescence will, of course, have depended upon the size of the Aralocaspian sea at the time being. The mere influx of this vast expanse will have been sufficient in itself to cause a great local glaciation. If the first universal glacial period began soon after this, the effect was doubled, if much later (when the lake was smaller) the transition must have been gradual, making the general glaciation appear as a continuation, slightly larger or even equal, of the local one. It will be one of the tasks of explorers to fit the advent of glaciation to the scale of desiccation.

[1] As our present temperatures are subjectively "normal" and as we imagine ourselves in an interglacial interval, the "normal heat" is that of the middle of an interglacial period.

As to Central Asia we had best say that it always stood in the same relation to the Duab as now, sharing its oceanic symptoms to a lesser degree. The Lop basin, the centre and extremest representative of continentality, was filled with water derived from the humidity spread by the Aralocaspian sea, but subsequent desiccation was even more rapid than in the outer belt. If my surmise is correct, the comparative extent of glaciation must have been much smaller than in the Duab. It will have shown itself more on the southern slopes of Tianshan than in Tibet separated from the sea of Lop by an enormous distance and presenting an analogy to the position of the Pamirs with regard to the Sea of Aral. I know very little of the Mongolian lakes on which Schwarz bases his hypothesis of the Flood and desiccation. Nobody can deny that Schwarz's idea is ingenious and beautiful as a whole, although slightly fantastic in detail. Price's remarks on the Dzungarian Gate seem to vindicate the bedrock of that great scientific romance. Price (quoted from diary in Carruthers, *North-west Mongolia*) gives an interesting description of the Dzungarian gate through which the Aralocaspian-Balkhash mediterranean ebbed and flowed, lapping the rim of Central Asia. As the sea retired from here first, and as general continentality is anyhow greatest in the middle, the extremes of the interior are easily understood through a cumulation of effects. Special topographic circumstances intensify the continental phenomenon of innermost Asia. A close ring of high mountains wards off the moisture bearing winds. Moreover these watersheds turn their shorter flanks towards the inside offering but little high lying surface to condensation. Thanks to its long, branching ranges the Duab has more glaciers and better irrigation. The vertical as well as the horizontal topography are combined against the inner circle. The accumulation of dry, heat storing sand within this narrow space also aggravates the climatic desert phenomenon.

This then is my idea of progressive desiccation and interrupted glaciations. I have kept them asunder for the sake of a diagrammatic exposition, without wishing to deny that an influx of outer humidity is thinkable at certain times, and can perhaps be supported by meteorological argument, especially in the Duab which is open in several directions. But these would be minor complications. Huntington speaks of pluvial periods. These one can only conceive as concurrent with glacial periods which must needs depress vertically, and also extend horizontally towards the coast, the rain belt of mountain slopes turned towards the sea. I fail to see how Central Asia can benefit by such a pluvial stimulus. Would it not make matters worse by allowing less vapour to cross the watersheds owing to a depression of the belt of thickest snow clouds? This agrees with my doubt—founded on morphological reasons—that renewed glaciation will not show itself most on the high passes of a mountain system, nor, of course, on its innermost chains (Pamirs, Eastern Caucasus), but in lower and outer zones. The reverse, a rise of universal temperature, may perhaps cause actual loss to the interior ring by abstraction of vapour, so that there is no gain either way. In a crude fashion the simile of a pot suggests itself. When heated it boils over, while cooling only gives the negative advantage of a saving and concentration of the water.

Having stated a general case, I shall now discuss at random and without prejudice various aspects of moisture. A favourite phrase is that of the excess of evaporation over precipitation which sounds very simple, though it may lead the unwary into mental traps. To it we must apply the limitation of place and time which I have hinted before. It is really an experimental idea meaning that in a dish exposed to the air more water can be evaporated during the year than rainfall would supply. In the steppes round Lake Aral the experimental evaporation is more than ten times as great as precipitation. For every inch of rain the air *could* absorb ten inches of water during the long, cloudless intervals. But that water is not there. So this relation may reveal much to us, or very little, according to topography and distribution of rain. If rainfall and evaporation are quite evenly spread over the year and on level ground, then a perfect balance without drainage will be established. The least drop more must run off. As it is, even the present poor rainfall would partly form rivers into Oxus and Aral, if it were crowded into a few days. Likewise no excess of evaporation, however great, can make the least difference to dry sand, where there is nothing to evaporate. Hence an overbearing evaporation without detailed morphological and seasonal comment gives no immediate clue to the hydrography and other conditions of a region. Unsatisfied evaporation cannot indemnify itself unless there is very much wind. It requires surface, and only wind will enable it to crowd itself into other surfaces, such as Lake Aral, where that air has to be removed which is already quenched. It has no sense to form averages and to speak of an excess all over the Duab which, strictly speaking, would mean that it contained no water at all. The excess of evaporation in a certain locality is another way of saying that much hot and dry air is produced there. The desert makes hot air and with its help can spread its influence according to the facility with which this desiccated air is mobilised against vital spots of condensation. What has to be remembered is, that a cube of air which can absorb a cubic inch of water at Petro-Alexandrovsk cannot do as much further north or higher up, its craving being dulled by cooling.

Wherever rainfall is collected or can seep away very quickly it has been rescued from potential evaporation and equally potential vegetation. This creates the contrast between bare slopes where evaporation has been cheated by withdrawal of water, and the rivers where it is disappointed by the small surface covering the annual cubic miles of running water. Above the permanent snow line there is an absolute excess of precipitation. Thus the flat desert with its conditions in favour of total evaporation is dovetailed into the network of moisture-saving mountains. At both ends, linked by running rivers, we have the old stores, the one preserved by cold, the other a surplus over evaporation in the lake. Between the two departments much crediting and debiting can go on without impairing the solvency of the partnership, only a reduction of the common fund denoting a loss to the whole of the Duab.

It will prove interesting to correlate the various data of rainfall, river discharge and evaporation. At the same time they will show how deficient our knowledge still is, how important it is that all our observations should be perfectly complete, equivalent and adequate, if guesswork is to be excluded. I begin by tabulating the most reliable figures from Berg, Hann, Ficker, and others. Some of them I have rounded off, very near calculations being defeated, in any case, by uncertainties and gaps. The only absolute basis is furnished by the long records of Russian observatories forming one of the best meteorological services in the world.

Sea of Aral.

L. Berg has made a special study of this interior lake and written an exhaustive monograph.

Surface (without islands) 63,000 sq. km.
Average depth 16 m. (greatest 68 m.)
Volume of water 1000 cb. km.
Sedimentation 34,000,000 cb. m. p. ann. (would fill in 30,000 years).
River discharge 1500 cb. m. p. sec., or 47 cb. km. p. ann., equal to 750 mm. over surface.
Precipitation: mean 110 mm. or 7 cb. km. (average between Kazalinsk-Petro-Alexandrovsk).
extreme maximum (wet years) 170 mm. or 11 cb. km.
extreme minimum (dry years) 65 mm. or 4 cb. km.
Rainydays 37; cloudless days 152; cloudy 61.
Evaporation: Kazalinsk 1060 mm. (corresponding rain in that year 100 mm.).
Nukus (delta) 1930 mm. (corresponding rain in that year 70 mm.).
Salinity 10·76 °/₀.

The common discharge of Amu and Sir of 47 cb. km. is probably an estimate, as it is not likely that the many distributaries of the Oxus delta have been gauged exactly during a year. As to evaporation on lakes, we know very little, and the safest plan is to determine it by minute control of inflow and outflow as Maurer did with Lake Zürich where he found an evaporation of 4·6 mm. in August 1911. Evaporation stands in some ratio to rainfall and cloudiness, but so far we have not found it, nor is it likely that we ever shall discover a reliable one, seeing that so much depends upon the seasonal distribution of various elements, especially of wind. Moreover there is a difference between evaporation on water and land, between small and large pools, ponds, lakes and swamps, sweetness and salinity. It will hardly ever form a trustworthy coefficient applicable to related conditions, such as other lakes in neighbouring regions. All meteorological and hydrographical measurements necessarily yield evaporation as a difference, and as these direct observations are so important in many directions, it is they which should be made, extended and improved. Compared to its volume the Aral lake is very shallow, thus offering a large surface to sun and air. The existence of a terminal lake is explained by the need of an evaporating surface for rivers stopped by configuration of the ground. The mirror always represents the area required for the establishment of a balance between influx (including rain on the spot) and evaporation. If evaporation increases the lake will shrink until it is so small that the surface cannot evaporate more than what comes in. Hence much depends upon the shape of the basin. The oscillations of a shallow lake with flat pools around its shores will be quicker and more erratic than those of a deep one sunk between steep rocks.

Amu-darya.

Length: total 2400 km.; from Aral to Surkhan 1300 km.
Basin (catchment area): 300,000 sq. km.
Rainfall of basin: mean 300 mm., or 90 cb. km.
extreme maximum 500 mm., or 150 cb. km.
Volume at Nukus (top of delta): 50 cb. km. p. ann. (doubtful).
Discharge into Aral: 32 cb. km.
Under cultivation in basin: 21,000 sq. km., or 7 %.
Irrigated plains below basin (including Khiva): about 6000 sq. km. with 2,000,000 population (?).

As we have no observations for Afghanistan, the rainfall over about one half of the basin is a guess. But it cannot be very far wrong unless areas of very much higher precipitation exist on the slopes of Hindukush.

Sir-darya.

Basin: 250,000 sq. km.
Rainfall of basin: mean 300 mm., or 75 cb. km. (Issik-kul and Ferghana average).
maximum 390 mm., or 96 cb. km.
minimum 156 mm., or 42 cb. km.
Volume at Parman-kurgan (near Khojent): 19 cb. km. p. ann.
Discharge into Aral: 15 cb. km.

The Oxus volume at Nukus cannot be accepted as a very reliable average being derived from a one year's observation in 1875/76. The measurement of the Sir (1899/1905) is good, but not immediately comparable owing to the different distances from the mouths and the great development of the Amu delta. I must therefore accept Berg's 47 cb. km. of total discharge into the lake as the best estimate at our disposal. Evaporation in the delta (sand, swamp, reeds) is naturally much higher than in the lake and might possibly amount to 8 cb. km., which would mean a volume of not more than 40 cb. km. at Nukus. This is more than enough compared to the 150 cb. km. of extreme rainfall over the basin. Taking a constant water level, the influx of 47 cb. km. into the Sea of Aral corresponds to an evaporation of 750 mm. to which must be added the mean rainfall of 110 mm., giving a total of 860 mm. Berg has shown a rise of the lake from 1880 to 1901 amounting to an average of 90 mm. per year. He uses this undeniable fact as an argument against progressive desiccation with which, of course, it need not have anything to do. So far it proves nothing but a fluctuation, and only another 1000 years or so of constant observation can disclose the continuance or arrest of desiccation. The known extremes of rainfall in the basins and over the lake itself suffice to explain changes of level by normal oscillations (Brückner cycles ?). The rise of level of 2000 mm. has taken place during 22 years, but the precipitation alone of about six very wet years would produce this amount, namely: rise of level by greater influx 250 mm.; more rain over lake 60 mm., over delta 10 mm.; total 320 mm. in a very wet year, according to records in hand. But as a considerable reduction of evaporation (in the lake, rivers and basins) has also to be considered, a couple of extremely wet and a number of moderately wetter years can easily account for the total increase of level. The delta (surface 20,000 sq. km.) has to be remembered as, owing to its nearness and facilities of drainage, it must send at least one half of the heavier rainfall into the lake. Speaking of the lakes in Nevada, Hann remarks, "There is no need of assuming an excessive augmentation of rain for a refilling of these lakes. A somewhat greater cloudiness coupled with a slightly lower temperature and a modicum of moisture would suffice in causing a considerable rise by slow degrees." Now the rise mentioned by Berg is a very rapid one, and as such speaks more against than for a change of climate, since we have become accustomed to judging the uniformity of a climate by centuries, not by years and decades. It is interesting, by the way, to take a note of the fact that a rise of 4 m. above the 1901 level will restore the outflow through the Uzboi into the Caspian. This adds but another complication to the vagaries of the Aralocaspian inundation at various stages of shrinkage.

By some of the preceding attempts at calculation I merely wish to point out how intricate these problems are, and how deficient the data at our command. The only solid basis is formed by the meteorological records. They alone, so far, give a true picture of the various sensitive as well as energetic climatological elements, and only their continuation can reveal small secular changes. As Mill (*Geogr. Jour.* June, 1906) once aptly remarked, "I can only hope, in the course of another century or so, with the development of the very excellent meteorological system which the Russians have introduced in the heart of Asia, there will be some more definite evidence on the subject to bring before the Research Committee." A century is but a small fraction of a glacial, still more of the post-tertiary period, but a hundred years of very exact instrumental records will demonstrate the shorter cycles and perhaps disclose the direction of a wider tendency. Our present and immediate thirst for knowledge can be greatly supplemented by geology and morphology, and by accurate statistics of glaciers, lakes, river volume, desert encroachment, population, etc. Lake level alone is the concentrated result of an indescribable mixture of causes distributed over wide areas including the locality and shape of the lake itself. Only a vast geological shrinkage from levels, to which there was no return, is evidence of protracted desiccation.

To students of physiography I would suggest the working out of interesting comparisons, such as that between the Oxus and the Nile.

Nile.

Length: total 6000 km.; to Atbara 3000 km.
Basin: 2,800,000 sq. km.
Rainfall of basin: 3720 cb. km. (Mill), or 1350 mm.
Discharge: 100 cb. km.
Area under cultivation in Egypt: 20,000 sq. km., but 26,000 sq. km. of crops (i.e. several harvests).
Population of Egypt 11,000,000: exports £25,000,000; imports £25,000,000; etc. etc.

It will be seen that the Nile discharges only 3 °/₀ of the rainfall of its catchment area, whereas Amu and Sir save a much larger proportion. Apart from the length of the river this discrepancy is chiefly due to topography and vegetation. The waters of the Oxus basin run together very quickly into their steep drainage channels. The Nile basin is flat and full of swamps and forests, facts to be connected with the questions of interior vapour circulation and the influence of forests. The quantities lost by evaporation below the last tributary do not materially change the proportion between Oxus and Nile as regards their relations of rainfall to discharge.

Huntington has given us a wonderful description of the dead cities of Lop where thriving oases of 50 square miles existed in the Takla-makan desert. On them, in conjunction with other phenomena, he bases his theory of climatic change during historical times. To this one cannot take exception, though I would slightly modify his conclusions. Stein has an idea of his own which may reveal a contributory cause. He says that continual irrigation must raise the level of the cultivated ground, the fields being analogous to a delta where silt is deposited by the checking of the current. In the course of centuries the land will rise above the rim of the ariks, so that finally the water cannot overflow into the plots. A new canal was then made to tap the river higher up. Fraser supplements this view by suggesting that in this way the garden towns travelled upstream towards the mountains. I have already said in the case of Agalik (Chap. VII) that villages seem to travel upstream. Much water in the Lop basin is simply wasted because it sinks into the gravel belt around the foothills, coming out further away charged with salts and therefore useless for agriculture.

Here then a surplus could be utilised if capting were possible. Before the recent desiccation only a portion seeped away, the rest remaining on the surface as, indeed, it does now during spring. Hedin and others speak of suffocation by sand as the prime factor. This cannot be positively denied either. But apart from the fact that the blowing in of sand can also be interpreted as a symptom of climatic variation, it may be that a line of advancing dunes saw its opportunity during a temporary disablement of the irrigation service through some human or natural calamity, such as wars or earthquakes. This must have happened at a time when the various effects of some period of desiccation were already taxing the resistance of the oases to the utmost, at a time, let us say, when the problem of the strictest economy of water was combined with a fight against drifting sand. Berg (*Izv. Imp. Russ. G. S.* 1905) attributes to wars the destruction of oases and cities. As Schwarz already pointed out, it seems more reasonable to argue the other way round, namely that more often want of water or wish for water and irrigated land led to Asiatic wars. Wars may destroy a high civilisation, institutions, monuments and races, but hardly ever the maximum of agricultural life which replenishes itself at once as long as the water is there. As Huntington says, the size of the communities is directly proportionate to that of the streams. It seems to me that floating humanity must refill the oases automatically with inhabitants, just as microbes soon crowd every spot of moisture on which a few of them have settled from the air. If the spot dries wholly or partly there must be death and dispersion. We must however steer a middle course and make some allowance for the disturbing influence of wars, especially those waged by nomads against settlers.

The point at which a landlocked river ends and becomes useless for irrigation we may call the point of desiccation. Likewise we may call level of precipitation the vague zone where most of the snow and rain is concentrated, and whence always issues an excess of running water. When streams unite they reduce their evaporating surface and, other things being equal, they depress the point of desiccation below the level at which they would have it severally. The Duab rivers are protuberances from the level of precipitation into the desert, and agricultural desiccation is the shortening of these outshoots. Any reduction of river volume or increase of evaporation must bring up the point of desiccation. The nearer an oasis lies to this point the sooner it will be affected by a change. Hence the nearer the mountain the better. Viewing desiccation in the light of historical colonisation, it is the oases at the fringe which will quickly respond to a variation of climate, while those at the foothills, near the issue of larger rivers, will practically remain undisturbed for ever. A comparatively small change may lead to the abandonment of the lowest settlements, while a return of former conditions need not necessarily bring about a resuscitation, because in the meantime the old ground may have got badly sanded up. From our economic point of view the loss of a thriving centre is something terrible, but morphologically it may be a very small affair, nothing but a slight shifting of the desert fringe. If for some reason or other, such as the lack of suitable ground, there are no considerable outer fields and gardens, the hill cities (e.g. those of Middle Bokhara) will always remain stable, and nobody would know anything of desiccation.

The Duab is composed of two meteorological and hydrographic portions, the productive and the wasting, or the mountains and the desert. Naturally the boundaries of two extreme regions are always shifting a little to and fro without causing a change of comparative areas worth speaking of. But such a trifling dislocation may cause a great human catastrophe thereby suggesting the enormous and rapid advance of certain conditions, thus seen through the spectacles of our vital interests. The risks of the fringe and competition will drive the population uphill towards the origin of the waters. This is perhaps one of the reasons why the mountains are so densely inhabited as compared to the Caucasus.

Between the point of desiccation and the altitude limit of vegetation a river system will be able to support a certain amount of life. Supposing that the ground remains equally suitable, villages can be nourished anywhere below the upper level of cereals. Whether the possible maximum of human settlement is attainable or attained, depends upon the way in which the water is tapped, which again depends upon the formation of the ground. The more scientific the system of irrigation, the nearer the maximum is the quantity of human life thriving on a given unit of water. The idea is to let the water run as short a distance as possible without being used, i.e. subject it as little as possible to direct evaporation before it touches the roots of the plants. The more concentrated the oasis, and the nearer the mountains, the better this condition is fulfilled. One big, connected area of gardens is more advantageous than a long string of settlements between which the river is exposed to unproductive waste. In the single and compact oasis much also depends upon its shape, the best plan being that of the distributed and reassembled network of arteries and veins, the garden expanding at first and then narrowing down again in the manner of the basin of Samarkand. Not so good is the system which branches out like a tree, especially when the main ariks have to be very long before the fields are reached. Better than this is the elongated shape disposed along the river in a strip with herring-bone channels on each side. It ensures the same minimum loss by evaporation as the first or lozenge shaped (reassembled) oasis, but shares with the tree pattern the disadvantage of blind endings in the desert, whereby progressive salinity is caused on the outer rim of the oasis, whereas the circulation type drains off its salts far away. The worst arrangement is that of straggling plots exposing the water travelling between them. The lowest villages, such as those of Karakul and Bokhara, will be very sensitive and the first to suffer from a change in rainfall.

But as far as human agency is concerned we may take it for granted that in recent times, and certainly up to the Russian conquest, the size and distribution of irrigated surfaces had become pretty stable, that no more wolves could drink above the lambs, because all upstream ground is already occupied as far as it can be turned to account by the traditional system of tapping and irrigation. The Samarkand basin, for instance, seems to be crowded, so that cultivation cannot be enlarged unless the belt of higher steppe is watered by modern works of engineering. Hence it is not likely that a lower town will be made to suffer by an increase of population in the higher reaches, as the people of Middle Asia have anyhow the tendency of using all upstream land (irrigable by their methods) to the last square inch. A growth of population will lead to distress which will be balanced again in some way or other, by emigration, war, higher mortality, etc. The natives remain wherever there is water, and the last family of peasants will cling to a place as long as there is sufficient moisture left as shown by numerous one-family settlements everywhere. The complete abandonment of a site is therefore tantamount to complete desiccation or excessive salinity of the spot.

When a settlement has attained full size, or water is getting scarce we may assume that the town will now begin to wander upstream (in olden times), people giving up fields at the lower end and making new ones above. In this way there would be a gradual headward crawl of the garden city, every new generation expanding above its fathers. The dead cities of Lop seem to have been abandoned rather suddenly judging from the aspect of the ruins and their archaeological treasures. A war may have been the initiating cause. In that case people felt the effect of desiccation acutely owing to the destruction and neglect of ariks and administration. Until then they had not noticed that they were fighting against a secular advance. Coming back they found it more convenient to begin higher up at once. Hence the gap. Lessening of water supply is probably the fundamental cause, while political disaster may have accentuated it more sharply by an interruption of continuity. As we may cling to the axiom that in these regions population is directly proportionate to water supply, the question can be clinched by knowing the true size and probable number of inhabitants of the ancient oases of Lop, comparing them to the modern ones. If it could be shown that the present cultivated sites are just as old as the abandoned ones, the solution would be simpler and more convincing still. Looking backwards, a progressive desiccation attacking the irrigation fringe seems fairly certain and need not have been very great in order to affect many thousands of people. It was more extreme and rapid in the Lop basin than in the Duab. Meteorological considerations do not favour the idea that precipitation in the higher mountains can be reduced much further otherwise than by morphological changes. As long as the Duab mountains remain as high as they are (within a margin of a few hundred feet in the next ten thousand years or so) there will be glaciers to feed the present number of inhabitants in the plains.

One cannot seriously connect *the* desiccation with questions of practical modern agriculture. Humanity does not really bother itself much about what may happen in a hundred years, still less, what may be the conditions a thousand years hence. I shall merely ask what are the prospects on the basis of the present water supply. Theoretically the surface of crops can be increased as long as the Amu-darya and Sir-darya reach Lake Aral during the vegetative season, save a certain allowance of water for carrying off salinity. But only irrigation works on a huge scale will enable us to utilise the nominal excess, and it remains doubtful if they will pay. Tapping the Oxus, say at Charjui, will be costly in itself, not to mention the enormous works of engineering necessary elsewhere as a direct consequence. The lowering water level of the river would fail to reach the capting ariks of Khiva for whose benefit a barrage would have to be built across the sandy and muddy river. Much may possibly be done by more intense methods of farming and a scientific system of water saving. Such methods, if feasible with the natives, will however leave a very narrow margin causing a strong reaction to every fluctuation of rainfall, unless the river volume of the driest years is taken as the limit. Hence there seems not much chance of getting more food out of the Duab. Something may be gained by the substitution of more valuable crops, but already now cotton has taken away grain land, and it remains to be seen if the export of cotton leaves a good profit, to the native population as a whole, over the import of wheat. This makes me think that rapid economic improvement lies chiefly in the direction of railway improvements and the lowering of transport rates.

SAND AND LOESS.

Sand, loess and ice are the most typical climatic rocks[1]. Sand symbolises the heat desert, ice the cold desert; loess is the characteristic non-forest soil as contrasted to mould and peat.

[1] Geologists are in the habit of summarising their objects as "rocks." Exceptions are suggested by feeling, although the logical definition ought to embrace all dead matter. It may seem absurd to classify a carcase as a rock, but a habitual dump of dead animals is most certainly a geological deposit or rock. Decomposition and change, to which even granite is subject, are a question of time and one would have to state how long an accumulation must remain physically and chemically stable before being entered in the list of stratigraphic occurrences. Glacier ice, for instance, which is a rock, remains structurally but not materially identical. Cities are rock formations by the same right as coral reefs. Quarrels over definitions are usually sterile as regards the matter of fact in itself, the aspect of which is mostly determined by what we wish to know, that is to say by precedent, by the purpose of investigation and the instrument of observation. But the value of discussion lies in showing the possible application of facts to all the different branches of science and in training the mental eye for a universal perspective. The methods of generalisation convince us of the identity of the laws of nature and the logical processes of the human mind.

Although they may generally be cousins to a common cause, more or less removed, the atmospheric dryness of a desert tract and desiccation of its rivers must at first be considered apart. The desert condition can be quite independent of the rivers originating outside. A complete desert cannot be made worse, just as little as a dry lake basin or an extinct oasis. It can be traversed by great streams which do not make any difference to it whatever their size. The desert is mainly an area of unfavourable rainfall, often aggravated by altitude, land-locked drainage, salinity and percolation. Its surface formation and petrographic character are the result of its own and surrounding climate, topography, geology and hydrography. The state of the rivers is viewed in the light of morphology and climate from watershed to mouth. Oases and desert lakes are the outcome of the desert and general drainage. The Duab Kizilkum is partly due to the accumulation of sand, and the accumulation of sand is partly due to desert, i.e. desert climate. Kizilkum is a sand dump sketched out by the neutralisation of various converging or conflicting physiographic agencies.

The close bond of theory between loess and glaciation (especially moraines) has accustomed us to the idea that the more sandy residue, being heavier, must lie nearer the glacial centre than the lighter sifting or loess. But in Turkestan we have two loess or clay belts, and a vast sand deposit in the middle. The present position of the desert must be partly ascribed to the difference between the sum total of wind effect at different seasons. The winds blow more frequently from the north than from the south. This would explain why the sands are nearer to the Alai-Pamirs. It is also a question of vertical level, for whereas there is no limit of distance on the flat, there is the different height to which various substances can rise in the air. Loess, as the lightest, will float highest, which is the reason generally advanced for its distribution on hillsides and upper valleys. But simple as this explanation sounds it forms but part of the truth, for what can prevent sand from travelling upwards by stages. Moreover loess contains a good deal of sand. On the whole the sand and loess areas represent the average effect of differential movements in many directions. The number of manifold crossings and overridings is very large though the visible end be charmingly simple.

To begin with we should remember that a partly aeolic origin is claimed for the northern (Kirghiz) steppes and the black earth of Russia, on the grounds of a physical mixture akin to that of the loess. Their major portion, not lying high and dry enough, cannot be genuine loess, because structure finally decides the definition. But as a genetic concept is also coupled with the name, we might concede the names of loess-clay (the valley-loess of several writers) and loess-mould to some of its modifications. These terms would convey that the loamy deposits of the lowlands diverge in their structure from the genuine, porous material, but that part of their substance may have been brought thither by air. Here they were mixed with water-borne matter, or themselves modified by water, and the further north we go, the greater the share of fluvial mud, until finally the possible percentage of original air dust becomes a mere speculation. One term however ought to disappear from geology, namely that of "secondary loess." Fluvially re-bedded loess cannot be loess any longer, neither genetically nor morphologically. To visit the birth of the father upon the nomenclature of succeeding generations is a dangerous practice where, as in this case, the word "loess" can leave a doubt as to whether it applies to the genesis of the deposit or its progenitor.

Admitting then that the loess-clay of the Kirghiz steppe is partly aeolic, we have to face the problem, why the sand is between two deposits in the formation of which wind played an important part. As dust is carried further than sand, the conclusion would be that the Duab loess came from the glacial centre of the north, and the loess-clay of the Kirghiz steppe from the Duab mountains, each being so to speak shot across the middle where the heavier sand was dropped. Otherwise we should much sooner have expected two sand belts and a loess and clay deposit in the middle. This is, of course, only a primitive way of stating the final result, as nobody will seriously expect a regular exchange of the aeolic material from opposing sides. It is merely a picture of horizontal distribution. We must figure to ourselves a great common fund of dust, sifted out of every imaginable kind of detritus, suspended in the atmosphere. Owing to a stronger wind pressure from the north the larger portion remained south of Lake Aral, the Duab thus retaining the most of its own contribution to the common fund, receiving in addition an extra share of which the Russian glacial plains had deprived themselves. The Kirghiz steppe must therefore contain a differential return of aeolic dust, even if the bulk of its earlier material was fluviatile or lacustrine. This sounds plausible, I hope, but as yet it is only an expression of total effect. We have still to ask why pure, or almost pure, sand should be collected in one place. Classified deposits of such magnitude can only occur where water or wind come to a dead stop, in a sea or in calms, or in relative calms produced by the balancing of currents. Otherwise transportation usually happens in mixtures and by stages, small sortings out appearing in quiet spots, eddies, and under lee sides. Thus the simple idea of the dropping out of the sand cannot suffice, even on differential grounds, for sand is ubiquitous, being able to travel almost anywhere by forward and upward steps. The Kizilkum is not exactly a region of calms, and if here wind energies were neutralised to perfection, the loess dust ought to accumulate just as well as the sand. Apart from the earlier river and lake deposition which prepared conditions and which again depended on purely topographic circumstances, we must therefore look upon the present precipitation and keeping together of the sand chiefly from a negative point of view. The true classifier is vegetation, and the desert belt of Turkestan is an aeolic enclave unable to retain dust owing to the absence of a close crop of grass. We know that water, moist surfaces, forest and pasture catch all air-borne matter, the coarse as well as the finest. Sand is everywhere, being contained in lake mud as well as in loess, of which latter it forms about one half. The sands of the Duab are partly loess without the dust. On a sea shore the surf has washed out the lighter stuff which is not allowed to settle within the disturbed zone. The Kizilkum is the shore of the steppes on either side, a desert island in the ocean of clay and loess.

The moraines of glacial periods, exposed lake bottoms, river alluvions, the weathering and corrasion of rocks, scree slopes, and the grinding to powder of the sand itself, have provided and are still providing a great quantity of dust. To this must be added the abrasion of the deserts, clay steppes and loess cliffs, whereby the original deposits are continually being stirred and re-bedded. This vast shifting mixture suspended in the air or travelling along the ground, is driven hither and thither, accumulating wherever the formation of the country will allow it to rest. Exposed ridges and steep summits form the absolute limit of aeolic formations. Within the zone of relative atmospheric calm, below the level of violent unrest which ever beats upon the mountain islands, plant life causes another sorting of the matter carried forwards and backwards by manifold currents. Wherever the deepest hollows of a land-locked drainage system were impregnated with mineral salts from the leaching of the heights, or where an old sea bottom was laid bare, vegetation was very poor from the first. Here then fine dust could not remain, being gradually blown out, or rather less and less came back every time, because the steppe every year caught and permanently held a portion. In other words, the everlasting turnover resulted in an increased saving of sand to the desert and a growing credit of dust to the loess steppe. The desert does contain just a little dust without which even the scanty vegetation of the sands would be well nigh impossible, but together with the powder continually floating about in the air, it is merely that last permanent residue or suspended balance which always remains equal, being neither augmented nor diminished in the general shaking to and fro. Neither sand nor atmosphere will hold more than this minimum in the course of time, whereas water and plants can bind all aeolic sediment. Thus the sandy stretches are not so much a deposit blown there, as a deposit left there.

Naturally there is also a constant adding to the yellow belt, for sand so to speak attracts sand. There must be, after all, a slight overweight of gravitation, inducing sand to favour the lower level, although this cannot be roughly expressed by the difference in weight between a grain of sand and a particle of dust; by saying that the dust is sent further away. This ultimate concentration works with a very fine margin in favour of the coarser fragments of quartz. The quantity of general mixed stuff drifted into and out of the desert always leaves a little more new sand there than is lost by old sand being blown out and retained elsewhere. Also the distance is less, for what comes out of the desert is promptly sent back by prevailing currents in company with dust and other matter. Moreover sand can only run and jump, while dust can float which gives it the possibility of settling down evenly. Thus the weight of sand also influences its gathering through the cumulative effect of a very small balance. Of course we know that the desert as a whole can travel, gaining inch by inch in the direction of prevailing winds, if not opposed by water or strong vegetation. But even so it remains, as a whole, an area of depression for aeolic transport, as well as one of residual condensation for aeolic sifting. The yellow sea attracts sand and repels dust. To resume, Kizilkum is a place where of all Duabic detritus only sand remains because it is the lowest land with the driest climate, where, owing to salinity and drought, vegetation is poor. No dust remains however much may be passed through it. It is also a depression or stagnation because a light breeze which will just carry sand downhill may not lift it uphill, weight thus acting as a prevailing factor in keeping the sand at some distance from the mountains. There also seems to be a prevalence of winds blowing into the desert. Theoretically this is very likely, especially in summer when the hot air rises, giving way to undercurrents. Ficker says, "We discover a great variability of the wind directions along the boundaries of mountain and steppe in the middle of the Duab. In the annual mean the winds along these boundaries veer around the country just as if a barometric minimum happened to be in the middle." Thus of all sand blown up and down, in and out, a tithe is claimed by the desert.

In Grund's opinion drifting sand near a loess belt is only a passing stage of destruction, the final geological product being loess which is the more permanent deposit. Loess depends upon the grip of vegetation, wherefore aeolic accumulation is generally a phenomenon on the outskirts of deserts. The friction and frittering of huge masses of sand must be an important source of dust, so that desert and loess are often closely associated. Dust can come from anywhere, and loess is a form of dust deposition favoured by certain climatic conditions. Grund is right in saying that the location of aeolic strata, with relation to their probable birth place, is no direct evidence of the direction whence blow the prevailing winds. The source of the material need not be sought in the quarter of the stronger or more frequent air currents, for the grassy surface of the loess can retain dust brought by weaker or rarer breezes, and need not suffer great loss from stronger but cleaner winds.

The case of sand is somewhat different. Once it has reached the lowlands or been exposed there from lake bottoms or sea shores, its boundaries are outlined by winds and slopes, as also by rivers, oceans, lakes and vegetation. The formation of loess is chiefly bound to the presence of grass, however much the dusty winds may be in the minority. It is therefore best to say that loess comes out of the air, and that the dust in the air comes from anywhere. From the genetic standpoint there is an important difference between loess and sand, it being wrong to call both aeolic products without discrimination. Subaeric sand is air-***bound***, being rounded up and fenced in by atmospheric play, whereas loess is exclusively and necessarily air-*born*. All loess must have come out of the atmosphere where it was suspended, and its origin is therefore unequivocal, always being a precipitation out of the same medium. But it is impossible to say how much of the substance of the barkhans of Kizilkum is marine, fluviatile or aeolic. Sand is a mixed product, although the life functions of the desert are aeolic. Loess is a stratigraphic, genetic and petrographic concept; sand is only petrographic. The drifting quartz is not so much an aeolic deposit, as an aeolic aspect of the many possible varieties of sand beds. As a whole sandy wastes have mostly been brought together by water, assuming their atmospheric features after exposure through manifold geological and physiographic processes. Thus sand is not necessarily aeolic, but dunes are; not the material but the living shape; not the sea, but its waves. It is well to insist upon this distinction nowadays, when aeolic theories excite so much interest, thus often leading us to search the air for the origin of sand.

Dunes are wanderers and their birthplace is sand gathered by various means, but rarely due to wind alone, as in some mountain valleys. Like a reservoir of water, sand is moveable within its boundaries. Between sand and water there is a strong antipathy revealed by sharp lines. Where the one loses the other gains. Waves are sucked up and evaporated by sand; dunes "evaporate" when lapping the fringe of lakes or oases. Most of the sand starts drifting after birth as a deposit, for owing to its weight and quick settling it is much better adapted to river transport. Loess as such has ceased to shift, and must first be transformed by water cutting canyons into its bulk. The metamorphosis and circulation of sand is slow; loess can be changed into mud or clay, and clay dust return to loess within the year.

The aeolic theory of loess established by Richthofen can be taken as proved. To one who has seen the shape and distribution of Asiatic loess it is impossible to conceive a fluvial or fluvioglacial origin. High loess reproduces the underlying contours and it is manifestly absurd to assume its sedimentation in enormous valleys, where only some mounds and slopes of the friable loess remained after all the surroundings had been eroded. Our knowledge can still be deepened by careful research into the relations between thickness of the deposits, windward and leeward sides of hills, vegetation, rainfall, etc. It is of course impossible to determine how much of the loess growth of to-day or of human history is due to aeolic re-bedding, and how much comes from original sources. Loczy has found 2 m. of loess on the tombs at Singanfu which are 2000 years old. Owing to its friability we can trace loess only as an interglacial and postglacial formation. Huntington has found it on the top of moraines. Steinmann has described the various layers of the Rhine loess separated by argilised strips representing moister periods, and corresponding to the set of glacial terraces. He also strongly suspects synchronism with South American (pampa) loess. As I have hinted before, the whole of the loess does not come out of moraines, but at the end of each glacial activity the great masses of boulder clay and morainic rock flour formed a store of ready material for immediate dispersion. German geologists believe that the loess of Middle Europe was blown down by northerly winds against the foothills and Alps. Nehring has found the bones of steppe mammals in German loess, so that we are led to infer that a drier climate than now reigned during glacial intervals. How such a postglacial "desiccation of Europe" came about, I do not profess to know; perhaps because, as Hann says, an oceanic climate is more sensitive to fluctuations of solar heat than a continental one. Evidently we have to read "climate" here in the sense of humidity. There being little absolute moisture in continental Asia its quantitative distribution cannot be so strongly affected by variations of sun temperature. The postglacial climatic stages of Europe may have been rather abrupt, and the interval of loess steppe may express contrast between the cold Skandinavian glaciers and the heated and windswept interior of the continent along the foot of the Alps. That would also present an analogy to the loess belt of the Duab. The very narrow strip of European loess is the relict of an older climate the aspect of which is still preserved locally on the Kaiserstuhl, at least to some extent, by the general effects of heat absorption, radiation, percolation, drainage, vegetation and the cultivation of the vine and walnut.

CLIMATE AS CAUSE AND EFFECT.

I have often spoken of the effects caused by the dry climate of the Duab. Now, although the expression is correctly applied on the whole, I wish to show that climate is a somewhat evasive concept, to be handled with due caution.

Let us first of all ask ourselves, "what is a dry climate?" To begin with, climate may be described as the average annual state of the outer sphere, that is to say the atmosphere combined with sun and cosmic space. For the sake of simplicity we shall understand atmosphere as including everything outside. Solar energy itself is a fairly simple factor as determined by latitude, but enormously varied by earth surface, altitude, etc. The sun's direct heat therefore remains self-understood and our chief concern is with its modifications by geographical dimensions and situations. These again determine the quality and movements of the air above them. Abstracting from the little difficulties and exceptions inherent to every definition, we may say that the atmosphere is the bearer of the climate, or, better still, climate is the average atmosphere. Hence a dry climate is one which we are pleased to define as dry, owing to certain qualities of the air compared to some standard, such as Northern Europe, Britain, or the Alps. One has only to be on one's guard against association with other elements, such as absolute temperature, seeing that a dry climate may be very cold, as well as very warm.

The definition of climate as a state is clear, it being in effect one thing. It is geographical conditions reflected by atmosphere, another word for surroundings in the widest sense. But matters become difficult when we begin to speak of climate as a cause, as something which makes other things. Here we have to remember that the condition of the atmosphere is the result of many factors, and that some of the features upon which it is supposed to work are really makers of climate. Whoever wishes to feel clear on the point had therefore better start by taking climate in the passive sense, i.e. the atmosphere as indicator of a complexity of causes. In that case the dry climate of the Duab as a cause of the physical conditions of the Duab, means nothing but the sum of those circumstances and agencies, producing the features of the Duab, expressed in terms of climate. Unless we become conscious of these distinctions we shall get lost in a maze of vicious circles when attempting to unravel the present combinations, past complications, or future probabilities. The character of the column of air above the Duab is part of the region and a result of its position (latitude, continentality) and relief (mountains). These one might call the major or determining climatic features. After them we have the minor climatic characteristics (texture, irrigation, vegetation, aeolism), namely those determined by the atmosphere, which thus forms the intermediary between the major and minor features. In the course of ages the resultant peculiarities may grow into major ones, as desiccation or accumulation may lead to an important rearrangement of morphological conditions. It is thinkable that a desert spreads through its own force. Thus the climate of a locality may change itself, so to speak, by setting up reactions from the topographical surface, whereby certain elements, such as humidity, may be lessened or increased.

Speaking in a rough and ready fashion we may say that one half of a climate comes from outside, is determined by the surroundings of the district, and that the other half is home-made or worked out by the formation of the region itself. It is one of the chief functions of climate to bring the influence of certain surfaces to other surfaces. Climate is a medium of conditions and exchanges, expressing the condition of the whole world as applied to, and modified by, the morphology of a particular place. The earth has a climate, the moon has not; or earth has a mild climate compared to the moon. There are climates of all sorts and degrees, some very definite, some exceedingly vague. Of hot sand deserts and of glaciers we can practically say that they *are* climates, being almost full and adequate expressions of the sum total of conditions regulating their atmospheres. We also have climatic zones and provinces described by the occurrence of certain means of temperature, humidity, etc. All these are fairly well defined because they delineate themselves. But the word is often used when no uniform condition is meant. The "climate of Europe" is nothing but the name of the catalogue of the many different conditions to be found in that area. As the term climate is usually suggestive of some sort of uniformity or average, the different sense in which it can be used may often be productive of wrong impressions. Strictly speaking only a floristic region has a flora, while to begin with it has simply vegetation. Only a climatic district has a climate, while there is meteorology everywhere. Rarely is there coincidence within political and topographical boundaries, as in the British Isles, the whole of which form part of a uniform climatic region. The Duab is a little more complex but ruled by the simple dualism of desert and glaciers, forming two climatic contrasts or spheres of influence. The only sweeping generalisation applies to the sway of continentality, chiefly expressed by annual extremes and aridity. Beyond this the meteorological physiognomy of the Duab must be described by an analysis of its parts. The mention of climate as a cause is best seen in the light of a first impression to be conveyed, as an introductory form of speech opening the discussion.

We can only say that an influence of continentality and dryness (compared to corresponding situations in other plains and mountains) pervades the entire Duab. But as the whole is composed of different local peculiarities, we would obscure these characteristic contrasts by pushing too far the formation of means. We have the mountains and the plains, the oases and deserts, drainage and adduction of rivers, deficiency of rainfall and correction by irrigation, want of water for agriculture, yet a waste flowing unused into the evaporating pan. Natural vegetation depends upon rainfall which may be excessive and badly distributed, or a well spread minimum. The strength and discharge of rivers depend upon the size and height of the ranges whether glaciated or not, the length of their course, interception by irrigation, and so on. The theoretical rate of evaporation in the desert, and the rivers running through it stand in no causative relationship, but only serve to demonstrate a contradiction. To say that the character of the steppes and the Pamirs is due to climate may often lead to misunderstandings, for in the first line it is these places that make their climates, at least partly. Thus one should take this mode of talking in the sense of "demonstrated by," i.e. climate is a special manner of showing the qualities of the Duab, without prejudice as to cause. Similarly, as to the past the favourite phrase of a "change of climate" is but a preliminary question in preparation of special inquiry.

We must look upon climate mainly as an indicator of causes, as an intermediary of changes. This refers especially to the old past whence we have no exact meteorological observations, without which a direct study of climate as a scientific object is impossible. Hence we must always ask what physical changes have taken place and how they have affected the atmospherical phenomena.

Helene Wiszwianski says that the desert is a function of climate. But this is already more or less stated by the definition, for in ordinary as well as in scientific language the desert is a fairly warm place bare of vegetation owing to insufficient rainfall. (N.B. The deficiency may be due to little rain or bad distribution, but also to percolation or salinity.) It is true that a knowledge of the climatic elements over a certain locality will allow us to expect a desert, because the condition of the atmosphere reflects the energies concentrated upon a locality, especially those revealed by things vital to man (food and health). A few days in summer are all that is needed to convince us that our desolate surroundings are not the outcome of cold or want of soil. Climate alone, scientifically described, does not quite unequivocally disclose the nature of the underlying country, for it is above all a study of effects. Given causes must have the same effect, but equal effects need not have identical causes. The behaviour of the atmosphere above a locality is the result of world surroundings and of the place itself, so that really we have to collect a vast amount of information of the past and present, before risking a mild guess at the old climate of some district. As regards modern times, the discrimination does not make any practical difference, seeing that the topography is already well known before meteorological observations begin, and that most climatological descriptions are accompanied by the geographical coordinates and other information.

When dealing with the past it behoves us to keep apart climate as a prime or exogenous factor (e.g. solar heat), as an intermediate factor (e.g. higher temperature due to a lowering of the land), and as an erroneous substitute (e.g. a change in rainfall may be fraudulently suggested by the formation of a lake through subsidence; by loss of water through altered drainage). All these possibilities have to be weighed carefully against each other. The first case, the real change of climate as a universal phenomenon is so far reaching as to demand the greatest caution. It is as likely as it is uncertain; likely, because no assumption seems easier; uncertain, because we do not know anything definite. The second likelihood, climate as a result of topographical features or changes, is the most frequent and offers the best chances of scientific proof. The third is rare and local, but may occasionally be important. Owing to the great length of geological periods, two of the leading possibilities are probably combined in most instances, weakening or strengthening their manifold effects, while the third may sometimes be added as a disturbing element. Hence the geographer is well advised in working out a number of tentative synthetic systems reaching backwards, and forwards, and in all directions of space, in order to show what can precede and follow the event and what is collateral to it in topography. In this manner he may avoid rash conclusions and find out whether to value more highly as a cause or as an effect, a phenomenon which he has endowed with a climatic function. He should handle very gingerly the suggestive idea of change of climate (as a cause) which forms a groove of thought too easily leading outside earth, where one only goes as a last resort.

One begins by separating the major and minor determinations. Latitude, hemisphere, seasons, continent or ocean, altitude and depression are the great makers of climate. But certain types of erosion and drainage, the finer texture of surfaces, and above all organic life are true effects and direct expressions of a specific climate without which they could not exist in this special form. Between them and the large topography the climate or atmosphere is interposed as an indispensable mediator. As a rule of thumb it is well to repeat to oneself that *topography goes before climate.* Glaciers depend upon a slope and a climate, certain plants upon land and a climate, etc. The natural life of plants, animals and primitive man are almost wholly subject to climate, so that here the intermediary rises supremely to the role of creator. Indeed it is mainly this, our own human and most personal interest, which so often tempts us to see climate in the character of first cause. Applied climatology is nearly quite a branch of biology and economics, while pure climatology is the application of topography to atmospheric dynamics. The habit of taking part for the whole is a fertile source of error, here as everywhere. We may be thinking of one particular phenomenon of a region unconsciously accepting it as a full equivalent of climate. Drastic examples are sand dunes (France and Takla-makan), woolly plants (Alps and steppes), fur clothes (Eskimo and Tibetan), snow mountains (Spitsbergen and East Africa), though the real danger lies in the more complex forms such as land-locked rivers, glaciers and certain types of erosion.

As an exact science climatology, being a study of effects in the atmosphere, depends upon modern instrumental records. The state of former climates we shall never know in an exact manner, being obliged to describe them in terms of physiography, glaciation, sedimentation, erosion, vegetation, from which widely divergent pictures may be drawn. Climate being a set of discrete causes is often used indefinitely like "force." It is a comfortable word, almost as general and easy as force which, being absolutely non-committal is also absolutely correct, for one can always say that this or that is due to force, without ever making a mistake. Climate has too many meanings; it may be a whole or something particular, a change or a stability, endogenous or exogenous dynamics, a time or a locality, a feeling or a set of meteorological tables. Vaguely stated a change of climate does not mean much more (or should not mean more) than a change pure and simple. Then it is a superfluous word, as in "change due to a change of force." But as nobody wishes to tell us that change is due to change, everybody has some specification at the back of his mind. It is doubtful if generalisations exist, except in the shape of mental concepts, so that a world climate or world force is nothing but the methodic bridge between particulars. Hence we cannot avoid the particulars of detailed measurement and description. A general force acting upon a particular thing becomes a particular force or phenomenon. We must therefore have complete knowledge either of the thing (ancient topography) and the constant principle (cosmic periods), or of the phenomena themselves (actual meteorological observation and experiment). The method of direct observation is evidently the simplest, and its prolongation into the future will also extend our retrospect of the past. This holds good of meteorology and geological morphology, for both combined must provide us with the picture of ancient climes. Meteorology is a branch of mathematical physical science using the features of earth as a huge experimental apparatus. Climatology is a branch of comparative geography. To-day speculation on old climates is still easy, because we cannot always disprove it, while exact reconstruction is exceedingly difficult owing to the paucity of data.

SPELLING AND PRONUNCIATION.

Native and Russian names are spelt according to the system of the Royal Geographical Society which everybody writing in English ought to obey. Exceptions are made in the case of such words as have become familiar to readers by long usage: Mecca, Calcutta, Jenghiz Khan, Kirghiz, Ferghana, etc.

Being neither philologist nor historian, but looking at maps from a purely topographic point of view, I have accepted without criticism the place names of the Russian ordnance survey. Only a vowel has been changed here and there (Bokhara for Bukhara; Zarafshan for Zerafshan) when confusion was out of the question. It must be kept in mind that all systems for the spelling of foreign words are merely approximate hints to those who do not know the language, an artificial geographical Esperanto, in fact. Even to the scholar such codes are nothing but conventional signs or aids to memory. No language has an absolutely fixed system, least of all for the people who speak it, they being more independent, through constant practice, of a rigid value of letters and syllables. We practically learn the whole of our mother tongue by heart, not a few rules. In English, for instance, we must often go by the shape of the whole word (enough, cough, through), while the pronunciation of some words is without parallel (gaol).

The place names of the Duab are either Turki or Persian, but both these language families are represented by a great number of dialects. Apart from the fact that the same place often has quite different names, there are many pronunciations for one and the same locality, according to the race of the speaker. As to the division or contraction of the parts of names I have been led by convenience only. It would be pedantic to write always "Surkh-ab," "Khing-ob," whereas Kalailiabiob simply shouts for dissection into "Kalai-liabi-ob."

a	as in *father*	Samarkánd, Hazrat-sultán
e	as in *benefit*; sometimes *ay* as in *stay*	Chashmé; Denáu
i	*ee* as in *beet*; sometimes *i* as in *sinner*	Bibi; Vadíf, háji
o	as in *mote*; or short as in *sobbing*	Kokán; onbashí
u	as in *flute*	Uratiubé, Saripúl
ai	as *i* in *ice*	Baizirék, Faizabád, sarái
au	sometimes *ow* as in *how*; but usually separate *a-u* i.e. *a-oo*	hauli, Denáu (howli, Denow); karaúl, Taumén, aúl (Kara-oól, Ta-oomén, a-oól).
g	always hard as in *get*	Yangi-bazár (old form: yanghi)
h	always pronounced, aspirate	hauli (howli)
kh	Scotch and German guttural *ch* as in *Loch*, *machen*	Bokhará, khaná, Shakhzindé
gh	soft guttural	Vashantágh, Yakabágh
y	always a consonant as in *yard*	daryá, Yakhsú, yuzbashí
x	*ks*, in Russian words	Alexandrovsk
zh	French *j*, or *s* as in *treasure*	Jizhik, Andizhán

ch (church); j (journal); s (less); sh (shame); v (veil); z (zeal) as in English
Khsh I have contracted into Ksh (Vaksh, Makshevat)

Accent. In nine cases out of ten the accent is on the last syllable or the terminal vowel: Amu-daryá, Kerminé, and as indicated by words accented above.

NATIVE WORDS.

In the Duab the Mongol and Aryan languages are thoroughly mixed, and even further west it is a common occurrence for names to be composed of two dialects (Eski-sher in Asia Minor). In the small dictionaries of Sart (Uzbeko-Sart) by Nalivkin, and Tajik by Khatimbayev words of both classes appear. Wherever I could trace the origin I have put (T) for Turki, embracing Mongol, Turkish, Kirghiz, Uzbek, etc., or (P) for Persian, including Tajik, Hindustani, Afghan, and Arabic. Arabic has had a great influence on the Persian language as well as on Turki, because the Mongoloid nomads received their civilisation through Persia. Hence most words connected with religion, the calendar, science and art are common to Turkish and Persian dialects, being generally derived from the common source of Arabic. As to etymology, some of the names are easily recognised, though sometimes one has to allow for inflection by declension (e.g. Puli-sangin). But the layman had better not venture too far in his interpretations, as even erudite scholars find enough to puzzle them. Persian is a very rich language with a vast number of words of similar sound, so that a name known by transcription only, often gives no safe clue to the orientalist, or rather misleads him. To this must be added the derivations from other dialects, not to speak of endless corruptions, whereby an original Turki word may assume the appearance of a perfectly self-evident Persian name. The possibilities are inexhaustible. Tashkent seems perfectly plain as "stone city," but it may (or may not) have been Shashkend or "six towns" (analogies, Hexapolis, Panjikent, Olti-shar, Yedti-shar). For these same reasons one must not expect uniformity of spelling and, still less, conformity between common words and "official" names on maps and in books. Topographers have different ears and different interpreters, although they may have the same system of transcription. It is impossible to reconcile conventional topography with philology, without perpetuating confusion into all eternity. For about one hundred of the words given below I am indebted to Knox's *Glossary of Geographical Terms.*

ab (P); water, river
abad (P); cultivated land, populous
achik (T); salt, bitter
ada (T); island
agach (T); tree
ak (T); white
al-sherif; noble
araba; cart
arabakesh; driver
aral (T); island
archa; juniper tree
arik (T); canal
art (T); pass, col
ashdahar (P); dragon
assia (P); mill
ata, ota, dada (T); father
at (T); horse
aul (T); camp
ayag (T); lower, under
bacha (P); boy
bag (T); village
bagh (P); garden
bai, boi (T); rich
baiga; game on horseback
bala (P); upper
bash (T); head
bazar (P); market
beg; high official, governor
bel (T); pass
bend (P); dike, dam
bibi (P); grandmother
bulak (T); spring
chai, choi; tea
chaikhana; tea house
chalma; turban
chap (T); left
chapan; male gown
charpoi (India); bedstead
chashma (P); spring
chil (T); chikor, partridge
chilim; water pipe
chukur (T); deep, gorge
chul (P); desert
chupan (P); shepherd
dagh, tagh, tau (T); mountain
dar (P); door, passage
darband (P); difficult pass
daroz (P); high, long
darvaza (P); gate
darya, dara (P); river
dasht (P); plain, steppe
dasturkhan (P); tablecloth
davan (T); pass
deh (P); village
dengiz (T); sea
dere, dara (T); valley
des, dās (P); plain
dushamba, dushanbe; Monday
frengi, feringi (P); Frank, Occidental
gardan (P); neck
garm (P); hot
gul (P); flower, rose
gur (T); tomb
guzar (P); ferry, passage
hauz (P); reservoir, pond
hissar (T); castle
issik (T); warm
jai, ir (T); place, spot
jilga (T); deep valley
ju (P); stream
juma; Friday
kabud (P); blue
kafir; unbeliever
kala, kila (P); fortress
kalan, kalon (P); great
kalta (T); short
kamarband (P); belt
kamish (T); reeds
kand, kend (P); city
kar (T); snow
kara (T); black
karaul (T); guard
kash (T); bank, border
katta (T); large
kazan (T); kettle
kazi (P); judge
ketmen; hoe
khan (P); inn
khana (P); house
kharam (T); dirty, disgusting
khatun (T); woman
khunuk (P); cold

khurjin; saddlebag
kichik, kchik (T); small, little
kishlak (T); village
kizil (T); red
koh, kugh (P); mountain
koi (T); sheep
kok, kuk, gok (T); blue, green
kol (T); ravine
kosh (T); camping place
kul, gol (T); lake
kum (T); sand
kurgan (T); hill, fort
kush (P); killer
kush (T); falcon
kutas (T); yak ox
kuyu (T); well, cistern
lak (P); place
langar (T); rest house
madrasa, medresse (Ar.); college
maidan (P); open place
mazar (T); saint's tomb
meshed, masjid (P); mosque
mihmon, meiman (T); guest, stranger
minar (Ar.); minaret
mingbashi (T); head of a thousand
mir (P); mountain
mirakhur; stable boy, equerry
mirza (P); scribe
mollah; priest
murg (P); fowl
muz (T); ice, snow
nau (P); new
onbashi (T); head of ten
ou, oi (T); water; hollow
ova (T); plain
palwan, palevan (P); hero, giant
panjshamba, peishamba; Thursday
pillau; dish of rice and meat
pul (P); bridge
pul (T); money
rabat, robat (T); resting place
rigistan (P); market place
rishta (T); thread
rud (P); river
safid, sefid (P); white
sai (T); ravine
sakal (T); beard
salaam; greeting
salla, sallia (P); turban
samovar (Russ.); hot water urn
sang (P); stone
sar, sir (P); head, summit
sarai (P); house
sard (P); yellow
sarik (T); yellow
sebz, siabz (P); green
sel; snowfield, ice
shahr (P); town
shaitan (P); devil
shakh (T); horn
shikari (P); hunter
shir (P); tiger; milk
shor, shur (P); saline, brackish
shutur (P); camel
sia (P); black
silau; present, bakshish
su (T); water
surkh (P); red
stan (P); place
tam (T); wall
tamasha; feast
tan (P); narrow
tang (P); defile
takht (P); throne
takhta (P); plank
tash (T); stone
tenga, tanka; a silver coin
tengi, tang; a path, road
tepe (T); hill
tik (T); steep
tilla; gold
timur, tumur, demir (T); iron
tiubiteka; cap
tugai (T); bushy place
tura; lord, master
turpak; clay, loess
tut (P); mulberry tree
ui (T); house
ulu, ulugh (T); great, high
uraza; fasting time
urta (T); middle
uzun (T); long, far
yaila, yailak, ailak, lailak (T); summer pasturage
yagh, yak (P); ice
yaka (T); boundary
yangi (T); new
yar (T); cliff, bluff
yashil, yeshil (T); yellow (also green?)
yelikbashi (T); head of fifty
yol (T); road
yurt (T); felt tent
yuzbashi (T); head of a hundred
zamin (P); earth
zindan (P); prison

P.		T.
yak	1	bir
du	2	iki
se	3	uch
chor	4	durt
panj	5	besh
shish	6	olti
haft	7	yiti
hasht	8	sakis
nu	9	tokus
da	10	on
bist	20	igirma
si	30	otus
chil	40	kirk
panjo	50	yilik
shast	60	oltmish
aftod	70	yitmish
ashtod	80	saksan
navad	90	toksan
sad	100	yuz
hazor	1000	ming

SCIENTIFIC GLOSSARY.

Besides explaining a few words to the general reader, this glossary states the sense in which I employ certain terms.

Abrasion. Scraping of broad surfaces, chiefly by the sea and ice. In connection with wind "deflation" is often used.

Aeolic. Originated by or acted upon by wind.

alpine and Alpine. With a capital "A" it refers to places and conditions in the Alps themselves, whereas "alpine" means akin to the type for which the Alps have furnished our first object lesson. It is a descriptive and comparative term conveying certain impressions of landscape not amenable to strict definition. But the distinction by spelling is not always made in literature. Likewise "Duabic" is in the Duab, while "duabic" may be anywhere in Asia. Also compare "Pamirs" and "pamir."

Breccia. A clastic (i.e. composed of fragments) rock of coarse *angular* débris. Breccias result from all sorts of geological processes: tectonic friction, volcanic, morainic and talus deposition, etc.

Convection. Mixing of layers by ascending currents.

Corrasion. Scraping by water, wind, etc.

Corrie or cirque. Characteristic, short valley in a mountain flank; of more or less half circular shape with vertical walls. Intimately connected with glaciation.

Corrosion. Chemical attack on surfaces.

Denudation. Destruction and lowering of the land by atmospheric agencies and transport by wind, rivers and glaciers.

Desquamation. Production of peels and scales by weathering.

Dolina. Funnel shaped sinkhole or subsidence of the Karst, q.v.

Endogenous and Exogenous. Acting from inside or outside. The sun and atmospheric agencies are exogenous, while the earth's own temperature (infinitesimal), movements of the crust and volcanic forces are endogenous. Gravity seen as attraction is endogenous, but viewed as falling matter (impact from above) it might pass as exogenous.

Erosion. Cutting down by water and ice.

Fluvioglacial. The collaboration of glaciers and rivers, but more especially the conversion of morainic material into alluvial pebble beds.

Fohn. A relatively warm wind of the Alps falling from over a mountain divide. Wrongly called "Scirocco" at Innsbruck, although it has nothing to do with an origin thereby suggested. The air, icy cold on the mountain ridge, is heated and desiccated by the descent into lower levels.

Glacial Periods. Pending further inquiry one may accept three for the Alps: the Mindel, Riss and Würm, called after the rivers where Penck studied glacial deposits (they follow in alphabetical order). A fourth and oldest, the Günz, may be added. In between many smaller ones are traceable. Speaking of the ice age I mean the whole set, unless "the last" is added. Whether the whole series was one (monoglacialism) or not (polyglacialism) can only be discussed on two alternative conditions. Either the glacial cause must be definitely known, or a standard level must be fixed, such as the present Alpine or Skandinavian snow line. Then monoglacialism is the continuation of the same (though fluctuating) cause, or the keeping of the snow lines within standard level, while polyglacialism would entail absolute interruptions of the cause, or occasional retreat of the snow line above standard level. But what's in a name? One might as well discuss, if the yearly shoots of a perennial are the same plant or not.

Halophytes. Plants able to thrive in saline soil.

Height. Absolute height is that above sealevel; relative height that above any other point, such as the plain at the foot of the mountain.

Humidity. The amount of water vapour in the air. Absolute humidity is the weight of water contained in a cube of air. Only the meteorologist making calculations or building theories is interested in absolute humidity. The geographer inquires after relative humidity (expressed in percentages) showing how far the air still is from saturation and condensation. This depends upon temperature, for hot air can hold more water than cold. Air can be wet at a cool temperature and very dry at a high temperature, although the actual quantity of water (i.e. absolute humidity) may have remained exactly the same.

Interior Mountains, Interior Heating. A short expression for an important phenomenon. Great mountain masses (Alpine groups, Monte Rosa, etc.) and plateau-like bulges of earth (Pamirs, etc.)—which are usually in the interior of a mountainous region—are more advantageous to heat storing than isolated summits or narrow ridges. Thus the mean temperature at say 4000 m. is higher on the Pamirs than it would be at the same height, in the foothills or outer ranges. This is the most important of the factors regulating the snow line, though in the Duab greater continentality and decrease of precipitation by interception have also to be considered.

Isoclinal, synclinal and anticlinal stratifications are obliquely parallel (slates on roof), converging downwards (gutter), and converging towards the zenith (gable or roof).

Karren. Peculiar limestone erosion of innumerable furrows, often very deep, with sharp ridges between.

Karst. Bare, deeply corroded and eroded limestone surfaces of the Adriatic regions, Istria, Dalmatia, etc.

Mesa. A table-land (Colorado).

Metrical (decimal) System:

1 km. (kilometre) = 0·62138 mile, or 3280·87 feet.
1 m. (metre) = 3·281 feet, or 39·37 inches.
1 cm. (centimetre) = 0·3937 inch, or 3·937 lines.
1 mm. (millimetre) = 0·03937 inch, or 0·3937 line.
1 km. = 1000 m.; 1 m. = 100 cm.; 1 cm. = 10 mm.
1 sq. km. (square km.) = 0·386 sq. mile, or 247·11 acres.
1 sq. m. = 10·764 sq. feet; 1 sq. cm. = 15·5 sq. lines.
1 cb. km. (cubic km.) = 0·24 cb. mile; 1 cb. m. = 35·316 cb. feet
1 l. (litre) = 1000 cb. cm. = 1·76 pint.
1 kg. (kilogramme) = 1000 g. (grammes)—2·2 lbs.; 1 g. = 0·35 oz.

Morphology. Short for geomorphology. Everything concerning the outward shape of the earth's features and their origin. Geology delves more deeply.

Névé. Snow fields of permanent, hardened snow. Usually applied to the larger expanses feeding a glacier.

Nunatak. Relatively small, rocky, peak protruding from a great expanse of glacier.

Orography, same as mountain topography.

Palaearctic. A region of zoogeography embracing a fauna of common characteristics. It covers the temperate zones of Europe and Asia from Iceland to Behring Strait, and from the Azores to Japan, including the northern Sahara, the greater part of Arabia, Persia, Afghanistan and Tibet.

Penitentes. A formation peculiar to the South American Andes. The snow-fields are cut up into rows of prisms or cones often several feet high.

Physiography. Stands for physical geography; see page 14.

Quaternary. The last great geological period, divided into pleistocene or diluvial; and recent, postglacial or alluvial. Although there is no fixed calendar, we may say that the last great glaciation (how great and where?) marks the end of pleistocene. One may also say that everything still forming (gravel beds of rivers, screes, fresh moraines) is recent. On the whole there is no sharp division and "alluvial or recent" simply means the present moment of geological times.

Rainfall. Used in the same sense as precipitation, i.e. including snow, dew, hoar frost, hail, etc.

Tertiary. Geological period preceding the quaternary, q.v.

Thermometer. The scientific scale is divided into centigrades or 100° (Celsius). 5° equal to 4° R(éaumur) or 9° F(ahrenheit). Centigrades and R. have the same freezing point of 0°, but 32° F. is the same as 0° or 0° R., so that after converting from or into F. one has to reckon from 32°. 41° F. = 5°; 23° F. = −5°.

Tor. A rocky hill, generally isolated, in the south-west of England.

Xerophilous plants or xerophytes are those specially protected against heat and excessive transpiration in dry regions.

Wadi. Dry (intermittent) water course or river bed in North Africa.

LITERATURE.

Owing to the composite character and wide range of the book, the compilation of this list was a matter of some difficulty. A complete bibliography is out of the question owing to want of space, nor can I mention everything I have read in the course of my studies on Middle Asia and physiography. For these reasons I have made what might be called a subjective and modern (mostly after 1890) list of those works and articles which interested me most, and which the reader has therefore a right of knowing. In addition to these there are the authors whom I must quote in connection with scientific data and theories. Bibliographical lists of reference are found in Curzon, Dmitriev-Mamonov, Friederichsen, Geiger, Huntington, Lansdell, Olufsen, Schwarz; the completest up to 1895 being Lidski's *Materials for a Bibliography of Middle Asia.* I have hardly ever mentioned the standard works of authors well-known to the public or to every geographer, and it must be taken for granted that I have read or used most of the physiographic textbooks and travel descriptions by Abercromby, Abruzzi, Aitken, Ball, Blanford, Bonvalot, Brückner, Conway, Credner, Davis, Deasy, Dutreuil de Rhins, Fedchenko, Forsyth, Freshfield,

Futterer, Geikie, Greely, Grenard, Grisebach, Günther, Hedin, Heim, Hess, Holdich, Humboldt, Huxley, Indian Government, Krahmer, Littledale, Longstaff, Lyell, Merzbacher, Michell, Mill, Moser, Mushketov, Neumair, Obruchev, Oshanin, Penck, Peschel, Przhevalski, Ratzel, Richter, Richthofen, Schlagintweit, Schuyler, Sievers, Stein, Supan, Tait, Tyndall, Vambery, Waddell, Walther, Wood, Younghusband, Yule, and the leading periodical journals on geography, glaciology, as well as British and German encyclopaedias.

The general reader desirous of gaining some insight into modern geography, which is practically physiography, will find himself rewarded by a study of one of three small and excellent elementary volumes:

HUGH ROBERT MILL, *The Realm of Nature.* London, John Murray, University Extension Manuals.

WILLIAM MORRIS DAVIS, *Physical Geography.* Boston, Ginn & Co.

RALPH S. TARR, *New Physical Geography.* New York, The Macmillan Company.

There are four books and one article in the English language to which my "Duab" may be considered as forming a complement. They are strongly recommended to whomsoever wishes to obtain a good survey over the whole of Inner Asia, without diving too deeply into several hundredweight of literature. Their perusal is easy and their contents are trustworthy:

GEORGE N. CURZON, "The Pamirs and the Source of the Oxus." *Geographical Journal,* Vol. VIII. London, 1896.

A most thorough review of everything related to the Pamirs. Very interesting.

DAVID FRASER, *The Marches of Hindustan,* the Record of a Journey in Tibet, Transhimalayan India, Chinese Turkestan, Russian Turkestan and Persia. Edinburgh & London, 1907, Blackwood.

This splendid book covers a wide range from India to Persia, describing the scenes and peoples along a route which one might call the historical and ethnological artery of Asia. Here the reader will find a long extension of my region in the most important directions. It is one of the most delightful books of travel I have ever enjoyed, some of its passages, full of poetry and humour, equalling many of the best things Kipling ever wrote.

ELLSWORTH HUNTINGTON, *The Pulse of Asia.* London, 1910, Constable.

A brilliant description of Central Asia proper, its physical features and conditions. It is full of lucid explanations and suggestive ideas. The great problem of desiccation is treated on a broad basis, and the dead cities of the Lop (or Tarim) basin are spread out to our gaze. Questions of climate, glaciation, the influence of surroundings, the geographic basis of history, provide food for thought.

O. OLUFSEN, *The Emir of Bokhara and his Country* (including a Journey to Khiva). London, 1911, Heinemann.

Valuable as a book of reference, containing a wealth of detailed information on architecture, archaeology, customs, habits, religion, dress, etc. together with a large number of excellent photographs of ethnographical objects, art, buildings, and racial types.

FRANCIS HENRY SKRINE and EDWARD DENISON ROSS, *The Heart of Asia.* London, 1899, Methuen.

The history of Central Asia, Bokhara and Russian Turkestan is here presented by two scholars and orientalists. This standard work forms the necessary historical supplement to the geographical books. We learn everything important concerning the early origins of various races, the Mongol invasions, the fates of dynasties, old trade routes, wars, and the march of the Russian conquest. To this the *Tarikh i Rashidi* by Elias and Ross (see below) forms a kind of complement.

G.J.—Geographical Journal; Monb.—Monatsberichte; Sb.—Sitzungsberichte; Abh.—Abhandlungen; P.M.—Petermanns Mitteilungen; M.—Mitteilungen; G.—Gesellschaft; Egh.—Ergänzungsheft; Zts.—Zeitschrift; Pr.—Proceedings; Jb.—Jahrbuch; (R.)—in the Russian language.

E. S. BALCH. Die Einteilung wellenförmiger Oberflächen. Berlin, 1900.

L. S. BERG. Journey to the Isfara Glaciers. Izv. Imp. Russ. Geog. Soc. Turkest. Div. 1907 (R.).

GEN. BERTHAUT. Topologie. Paris, 1909–10.

A. A. BOBRINSKI. Mountain Tribes of the Panj. Moscow, 1908.

A. BÖHM. Geschichte der Moränenkunde. Abh. K.K. Geog. G. Wien, 1901.

G. BOEHM. Reiseskizzen aus Transkaspien. Geog. Zts. Leipzig, 1899.

A. BOUTQUIN. L'Asie Centrale, la Question du Dessèchement du Globe. Paris, 1910.

E. BRÜCKNER. Klimaschwankungen seit 1700. Geog. Abh. Wien, 1890.

J. BRUNHES. Sur les Contradictions de l'Erosion Glaciaire. C.R. 142, 1906.

S. R. CAPPS. Rock Glaciers in Alaska, cited in G.J. from Journal of Geology, 1910.

G. CAPUS. A travers le Royaume de Tamerlane. Paris, 1892.

D. CARRUTHERS. Arpa and Aksai Plateaus. G.J. Nov. 1910.

C. CHELIUS und C. VOGEL. Zur Gliederung des Löss. N. Jb. Min. Geol. 1891.

R. CREDNER. Die Deltas. P.M. Egh. 56, 1878.

J. CVIJIC. Das Karstphänomen. Geog. Abh. Wien, 1895.

J. CVIJIC. Morphologische und Glaziale Studien aus Bosnien, etc. Abh. K.K. Geog. G. Wien, 1900–1901.

J. DAMIAN. Seestudien. Abh. K.K. Geog. G. Wien, 1899.

A. I. DMITRIEV-MAMONOV. Guide through Turkestan. Petersburg, 1903 (R.).

W. A. DOLGOROUKOFF. Guide à travers la Sibérie. Tomsk, 1899.

DUNMORE. Journeyings in the Pamirs. G.J. 1893.

W. R. ECKARDT. Einfluss des Waldes auf das Klima. Met. Jb. f. Aachen,. 1907.

M. ECKERT. Das Gottesackerplateau, Ein Karrenfeld. Innsbruck, 1902.

N. EKHOLM. On the Meteorological Conditions of the Pleistocene Epoch. Quart. J. Geol. Soc. 1902.

N. ELIAS and E. D. ROSS. Tarikh i Rashidi. London, 1895.

P. T. ETHERTON. Across the Roof of the World. London, 1911.

B. A. FEDCHENKO. Journey through the Pamirs. Yearbook V, Russ. Alpine Club. Moscow, 1906 (R.).

H. v. FICKER. Zur Meteorologie von West Turkestan. Denks. Ak. Wiss. Wien, 1908.

H. v. FICKER. Wolkenbildung in Alpentälern. Innsbruck, 1905.
W. FILCHNER. Ein Ritt über den Pamir. Berlin, 1903.
E. FISCHER. Eiszeittheorie. Heidelberg, 1902.
F. J. FISCHER. Meer- und Binnengewässer in Wechselwirkung. Abh. K.K. Geog. G. Wien, 1902.
H. FISCHER. Aequatorialgrenze des Schneefalles. M. Ver. Erdk. Leipzig, 1888.
F. A. FOREL. Handbuch der Seenkunde. Stuttgart, 1901.
M. FRIEDERICHSEN. Reisebriefe aus Russisch Zentralasien. M. Geog. G. Hamburg, 1902.
M. FRIEDERICHSEN. Die heutige Vergletscherung des Khan-Tengri Massivs, etc. Zts. Glk. 1908.
M. FRIEDERICHSEN. Morphologie des Tienschan. Berlin, 1898.
E. J. GARWOOD. Features of Alpine Scenery due to Glacial Protection. G.J. Sept. 1910.
W. GEIGER. Die Pamir Gebiete. Geog. Abh. Wien, 1887.
F. E. GEINITZ. Die Eiszeit. Braunschweig, 1906.
E. GOGARTEN. Alpine Randseen und Erosionsterrassen. P.M. Egh. 165.
Y. D. GOLOVNIN. To the Pamirs. Moscow, 1902 (R.).
I. G. GRANÖ and G. MERZBACHER. Die Eiszeitfrage i. d. Nordw. Mongolei. P.M. 1911.
J. W. GREGORY. The Terms "Denudation," "Erosion," etc. G.J. Feb. 1911.
A. GRUND. Probleme der Geomorphologie am Rande von Trockengebieten. Sb. Akad. Wien, 1906.
A. GRÜNWEDEL. Archäologische Arbeiten in Idikutschari. Abh. Ak. Wiss. München, 1906.
A. HEIM. Ueber Verwitterung im Gebirge. Basel, 1882.
M. E. HENRY. Sur le Rôle de la Forêt etc. Paris, 1902.
E. HUNTINGTON. The Rivers of Chinese Turkestan and the Desiccation of Asia. G.J. Oct. 1906.
A. E. KITSON and E. O. THIELE. Upper Waitaki Basin. G.J. Nov. 1910.
J. KLOOS. Entstehung d. lössartigen Lehmes. Zts. D. Geol. G. 44/324.
A. KNOX. Glossary of Geographical Terms. London, 1904.
A. v. KOENEN. Alter der Schotterterrassen. N. Jb. Min. Geol. 1891.
E. KOKEN. Die Eiszeit. Tübingen, 1896.
F. KÖNIG. Die Verteilung des Wassers über, auf und in der Erde. Jena, 1901.
N. KORZHENEVSKI. Along the Muksu River. Yearbook V, Russ. Alpine Club. Moscow, 1906 (R.).
KOSSIAKOF'S Journey in Karategin and Darwaz. Pr. R.G.S. 1886.
A. KOSTANKO. Turkestan. Petersburg, 1880 (R.).
K. KRAEPELIN. Zur Systematik der Solifugen. Hamburg, 1899.
KRAFFT. Reiseerinnerungen von der Russischen Transkaspibahn. Militär Wochbl. Berlin, 1898.
A. v. KRAFFT. Geologische Ergebnisse einer Reise durch das Chanat Bokhara. Ak. Wiss. Wien, 1900.
A. v. KRAFFT. Das Ostbokharische Goldgebiet. Zts. Prakt. Geol. Feb. 1899.
HUGUES KRAFFT. A travers le Turkestan Russe. Paris, 1902.
PRINCE KROPOTKIN. The Desiccation of Eur-Asia. G.J. June, 1904.
L. KUROWSKI. Die Höhe der Schneegrenze, etc. Geog. Abh. Wien, 1891.
M. A. LEVANEVSKI. Sketches in the Kirghiz Steppes. Zemlevedenie, 1894 (R.).
M. E. D. LEVAT. Turkestan et Boukharie. Paris, 1902.
M. E. D. LEVAT. Richesses Minérales en Asie Centrale. Paris, 1903.
V. I. LIPSKI. Mountainous Bokhara. III Vols. Petersburg, 1902 and 1905 (R.).
S. G. LITTLEDALE. Across the Pamir from North to South. Pr. R.G.S. 1892.
T. G. LONGSTAFF. Mountain Sickness. London, 1906.
W. LOZINSKI. Versuch einer Charakteristik der Canyontäler. Jb. K.K. Geol. Rchsanst. Wien, 1909.

F. MACHACEK. Ergebnisse etc. im westlichen Tianschan. M.K.K. Geog. G. Wien, 1912.
F. MACHACEK. Geomorphologische Studien aus dem Norwegischen Hochgebirge. Abh. K.K. Geog. G. Wien, 1908.
M. MANSON. The Evolution of Climates. Amer. Geol. 1898.
E. MARKOV. The Sea of Aral. G.J. Nov. 1911.
G. P. MERRIL. Treatise on Rocks, Rock Weathering, etc. New York, 1897.
H. MEYER. Die Eiszeit in den Tropen. Geog. Zts. X/11.
G. MERZBACHER. Zur Eiszeitfrage i. d. Nordw. Mongolei. P.M. 1911.
R. MICHELL. The Regions of the Upper Oxus. Pr. R.G.S. 1884.
A. v. MIDDENDORF. Einblicke in das Ferghana Thal. Petersburg, 1881.
H. MILLER. River Terracing. Pr. R. Phys. Soc. Edinburgh, 1883.
E. D. MORGAN. Steppe Routes from Karshi. Pr. R.G.S. 1881.
E. D. MORGAN. Geography of Central Asia from Russian Sources. Suppl. Paps. R.G.S. 1884.
F. W. K. MÜLLER. Neutestamentliche Bruchstücke in soghdischer Sprache. Sb. Ak. Wiss. Berlin, 1907.
P. S. NAZAROV. Travels in the Pamirs. Zemlevedenie, 1896 (R.).
A. NEHRING. Zur Steppenfrage. Globus, 1894.
A. NEVE. Journeys and some Factors of Himalayan Erosion. G.J. Oct. 1911.
A. NEVE. The Ranges of the Karakoram. G.J. Nov. 1910.
V. NIKOLSKI. Promenade through the Pamirs. Yearbook II, Russ. Alpine Club. Moscow, 1904 (R.).
F. NÖLKE. Die Entstehung der Eiszeiten. Deutsche Geog. Bll. 1909.
V. F. NOVITSKI. Travels into the Mts. of Peter the Great. Izv. Imp. Russ. Geog. Soc. XL/1, 2 (R.).
V. OBRUCHEV. Weathering and Deflation in Middle Asia. Zap. Imp. Russ. G.S. 1895 (R.).
V. OBRUCHEV. Transcaspia. Zap. Imp. Russ. G.S. 1890 (R.).
N. V. OSTROUMOV. Geography of Turkestan. Samarkand, 1891 (R.).
W. OSTWALD. Naturphilosophie. Leipzig, 1901.
V. PASCHINGER. Schneegrenze und Klima. P.M. 1911.
V. PASCHINGER. Die Schneegrenze. P.M. Egh. 173, 1912.
A. PHILIPPSON. Die Humusbildung. Geog. Zts. 1897.
N. V. POGGENPOHL. To the Sources of the Muksu across the Western Pamirs. Izv. I.R.G.S. 1908 (R.).
PONIATOVSKI. Rice Cultivation in Turkestan. Govt. Rept. Tashkent, 1905 (R.).
I. PRINZ. Die Vergletscherung des nördlichen Teiles des Tienschan Gebirges. M.K.K. Geog. G. Wien, 1909.
R. PUMPELLY. Explorations in Turkestan. Washington, 1908.
G. RADDE. Transkaspien und Nordchorassan. P.M. Egh. 126, 1898.
F. RATZEL. Zur Kritik der sogenannten Schneegrenze. Leopoldina, Halle, 1886.
E. RICHTER. Neue Ergebnisse und Probleme der Gletscherforschung. Abh. K.K. Geog. G. Wien, 1899.
E. ROCCA. Karategin et Darvaz. Revue de Géog. 36.
G. E. RODIONOV. Routes in Turkestan. Local Staff Rept. Tashkent, 1907 (R.).
The Russian Pamir Expedition. Pr. R.G.S. 1884.
E. v. SALZMANN. Im Sattel durch Zentralasien. Berlin, 1902.
A. SAUER. Stand der Lössfrage in Deutschland. Globus, 1891.
A. F. W. SCHIMPER. Pflanzen Geographie. Jena, 1898.
A. v. SCHULTZ. Vorläufiger Bericht ü.m. Pamirexpedition. P.M. 1910.
A. v. SCHULTZ. Volks- und Wirtschaftliche Studien im Pamir. P.M. 1910.
F. v. SCHWARZ. Turkestan. Freiburg i. B. 1900.
F. SCHWARZ. Astronomical etc. Observations in Bokhara, Darwaz, etc. Petersburg, 1893 (R.).

F. v. SCHWARZ. Sintfluth und Völkerwanderungen. Stuttgart, 1894.
H. H. SCHWEINITZ. Orientalische Wanderungen. Berlin, n.d. (1910).
R. SIEGER. Studien über Oberflächenformen der Gletscher. Wien, 1898.
H. SJÖGREN. Ueber das diluviale Aralokaspische Meer. Jb. K.K. Geol. Rchsanst. Wien, 1888.
N. A. SOKOLOV. Die Dünen. Berlin, 1894.
Z. SOLDERN. Bochara. Allg. Bauztg. Wien, 1899.
SOLGER, GRAEBNER, U.A. Dünenbuch. Stuttgart, 1910.
M. A. STEIN. Sand-buried Ruins of Khotan. London, 1903.
G. STEINMANN. Diluvium in Südamerika. Monb. D. Geol. G. 1906.
G. STEINMANN. Entwicklung des Diluviums in Südwestdeutschland. Zts. Deut. Geol. G. 1898.
J. STINY. Die Muren. Innsbruck, 1910.
A. SUPAN. Die Jährl. Niederschlagsmengen auf den Meeren. P.M. 1898.
E. R. TSIMMERMANN. In the Depths of Asia. Moscow, 1892 (R.).
P. A. TUTKOVSKI. Discussion on Mineral Deserts (Loess, etc.). Bull. Russ. Alpine Club, Oct. 1911 (R.).
W. ULE. Die Aufgabe geographischer Forschung an Seen. Abh. K.K. Geog. G. Wien, 1902.
W. ULE. Die Aufgabe geographischer Forschung an Flüssen. Abh. K.K. Geog. G. Wien, 1902.
H. VALLOT. Manuel de Topographie Alpine. Paris, 1904.
M. M. VIRSKI. Settled Places in the Samarkand District. Local Govt. Rept. Samarkand, 1906 (R.).
F. WAHNSCHAFFE. Beitrag zur Lössfrage. Jb. Kgl. Preuss. Geol. Ldsanst. 1889.
J. WALTHER. Das Oxusproblem. P.M. 1898.
J. WALTHER. Das Gesetz der Wüstenbildung. Berlin, 1900.
E. WARMING. Oekologische Pflanzengeographie. Berlin, 1896.
E. WISOTZKI. Zeitströmungen in der Geographie. Leipzig, 1897.
H. WISZWIANSKI. Die Faktoren der Wüstenbildung. Veröff. Inst. f. Meeresk. Berlin, 1906.
A. WOEIKOF. Einfluss der Schneedecke auf Boden, etc. Geog. Abh. Wien, 1889.
A. WOEIKOW. Der Aralsee. P.M. 1909/4.
A. C. YATE. Afghan Boundary Commission. London, 1887.
C. E. YATE. Northern Afghanistan. London, 1888.
Y. L. YAVORSKI. Travels into the Mountains of Bokhara etc. Zemlevedenie, 1895 (R.).

INDEX

The first part of this index is a classification of subjects with numerous cross-references. It is followed by an alphabetical list consisting chiefly of proper names. The Lists of Contents, Illustrations and Literature, and the Native and Scientific Glossaries should also be referred to, none of which are indexed.

I. SUBJECT INDEX.

A. General Topography and Mountains; Geology.

B. General Morphology and Physiographic Agencies.

C. Water; Rivers and Valleys; Lakes.

D. Snow; Glaciers and Moraines.

E. Steppe and Desert; Loess and Sand.

H. Plants and Animals.

I. Man and Surroundings.

K. Native Life.

L. Folklore and Religion; Monuments and Art.

M. History and Politics.

N. Trade and Industry; Traffic and Travel.

O. Landscapes, Scenes, Adventures.

P. Reflections and Discussions.

II. ALPHABETICAL INDEX.

This part is almost entirely restricted to proper names. Not included—with a few exceptions—are the English or Latin names of plants and animals, nor most of the meteorological stations and localities of the Appendix.

UNPUBLISHED PHOTOGRAPHS IN AUTHOR'S COLLECTION

Makhan-kul district, dunes, reeds, vegetation, etc.

Bokhara and Samarkand, many views of loess, buildings, architecture, gardens, bazars, avenues, Afrosiab, natives.

Agalik valley and Kemkutan mountains, formation, surface, vegetation.

Zarafshan valley, landscapes along the whole course of the river—Sundry gorges, conglomerate cliffs, terraces, villages, screes, side valleys, etc.—Cracked and sliding slopes near Sasun—Dissected terrace at Dangari-yoz—Cone near Yarom—Lateral gorges at Taumen and Yarom—Mingbashi's house at Varziminar—Old river bed near Vardachit—Talus near Madrushkat—Houses of Paldorak—Terraces near Pakshif—Mudspate near Dikhab—Summer village—Slipping moraine—Rama glacier—Moraines, pocket lakes—Mount Bielaya—Eastward view from Macha pass—Zarafshan pass.

Fan river—Puli-mulla—Village of Piti—Talus at Piti—Castle of Zarvadan—Fan gorge above Piti, with terrace strip and benches—Vashantagh from Laudan pass—Sundry views of Lailak Chapdara and moraines in Pasrud valley—Khontagh—Third lake—Mt Chapdara from Laudan pass—Twin Pasrud glacier and wash plain (sandr)—Scree cone of Surkhab valley—Surkhab glacier—Boulderclay of Iskander valley—Views of Iskander-kul and neighbourhood—Great juniper trees of Iskander-kul—Talus steppe below Khoja Isak.

Pakshif pass—Mt Yangi-sabak from Pakshif pass—Twin Bodravak glacier—Mazar and other views of Gorif—Garm—Surkhab landscapes—Kalai-liabi-ob and Ganishau—Giant plane tree—Sagunaki mountains—Liulikharvi pass and peak—Farkikush terraces—Yughan and rose scrub—Welted furrows on scree—Khingob valley—Walnut trees, cave, mountains near Liangar—Erosion by snow (protection?)—Mountains around Gardani-kaftar.

Tupchek, many views of the mountains, moraines, glaciers, screes, steppes, etc.—Residual rock strips—Dry bed—Meanders and oxbows of Karashura.

Mountains and valleys between the Khingob valley and Sagirdasht—Village of Sagirdasht—Ridge dichotomy—Landslide in Khumbau valley—Morainic and torrential side valley of Khumbau—Mudspate near Seu—General view of Kalaikhumb and the Panj—Lateral valley of Panj and welted mudspate track—Mulberry trees and stone walls—Soldiers' gardens—"Gallic" wall and tower with machicolations at Kalaikhumb—Hillscape and snow crescents on Hazrat-ishan pass—View northwards.

Yakhsu district, many views of landscapes, conglomerate formations, villages, etc.—Hamlet with haystacks—Erosion of conglomerate peneplain (or subdued hills)—Tilted strata at Talbar—Gorge and near detail of conglomerate rock—Fields and scrub.

Baljuan to Samarkand, views of landscapes, loess formations, river plains, structure—Panorama of Turpak pass—Nomad camp—Savanna of Middle Bokhara—Karatagh and surroundings—Groves below Mura pass—Old towers of Sariassia—*Lasiagrostis splendens*—Iron Gate and the hills through which it is cut—Steppe at Yakasarai—Sandstone, bazars, villages, etc. etc.

Some day I hope to be able to make an album of extra illustrations (prints from negatives) to be deposited with the Royal Geographical Society. W. R. R.

CAMBRIDGE: PRINTED BY JOHN CLAY, M.A. AT THE UNIVERSITY PRESS

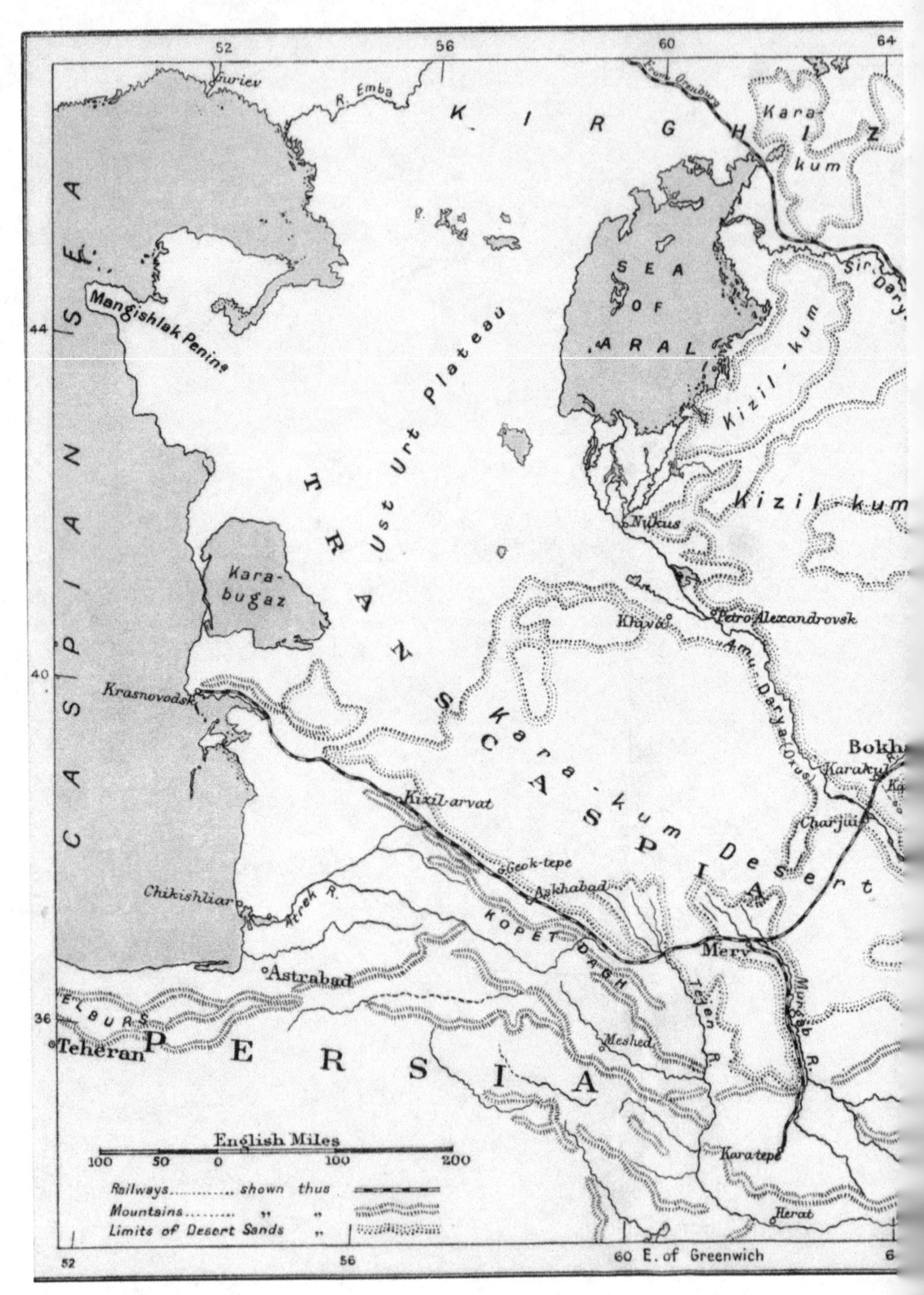
KIRGHIZ
Kara-kum
SEA OF ARAL
Sir Darya
Kizil-kum
Kizil-kum
Ust Urt Plateau
R. Emba
Guriev
Mangishlak Penin.
CASPIAN SEA
Kara-bugaz
TRANSCASPIA
Nukus
Khiva
Petro Alexandrovsk
Amu Darya (Oxus)
Krasnovodsk
Kara-kum Desert
Kizil-arvat
Geok-tepe
Askhabad
Charjui
Karakul
Bokh
Merv
Chikishliar
Atrek R.
KOPET DAGH
Astrabad
ELBURS
Teheran
PERSIA
Meshed
Tejen R.
Murgab R.
Karatepe
Herat
English Miles
100 50 0 100 200
Railways............ shown thus
Mountains.........
Limits of Desert Sands
52
56
60
64
44
40
36
60 E. of Greenwich

Sketch Ma

Map I.

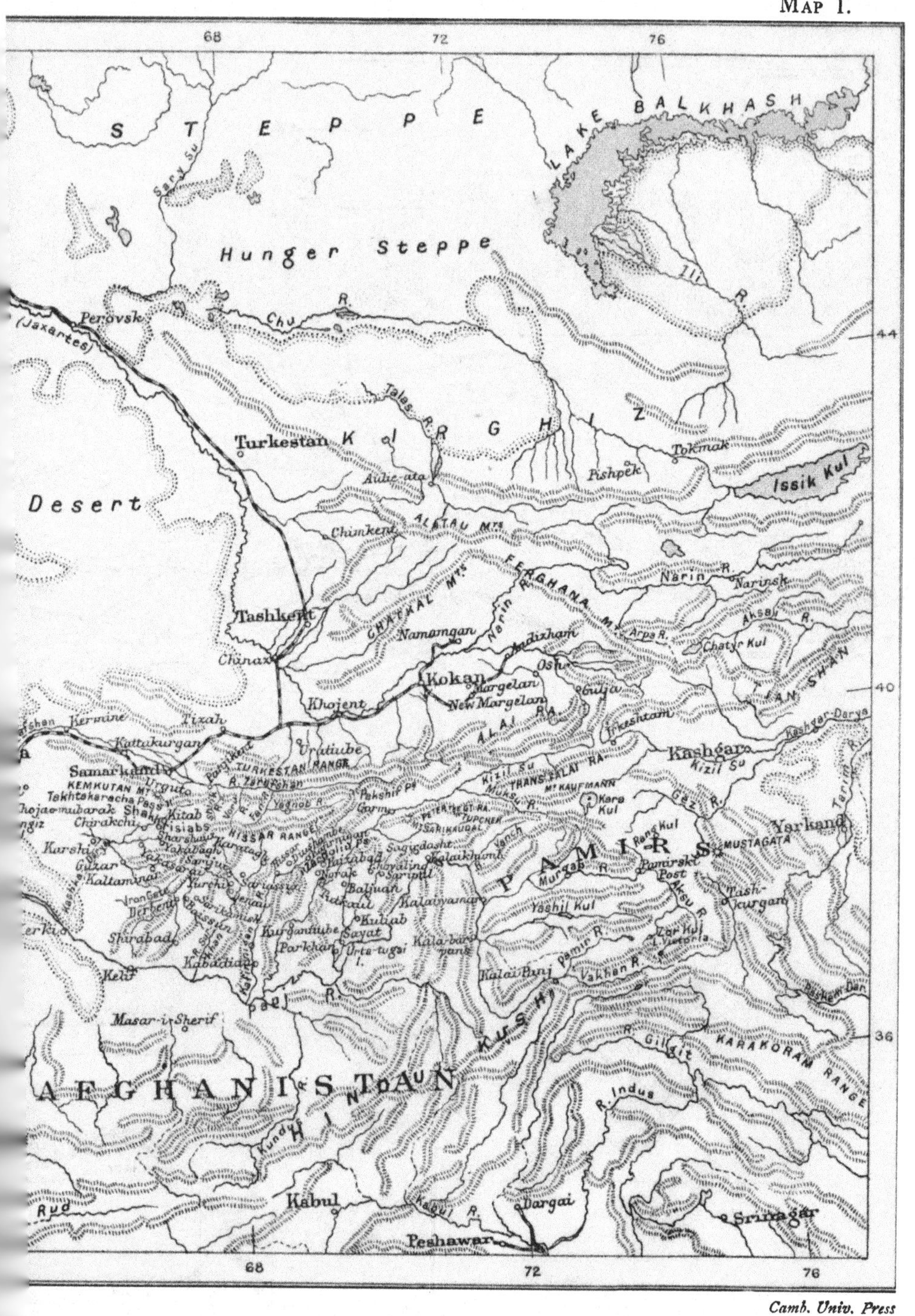

Camb. Univ. Press

Middle Asia.

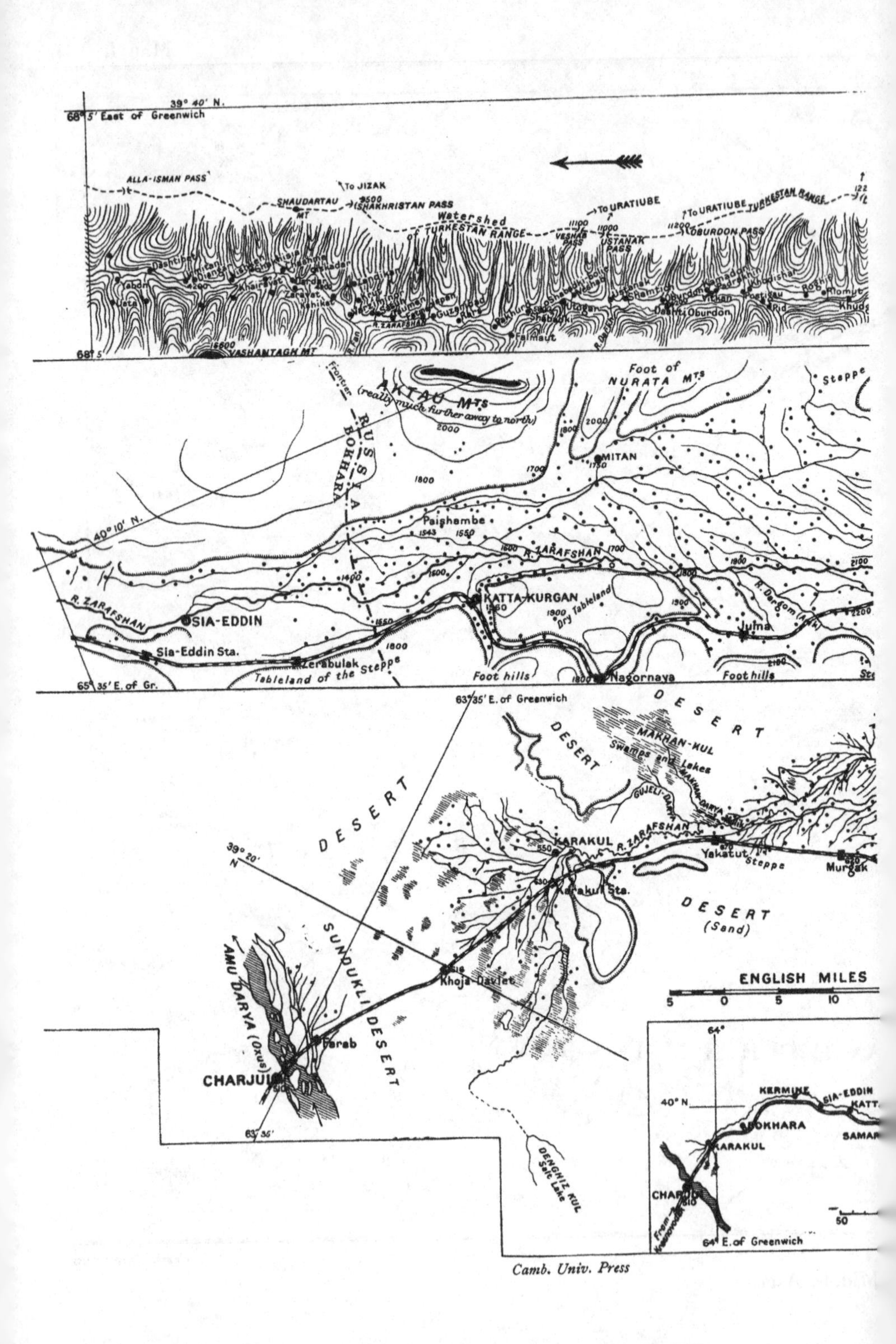

Camb. Univ. Press

MAP II.

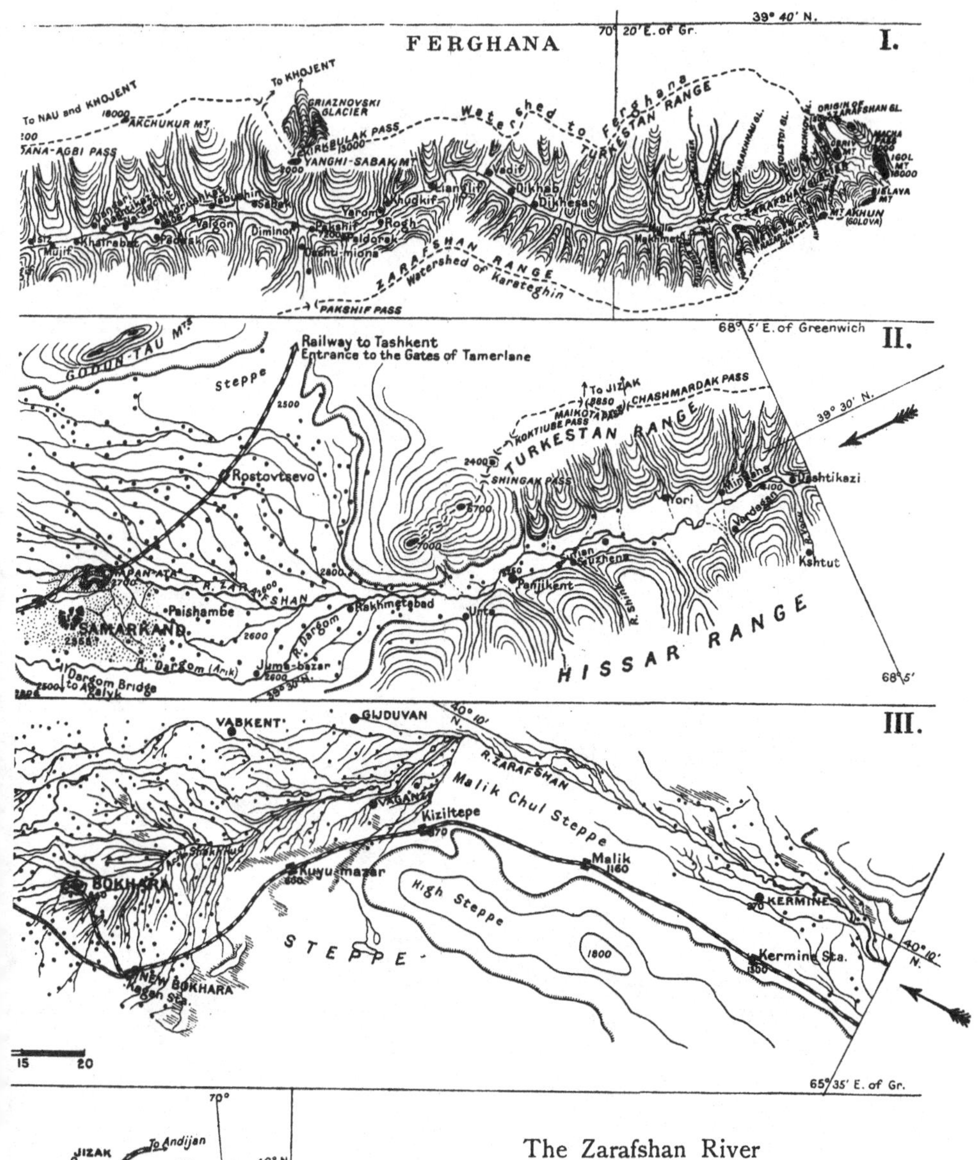

The Zarafshan River
from end to end.

Only rivers and irrigation canals are shown, but no roads.
The dots represent villages of 200 to 2000 inhabitants.

Printed in the United States
By Bookmasters